Lecture Notes in Computer Science 9108

Commenced Publication in 1973
Founding and Former Series Editors:
Gerhard Goos, Juris Hartmanis, and Jan van Leeuwen

Editorial Board

More information about this series at http://www.springer.com/series/7407

José Manuel Ferrández Vicente
José Ramón Álvarez-Sánchez
Félix de la Paz López
Fco. Javier Toledo-Moreo
Hojjat Adeli (Eds.)

Bioinspired Computation in Artificial Systems

International Work-Conference on the Interplay
Between Natural and Artificial Computation, IWINAC 2015,
Elche, Spain, June 1–5, 2015, Proceedings, Part II

 Springer

Editors
José Manuel Ferrández Vicente
Universidad Politécnica de Cartagena
Cartagena
Spain

José Ramón Álvarez-Sánchez
Universidad Nacional de Educación
   a Distancia
Madrid
Spain

Félix de la Paz López
Universidad Nacional de Educación
   a Distancia
Madrid
Spain

Fco. Javier Toledo-Moreo
Universidad Politécnica de Cartagena
Cartagena
Spain

Hojjat Adeli
Ohio State University
Columbus, Ohio
USA

ISSN 0302-9743   ISSN 1611-3349   (electronic)
Lecture Notes in Computer Science
ISBN 978-3-319-18832-4   ISBN 978-3-319-18833-1   (eBook)
DOI 10.1007/978-3-319-18833-1

Library of Congress Control Number: 2015938321

LNCS Sublibrary: SL1 – Theoretical Computer Science and General Issues

Springer Cham Heidelberg New York Dordrecht London

Printed on acid-free paper

Springer International Publishing AG Switzerland is part of Springer Science+Business Media
(www.springer.com)

Preface

The computational paradigm considered here is a conceptual, theoretical, and formal framework situated above machines and living creatures (two instantiations), sufficiently solid, and still non exclusive, that allows us:

1. to help neuroscientists to formulate intentions, questions, experiments, methods, and explanation mechanisms assuming that neural circuits are the psychological support of calculus;
2. to help scientists and engineers from the fields of artificial intelligence (AI) and knowledge engineering (KE) to model, formalize, and program the computable part of human knowledge;
3. to establish an interaction framework between natural system computation and artificial system computation in both directions, from Artificial to Natural and from Natural to Artificial.

With these global purposes, Prof. José Mira organized the 1st International Work Conference on the Interplay between Natural and Artificial Computation, which took place in Las Palmas de Gran Canaria, Canary Islands (Spain), 10 years ago, trying to contribute to both directions of the interplay.

Today, the hybridization between social sciences and social behaviors with robotics, neurobiology and computing, ethics and neuroprosthetics, cognitive sciences and neurocomputing, neurophysiology and marketing is giving rise to new concepts and tools that can be applied to ICT systems, as well as to natural science fields. Through IWINAC we provide a forum in which research in different fields can converge to create new computational paradigms that are on the frontier between Natural sciences and Information technologies.

As a multidisciplinary forum, IWINAC is open to any established institutions and research laboratories actively working in the field of this interplay. But beyond achieving cooperation between different research realms, we wish to actively encourage cooperation with the private sector, particularly SMEs, as a way of bridging the gap between frontier science and societal impact, and young researchers in order to promote this scientific field.

In this edition, four main themes outline the conference topics: gerontechnology and e-therapy, Brain–Computer Interfaces, Biomedical imaging applications for health, and artificial vision and robotics.

Gerontechnology is an interdisciplinary field combining gerontology and technology. Gerontechnology aims at matching systems to health, housing, mobility, communication, leisure, and work of the elderly. The development of computing systems for gerontechnology has turned into a challenging activity requiring disciplines as diverse as artificial intelligence, human–computer interaction, and wireless sensor networks to work together in order to provide solutions able to satisfy this growing societal demand.

Brain–Computer Interfaces implement a new paradigm in communication networks, namely Brain Area Networks. In this paradigm, our brain inputs data (external stimuli), performs multiple media-access control by means of cognitive tasks (selective attention), processes the information (perception), takes a decision (cognition) and, eventually, transmits data back to the source (by means of a BCI), thus closing the communication loop. The objectives include neuro-technologies (e.g. innovative EEG/ECG/fNIRS headsets, integrated stimulation-acquisition devices, etc.), Tele-services (e.g. applications in Telemedicine, tele-rehabilitation programs, tele-control, mobile applications, etc.), innovative biosignal processing algorithms, training techniques, and novel emerging paradigms.

Image understanding is a research area involving both feature extraction and object identification within images from a scene, and a posterior treatment of this information in order to establish relationships between these objects with a specific goal. In biomedical and industrial scenarios, the main purpose of this discipline is, given a visual problem, to manage all aspects of prior knowledge, from study start-up and initiation through data collection, quality control, expert independent interpretation, to design and development of systems involving image processing capable of tackling with these tasks. Brain imaging using EEG techniques or different MRI systems can help in some neural disorders, like epilepsy, Alzheimer, etc.

Over the last decades there has been an increasing interest in using machine learning methods combined with computer vision techniques to create autonomous systems that solve vision problems in different fields. This research involves algorithms and architectures for real-time applications in the areas of computer vision, image processing, biometrics, virtual and augmented reality, neural networks, intelligent interfaces, and biomimetic object-vision recognition. Autonomous robot navigation sets out enormous theoretical and applied challenges to advanced robotic systems using these techniques.

Ten years after the birth of IWINAC meetings these ideas maintain the visionary objectives of Prof. Mira. This wider view of the computational paradigm gives us more elbow room to accommodate the results of the interplay between nature and computation. The IWINAC forum thus becomes a methodological approximation (set of intentions, questions, experiments, models, algorithms, mechanisms, explanation procedures, and engineering and computational methods) to the natural and artificial perspectives of the mind embodiment problem, both in humans and in artifacts. This is the philosophy that continues in IWINAC meetings, the "interplay" movement between the natural and the artificial, facing this same problem every two years. This synergistic approach will permit us not only to build new computational systems based on the natural measurable phenomena, but also to understand many of the observable behaviors inherent to natural systems.

The difficulty of building bridges between natural and artificial computation is one of the main motivations for the organization of IWINAC 2015. The IWINAC 2015 proceedings contain the works selected by the Scientific Committee from more than 190 submissions, after the refereeing process. The first volume, entitled Artificial Computation in Biology and Medicine, includes all the contributions mainly related to the methodological, conceptual, formal, and experimental developments in the fields of neural sciences and health. The second volume, entitled Bioinspired Computation in Artificial Systems, contains the papers related to bioinspired programming strategies and all the contributions related to the computational solutions to engineering problems in different application domains.

An event of the nature of IWINAC 2015 cannot be organized without the collaboration of a group of institutions and people who we would like to thank now, starting with UNED and Universidad Politécnica de Cartagena. The collaboration of the UNED Associated Center in Elche was crucial, as was the efficient work of the Local Organizing Committee, chaired by Eduardo Fernández with the close collaboration of the Universidad Miguel Hernández de Elche. In addition to our universities, we received financial support from the Spanish CYTED, Red Nacional en Computación Natural y Artificial and Apliquem Microones 21 s.l.

We want to express our gratefulness to our invited speakers Prof. Hojjat Adeli, Ohio State University (USA), Prof. Marc de Kamps, University of Leeds (UK), Prof. Richard Duro, University of A Coruña (Spain), and Prof. Luis Miguel Martínez Otero, University Miguel Hernández (Spain) for accepting our invitation and for their magnificent plenary talks.

We would also like to thank the authors for their interest in our call and the effort in preparing the papers, condition sine qua non for these proceedings. We thank the Scientific and Organizing Committees, in particular the members of these committees who acted as effective and efficient referees and as promoters and managers of preorganized sessions on autonomous and relevant topics under the IWINAC global scope.

Our sincere gratitude goes also to Springer and to Alfred Hofmann and his collaborators, Anna Kramer and Christine Reiss, for the continuous receptivity, help efforts, and collaboration in all our joint editorial ventures on the interplay between neuroscience and computation.

Finally, we want to express our special thanks to Viajes Hispania, our technical secretariat, and to Chari García and Beatriz Baeza, for making this meeting possible, and for arranging all the details that comprise the organization of this kind of event. We want to dedicate these two volumes of the IWINAC proceedings to the memory of Professor Mira, whose challenging and inquiring spirit is in all of us. We greatly miss him.

June 2015

José Manuel Ferrández Vicente
José Ramón Álvarez-Sánchez
Félix de la Paz López
Fco. Javier Toledo-Moreo
Hojjat Adeli

Organization

General Chairman

José Manuel Ferrández Vicente   Universidad Politécnica de Cartagena, Spain

Organizing Committee

José Ramón Álvarez-Sánchez	Universidad Nacional de Educación a Distancia, Spain
Félix de la Paz López	Universidad Nacional de Educación a Distancia, Spain
Fco. Javier Toledo-Moreo	Universidad Politécnica de Cartagena, Spain

Honorary Chairs

Rodolfo Llinás	New York University, USA
Hojjat Adeli	Ohio State University, USA
Zhou Changjiu	Singapore Polytechnic, Singapore

Local Organizing Committee

Eduardo Fernández Jover	Universidad Miguel Hernández, Spain
Arantxa Alfaro Sáez	Universidad Miguel Hernández, Spain
Ariadna Díaz Tahoces	Universidad Miguel Hernández, Spain
Nicolás García Aracil	Universidad Miguel Hernández, Spain
Alejandro García Moll	Universidad Miguel Hernández, Spain
Lawrence Humphreys	Universidad Miguel Hernández, Spain
Carlos Pérez Vidal	Universidad Miguel Hernández, Spain
José María Sabater Navarro	Universidad Miguel Hernández, Spain
Cristina Soto-Sánchez	Universidad Miguel Hernández, Spain

Invited Speakers

Hojjat Adeli	Ohio State University, USA
Marc de Kamps	University of Leeds, UK
Richard Duro	University of A Coruña, Spain
Luis Miguel Martínez Otero	University Miguel Hernández, Spain

Field Editors

Diego Andina	Spain
Jorge Azorín-López	Spain
Rafael Berenguer Vidal	Spain
Germán Castellanos-Dominguez	Spain
Miguel Cazorla	Spain
Antonio Fernández-Caballero	Spain
José Garcia-Rodriguez	Spain
Pascual González	Spain
Álvar Ginés Legaz Aparicio	Spain
Javier de Lope Asiaín	Spain
Miguel Ángel López Gordo	Spain
Darío Maravall Gómez-Allende	Spain
Rafael Martínez Tomás	Spain
Elena Navarro	Spain
Pablo Padilla de la Torre	Spain
Daniel Ruiz Fernández	Spain
Antonio J. Tallón-Ballesteros	Spain
Hujun Yin	UK

International Scientific Committee

Andy Adamatzky	UK
Michael Affenzeller	Austria
Abraham Ajith	Norway
José Ramón Álvarez-Sánchez	Spain
Antonio Anaya	Spain
Diego Andina	Spain
Davide Anguita	Italy
Margarita Bachiller Mayoral	Spain
Dana Ballard	USA
Emilia I. Barakova	The Netherlands
Francisco Bellas	Spain
Guido Bologna	Switzerland
María Paula Bonomini	Argentina
François Bremond	France
Giorgio Cannata	Italy
Enrique J. Carmona Suarez	Spain
German Castellanos-Dominguez	Colombia
Jose Carlos Castillo	Spain
Antonio Chella	Italy

Santi Chillemi	Italy
Ricardo Contreras	Chile
Carlos Cotta	Spain
Erzsebet Csuhaj-Varju	Hungary
Felix de la Paz Lopez	Spain
Javier de Lope Asiaín	Spain
Erik De Schutter	Belgium
Ana E. Delgado García	Spain
Gerard Dreyfus	France
Richard Duro	Spain
Reinhard Eckhorn	Germany
Patrizia Fattori	Italy
Antonio Fernández-Caballero	Spain
Miguel A. Fernández-Graciani	Spain
José Manuel Ferrandez	Spain
Kunihiko Fukushima	Japan
Francisco J. Garrigos Guerrero	Spain
Tom D. Gedeon	Australia
Charlotte Gerritsen	The Netherlands
Marian Gheorghe	UK
Pedro Gomez Vilda	Spain
Juan M. Gorriz	Spain
Manuel Graña	Spain
John Hallam	Denmark
Tom Heskes	The Netherlands
Roberto Iglesias	Spain
Igor Aleksaner	UK
Joost N. Kok	The Netherlands
Kostadin Koroutchev	Spain
Elka Korutcheva	Spain
Yasuo Kuniyoshi	Japan
Ryo Kurazume	Japan
Petr Lansky	Czech Rep.
Jerome Leboeuf	Mexico
Maria Longobardi	Italy
Maria Teresa Lopez Bonal	Spain
Tino Lourens	The Netherlands
Max Lungarella	Japan
Manuel Luque	Spain
George Maistros	UK
Vincenzo Manca	Italy
Daniel Mange	Switzerland
Riccardo Manzotti	Italy

Rafael Martinez Tomas	Spain
Antonio Martinez-Alvarez	Spain
Jose Javier Martinez-Alvarez	Spain
Jose del R. Millan	Switzerland
Taishin Y. Nishida	Japan
Richard A. Normann	USA
Lucas Paletta	Austria
Juan Pantrigo	Spain
Alvaro Pascual-Leone	USA
Gheorghe Paun	Spain
Francisco Peláez	Brazil
Franz Pichler	Austria
Maria Pinninghoff	Chile
Andonie Razvan	USA
Mariano Rincon Zamorano	Spain
Victoria Rodellar	Spain
Camino Rodriguez Vela	Spain
Daniel Ruiz	Spain
Ramon Ruiz Merino	Spain
José María Sabater Navarro	Spain
Diego Salas-Gonzalez	Spain
Pedro Salcedo Lagos	Chile
Angel Sanchez	Spain
Eduardo Sánchez Vila	Spain
Jose Santos Reyes	Spain
Shunsuke Sato	Japan
Andreas Schierwagen	Germany
Guido Sciavicco	Spain
Radu Serban	The Netherlands
Igor A. Shevelev	Russia
Shun-ichi Amari	Japan
Settimo Termini	Italy
Javier Toledo-Moreo	Spain
Jan Treur	The Netherlands
Ramiro Varela Arias	Spain
Marley Vellasco	Brazil
Lipo Wang	Singapore
Stefan Wermter	UK
Hujun Yin	UK
Juan Zapata	Spain
Changjiu Zhou	Singapore

Contents – Part II

Contents – Part I

Online Control of Enumeration Strategies via Bat-Inspired Optimization

Ricardo Soto[1,2,3], Broderick Crawford[1,4,5], Rodrigo Olivares[1(✉)],
Franklin Johnson[6], and Fernando Paredes[7]

[1] Pontificia Universidad Católica de Valparaíso, Chile
rodrigo.olivares@uv.cl
[2] Universidad Autónoma de Chile, Chile
[3] Universidad Central de Chile, Chile
[4] Universidad Finis Terrae, Chile
[5] Universidad San Sebastián, Chile
[6] Universidad de Playa Ancha, Chile
[7] Escuela de Ingeniería Industrial, Universidad Diego Portales, Chile
{ricardo.soto,broderick.crawford}@ucv.cl,
franklin.johnson@upla.cl, fernando.paredes@udp.cl

Abstract. Constraint programming allows to solve constraint satisfaction and optimization problems by building and then exploring a search tree of potential solutions. Potential solutions are generated by firstly selecting a variable and then a value from the given problem. The enumeration strategy is responsible for selecting the order in which those variables and values are selected to produce a potential solution. There exist different ways to perform this selection, and depending on the quality of this decision, the efficiency of the solving process may dramatically vary. A modern idea to handle this concern, is to interleave during solving time a set of different strategies instead of using a single one. The strategies are evaluated according to process indicators in order to use the most promising one on each part of the search process. This process is known as online control of enumeration strategies and its correct configuration can be seen itself as an optimization problem. In this paper, we present a new system for online control of enumeration strategies based on bat-inspired optimization. The bat algorithm is a relatively modern metaheuristic based on the location behavior of bats that employ echoes to identify the objects in their surrounding area. We illustrate, promising results where the proposed bat algorithm is able to outperform previously reported metaheuristic-based approaches for online control of enumeration strategies.

Keywords: Constraint Programming · Constraint Satisfaction Problems · Swarm Intelligence · Bat Algorithm

1   Introduction

Constraint programming is an efficient technology for solving constraint satisfaction and optimization problems. Under this framework, problems are formulated

© Springer International Publishing Switzerland 2015
J.M. Ferrández Vicente et al. (Eds.): IWINAC 2015, Part II, LNCS 9108, pp. 1–10, 2015.
DOI: 10.1007/978-3-319-18833-1_1

as a set of variables and constraints. The variables represent the unknowns of the problem and are linked to a non-empty domain of possible values, while the constraints define relations among those variables. A solution to the problem is defined by an assignment of values to variables that does not violate any constraints. The resolution process is carried out by a search engine, commonly called solver, which attempt to reach a result by building and exploring a search tree of potential solutions.

The performance and behavior of most solvers noticeably depends on the quality of their configuration. Particularly, the enumeration strategy is a key configuration component that may dramatically influence the performance of constraint programming solvers. The enumeration strategy is responsible for selecting the order in which variables and values are selected to build the potential solutions of the problem. However, predicting the correct strategy in advance is often unfeasible as its behavior is commonly unpredictable. A recent idea to handle this concern is about interleaving a set of different strategies along the search instead of using a single one. In this way, enumeration strategies are replaced during search depending on the performance exhibited along the solving process. The performance of strategies is evaluated according to process indicators in order to use the most promising one on each part of the search process. This form of search is called online control within the autonomous search (AS) framework [10] and its correct configuration can be seen itself as an optimization problem. Indeed, we are in the presence of an hyper-optimization [17] problem, which in this case is the optimization of the process for solving an optimization (satisfaction) problem.

In this paper, we present a new system for online control of enumeration strategies based on bat-inspired optimization. The bat algorithm is a recent metaheuristic based on the echolocation behavior of bats that identify positions and objects by emitting pulses of sound and retrieving the corresponding produced echoes. We illustrate, promising results where the proposed bat algorithm is able to outperform previously reported metaheuristic-based approaches for online control of enumeration strategies. The remainder of this paper is organized as follows. The related work is given in Section 2 followed by an overview of constraint programming and a description of the hyper-optimization problem. The bat algorithm is illustrated next followed by the experimental evaluation. Finally, we give conclusion and some future work.

2   Related Work

The integration of control methods in constraint programming is a recent trend. Preliminary approaches proposed to sample and learn good strategies after solving a problem, such as the works reported in [8,7] and [15]. Another idea following the same goal proposes to associate weights to constraints [3], which are incremented once they conduct the search to domain wipeouts. In this way, variables that participate within heavier constraints may be selected first as they are more likely to cause failures. A variation of this approach is reported in [9,14], which argue that initial choices are often more important, suggesting to perform an a priori sampling phase for a better initial selection. In this paper, we focus on online control, based on the work done in [1] we employ an online method to adapt the enumeration strategies during solving time. The goal is to replace the strategies depending on the performance exhibited along the solving process. In this way, the most promising ones are used on each part of the search process. In this context, two approaches have been proposed to online control, one based on genetic algorithms [5] and a second one based on particle swarm optimization [6].

3 Constraint Programming and the Hyper-Optimization Problem

A constraint satisfaction problem (CSP) $\mathcal{P}$ is defined by a triple $\mathcal{P} = \langle \mathcal{X}, D, C \rangle$ where $\mathcal{X}$ is an n-tuple of variables $\mathcal{X} = \langle x_1, x_2, \ldots, x_n \rangle$. $\mathcal{D}$ is a corresponding n-tuple of domains $\mathcal{D} = \langle d_1, d_2, \ldots, d_n \rangle$ such that $x_i \in d_i$, and d_i is a set of values, for $i = 1, \ldots, n$; and $\mathcal{C}$ is an m-tuple of constraints $\mathcal{C} = \langle c_1, c_2, \ldots, c_m \rangle$, and a constraint c_j is defined as a subset of the Cartesian product of domains $d_{j_1} \times \cdots \times d_{j_{n_j}}$, for $j = 1, \ldots, m$. A solution to a CSP is an assignment $\{x_1 \rightarrow a_1, \ldots, x_n \rightarrow a_n\}$ such that $a_i \in d_i$ for $i = 1, \ldots, n$ and $(a_{j_1}, \ldots, a_{j_{n_j}}) \in c_j$, for $j = 1, \ldots, m$.

CSPs are usually solved by combining enumeration and propagation phases. The enumeration strategy decides the order in which variables and values are selected by means of the variable and value ordering heuristics, respectively. The propagation phase tries to delete from domain the values that do not drive the search process to a feasible solution.

In this work, we aim at online controlling a set of enumeration strategies which are dynamically interleaved during solving time. Our purpose is to select the most promising one for each part of the search tree. To this end we need to evaluate the strategies by penalizing the ones exhibiting poor performances and giving more credits to better ones. Based on the work done on [6], we can evaluate the performance of strategies via a set of indicators that are able to measure the quality of the search. The performance is evaluated via a weighted sum model (WSM), which is a well-known decision making method from multi-criteria decision analysis for evaluating alternatives in terms of decision criteria. Formally, we define a weighted sum model $A_t(S_j)^{WSM-score}$ that evaluates a strategy S_j in time t as follows:

$$A_t(S_j)^{WSM-score} = \sum_{i=1}^{IN} w_i a_{it}(S_j) \tag{1}$$

Where IN corresponds to the indicator set, w_i is a weight that controls the importance of the ith-indicator within the WSM and $a_{it}(S_j)$ is the score of the ith-indicator for the strategy S_j in time t. A main component of this model are the weights, which must be finely tuned by an optimizer. This is done by carrying out a sampling phase where the CSP is partially solved to a given cutoff.

The performance information gathered in this phase via the indicators is used as input data of the optimizer, which attempt to determine the most successful weight set for the WSM. This tuning process is very important as the correct configuration of the WSM may have essential effects on the ability of the solver to properly solve specific CSPs. Parameter (weights) tuning is hard to achieve as parameters are problem-dependent and their best configuration is not stable along the search [12]. As previously mentioned, we are in the presence of an hyper-optimization problem, which in practice is the optimization of the process (the optimal configuration of the WSM) for solving an optimization (satisfaction) problem. In Section 3.1, we present the approach used to optimize the WSM.

3.1   Bat Algorithm

As previously mentioned, to determine the most successful weight set for performance indicators, the WSM must be finely tuned by an optimizer. To this end we propose the use of a bat algorithm, which is a recent metaheuristic [19] inspired on the echolocation behavior of bats. Bats and particularly micro-bats are able to identify objects in their surrounding areas by emitting pulses of sound and retrieving the corresponding produced echoes. This advanced capability allows bats even to distinguish obstacles from preys, being able to hunt in complete darkness. The bat algorithm as been developed following three rules: (1) It is assumed that all bats use echolocation to determine distances, and all of them are able to distinguish food, prey, and background barriers. (2) A bat b_i searches for a prey with a position x_i and a velocity v_i. The pulses of sound emitted have the following features: a frequency f_{min}, (or varying λ), a varying wavelength λ (or frequency f), loudness A_0, and a rate of pulse emission $r \in [0, 1]$. All sound features can be automatically adjusted depending on their target proximity. (3) Although the loudness can vary in many ways, it is assumed that the loudness varies from a large (positive) A_0 to a minimum constant value A_{min}.

Algorithm 1 illustrates the pseudo-code for bat optimization. At the beginning, a population of m bats is initialized with position x_i and velocity v_i. Then, the frequency f_i at position x_i is set followed by pulse rates and loudness. Then, a while loop encloses a set of actions to be performed t times until the fixed number T of iterations if reached. The first action to be done between the loop is the movement of bats according to Eq. 2, 3, and 4.

$$f_i = f_{min} + (f_{max} - f_{min})\beta \tag{2}$$

$$v_i^j(t) = v_i^j(t-1) + [(\hat{x})^j - x_i^j(t-1)]f_i \tag{3}$$

$$x_i^j(t) = x_i^j(t-1) + v_i^j(t) \tag{4}$$

Eq. 2 is used to control the pace and range of bats movements, where β is a randomly generated number drawn from a uniform distribution within the interval $[0, 1]$. Eq. 3 defines the velocity of decision variable j held by bat i in time t, where $(\hat{x})^j$ represents the current global best position encountered from the m bats for decision variable j; and finally Eq. 4 defines the position of

---

Algorithm 1. Bat algorithm

Objective function $f(x), x = x^1, ..., x^n$.
Initialize the bat population x_i and velocity v_i, $i = 1, 2, ..., m$.
Define pulse frequency f_i, at x_i, $i = 1, 2, ..., m$.
Initialize pulse rates r_i and the loudness A_i, $i = 1, 2, ..., m$.

1: **while** $t < T$ **do**
2:    **for all** b_i **do**
3:       *Generate new solutions through Eq.(2), (3) and (4).*
4:       **if** $rand > r_i$ **then**
5:          *Select a solution among the best solutions.*
6:          *Generate a local solution around the best solution.*
7:       **end if**
8:       **if** $rand < A_i$ and $f(x_i) < f(\hat{x})$ **then**
9:          *Accept the new solutions.*
10:          *Increase r_i and reduce A_i.*
11:       **end if**
12:    **end for**
13: **end while**
14: *Rank the bats and find the current best $\hat{x}$.*

---

decision variable j held by bat i in time t. Then, at line 4, a condition handles the variability of the possible solutions. Firstly, a solution is selected among the current best solutions, and a new solution is generated via random walks as proposed in [20]. Next, at line 8, a second condition is responsible for accepting the new best solution and for updating r_i and A_i according to Eq. 5 and 6, where α and γ are ad-hoc constants between with $0 < \alpha < 1$ and $\gamma > 0$. Finally, the bats are ranked in order to find $\hat{x}$.

For the experiments, we employ the following bat configuration as suggested in [18]: $\alpha = \gamma = 0.9$; $f_{min} = 0.75$; and $f_{max} = 1.25$. At the beginning, $r_i(0)$ and $A_i(0)$ are selected randomly, with $A_i(0) \in [1, 2]$ and $r_i(0) \in [0, 1]$.

$$A_i(t + 1) = \alpha A_i(t) \qquad (5) \qquad r_i(t + 1) = r_i(0)[1 - exp(-\gamma t)], \quad (6)$$

4   Experimental Results

We have performed an experimental evaluation of the proposed approach on different instances of classic constraint satisfaction problems: Magic Square with N={3, 4, 5}, Latin Square with N={4, 5, 6, 7, 8}, Knight's Tour with N={5, 6}, Quasigroup with N={5, 6, 7}, and Langford with $M = 2$ and N={8, 12, 16, 20, 23}.

The adaptive enumeration component has been implemented on the Ecl^ips^e Constraint logic Programming Solver v6.10, and the bat-optimizer has been developed in Java. The experiments have been launched on a 3.3GHz Intel Core i3 with 8Gb RAM running Ubuntu Desktop 12.04.4 LTS. The instances are solved to a maximum number of 65535 steps as equally done in previous work [4]. If no solution is found at this point the problem is set to t.o. (time-out). Let us recall that a step refers to a request of the solver to instantiate a variable by enumeration. The adaptive enumeration uses a portfolio of twenty four enumeration strategies, which is detailed in table 1. We employ the following WSM for the experiments: $w_1 SB + w_2 In1 + w_3 In2$, where SB is the number of shallow backtracks [2] (SB), $In1 = CurrentMaximumDepth - PreviousMaximumDepth$, and $In2 = CurrentDepth - PreviousDepth$, where $Depth$ refers to the depth reached within the search tree. This WSM was the best performing one after the corresponding training phase of the algorithm.

Table 1. Portfolio of the enumerations strategies used

Ordering Heuristics								
Id	Variable	Value	Id	Variable	Value	Id	Variable	Value
S_1	First		S_9	First		S_{17}	First	
S_2	MRV		S_{10}	MRV		S_{18}	MRV	
S_3	AMRV		S_{11}	AMRV		S_{19}	AMRV	
S_4	O	Min	S_{12}	O	Mid	S_{20}	O	Max
S_5	S		S_{13}	S		S_{21}	S	
S_6	L		S_{14}	L		S_{22}	L	
S_7	MC		S_{15}	MC		S_{23}	MC	
S_8	MR		S_{16}	MR		S_{24}	MR	

First: the first entry in the list is selected.
MRV: the entry with the smallest domain size is selected.
AMRV: the entry with the largest domain size is selected.
S: the entry with the smallest value in the domain is selected.
L: the entry with the largest value in the domain is selected.
O: the entry with the largest number of attached constraints is selected.
MC: the entry with the smallest domain size is selected.
MR: the entry with the largest difference between the smallest and second smallest value in the domain is selected.

Min Values are tried in increasing order.
Mid Values are tried beginning from the middle of the domain.
Max Values are tried in decreasing order.

We compare the proposed approach based on bat-optimization (BAT) with the two previously reported optimized online control systems, one based on genetic algorithm (GA) [5] and the other one based on particle swarm optimization (PSO) [6]. We consider for the comparison as well, the results obtained with no online control, i.e., a single enumeration strategy (S_1 to S_{24}) is employed for the complete CSP resolution. At the end, we also include a random selection strategy. For the evaluation, we consider number of backtracks and runtime needed to reach a solution, both being widely employed indicators of search performance.

The results in terms of backtracks (see tables 2 and 3) illustrate that the proposed approach is able to compete with previous employed optimizers as well as with single strategies. For instance, taking into account magic squares problems, for small instances ($n = 3, 4$), the enumeration strategies require similar backtrack calls than BAT. However, for $n = 5, 6, 7$; BAT outperforms PSO, GA and the use of single strategies. Now, evaluating the knight's tour

Table 2. Magic Square and Latin Square results, in terms of backtracks

S_j	Magic Square					$\bar{X}_{MS}$	Latin Square					$\bar{X}_{LS}$
	3	4	5	6	7		4	5	6	7	8	
S_1	0	12	910	>177021	>170762	>177021	0	0	0	12	0	2.4
S_2	4	1191	>191240	>247013	>39181	>247013	0	9	163	>99332	0	>99332
S_3	0	3	185	>173930	>169053	>173930	0	0	0	0	0	0
S_4	0	10	5231	>187630	>127273	>187630	0	0	0	14	0	2.8
S_5	0	22	>153410	>178895	>79041	>178895	0	7	61	>99403	0	>99403
S_6	4	992	>204361	>250986	>230408	>250986	0	0	0	71	0	14.2
S_7	0	3	193	>202927	>166861	>202927	0	0	0	0	0	0
S_8	0	13	854	>190877	>202632	>202632	0	0	0	0	0	0
S_9	0	12	910	>177174	>170762	>177174	0	0	0	9	0	1.8
S_{10}	4	1191	>191240	>247013	>63806	>247013	0	9	163	>99332	0	>99332
S_{11}	0	3	185	>174068	>169053	>174068	0	0	0	0	0	0
S_{12}	0	10	5231	>187777	>153768	>187777	0	0	0	9	0	1.8
S_{13}	0	22	>153410	>179026	>77935	>179026	0	7	61	>99539	0	>99539
S_{14}	4	992	>204361	>251193	>230408	>251193	0	0	0	71	0	14.2
S_{15}	0	3	193	>203089	>96730	>203089	0	0	0	0	0	0
S_{16}	0	13	854	>191042	>247427	>247427	0	0	0	0	0	0
S_{17}	1	51	>204089	>237428	>229035	>237428	0	0	0	9	0	1.8
S_{18}	0	42	>176414	>176535	>116846	>176535	0	9	163	>99481	0	>99481
S_{19}	1	3	>197512	>231600	>185681	>231600	0	0	0	0	0	0
S_{20}	1	29	74063	>190822	>187067	>190822	0	0	0	9	0	1.8
S_{21}	1	95	>201698	>239305	>249686	>249686	0	0	0	71	0	14.2
S_{22}	0	46	74711	>204425	>46233	>204425	0	7	61	>99539	0	>99539
S_{23}	1	96	>190692	>204119	>130196	>204119	0	0	0	0	0	0
S_{24}	1	47	>183580	>214287	>219116	>219116	0	0	0	0	1	0.2
PSO	0	0	14	>47209	>56342	>56342	0	0	0	0	0	0
GA	0	42	198	>176518	>213299	>213299	0	0	0	0	0	0
BAT	0	3	99	733	1445	456	0	0	0	0	0	0

Table 3. Knight's Tour, Quasigroup and Langford results, in terms of backtracks

S_j	Knight's Tour		$\bar{X}_{KT}$	Quasigroup					$\bar{X}_{QG}$	Langford $L_{m=2}$					$\bar{X}_L$
	5	6		1	3	5	6	7		8	12	16	20	23	
S_1	767	37695	19231	0	0	>145662	30	349	>145662	2	16	39	77	26	32
S_2	>179097	>177103	>179097	0	0	>103603	>176613	3475	>176613	15	223	24310	>68157	>97621	>97621
S_3	767	37695	19231	0	0	8343	0	1	1668.8	2	1	97	172	64	67.2
S_4	>97176	35059	>97176	0	0	>145656	30	349	>145656	2	16	39	77	26	32
S_5	>228316	>239427	>239427	1	0	>92253	>83087	4417	>92253	2	29	599	26314	29805	11349.8
S_6	>178970	>176668	>178970	0	0	>114550	965	4417	>114550	1	22	210	1	3	47.4
S_7	>73253	14988	>73253	0	0	8343	0	1	1668.8	2	1	97	172	64	67.2
S_8	>190126	>194116	>194116	1	0	>93315	>96367	4	>96367	0	12	0	64	7	16.6
S_9	767	37695	19231	1	0	>145835	30	349	>145835	2	16	39	77	26	32
S_{10}	>179126	>177129	>179126	0	0	>103663	>176613	3475	>176613	15	223	24310	>68157	>97621	>97621
S_{11}	767	37695	19231	0	0	8343	0	1	1668.8	2	1	97	172	64	67.2
S_{12}	>97176	35059	>97176	0	0	>145830	30	349	>145830	2	16	39	77	26	32
S_{13}	>228316	>239427	>239427	1	0	>92355	>83087	583	>92355	2	29	599	26314	29805	11349.8
S_{14}	>178970	>176668	>178970	0	0	>114550	965	4417	>114550	1	22	210	1	3	47.4
S_{15}	>73253	14998	>73253	0	0	8343	0	1	1668.8	2	1	97	172	64	67.2
S_{16}	>190116	>194116	>194116	1	0	>93315	>93820	4	>93820	0	12	0	64	7	16.6
S_{17}	767	37695	19231	0	0	7743	2009	3	1951	2	16	39	77	26	32
S_{18}	>179126	>177129	>179126	0	0	>130635	>75475	845	>130635	15	223	24592	>68157	>97621	>97621
S_{19}	767	37695	19231	0	0	0	89	1	18	2	1	98	172	64	67.4
S_{20}	>97178	35059	>97178	0	0	7763	2009	3	1955	2	16	39	77	26	32
S_{21}	>178970	>176668	>178970	0	0	>96083	>108987	773	>108987	1	22	210	1	3	47.4
S_{22}	>228316	>239427	>239427	0	0	>94426	>124523	1	>124523	2	29	599	26314	29805	11349.8
S_{23}	>73253	14998	>73253	0	0	0	89	1	18	2	1	98	172	64	67.4
S_{24}	>190116	>160789	>190116	0	0	>95406	>89888	1	>95406	4	6	239	4521	0	954
PSO	106	12952	6529	0	0	0	0	0	0	0	1	0	1	3	1
GA	5615	86928	46271.5	0	0	7763	0	4	1553.4	15	223	97	64	0	79.8
BAT	7	1499	753	0	0	0	0	0	0	0	1	0	1	0	0.4

problem for $n = 5$, BAT and PSO clearly fail less than GA; and for the hardest instance, BAT takes the first place. In the smaller instances of quasigroup problem ($n = 1, 3$), BAT behaves similar than enumeration strategies, but for the hard instances ($n = 5, 6, 7$) BAT again outperforms PSO, GA and the use of single strategies. The same applies to small instances ($n = 8, 12$) and hard

Table 4. Magic Square and Latin Square results, in terms of Solving Time

S_j	Magic Square					$\bar{X}_{MS}$	Latin Square					$\bar{X}_{LS}$
	3	4	5	6	7		4	5	6	7	8	
S_1	1	14	1544	t.o.	t.o.	t.o.	2	3	5	9	11	6
S_2	5	2340	t.o.	t.o.	t.o.	t.o.	2	11	102	t.o.	14	t.o.
S_3	1	6	296	t.o.	t.o.	t.o.	1	3	5	7	12	5.6
S_4	1	21	6490	t.o.	t.o.	t.o.	1	3	5	9	12	6
S_5	1	21	t.o.	t.o.	t.o.	t.o.	2	8	60	t.o.	14	t.o.
S_6	4	1500	t.o.	t.o.	t.o.	t.o.	2	4	7	88	12	22.6
S_7	1	6	203	t.o.	t.o.	t.o.	1	2	4	7	12	5.2
S_8	1	11	1669	t.o.	t.o.	t.o.	1	2	5	7	12	5.4
S_9	1	13	1498	t.o.	t.o.	t.o.	2	3	5	13	11	6.8
S_{10}	4	2366	t.o.	t.o.	t.o.	t.o.	1	11	103	t.o.	14	t.o.
S_{11}	1	6	297	t.o.	t.o.	t.o.	2	3	6	8	12	6.2
S_{12}	1	21	6053	t.o.	t.o.	t.o.	2	3	6	14	13	7.6
S_{13}	1	21	t.o.	t.o.	t.o.	t.o.	2	8	62	t.o.	15	t.o.
S_{14}	4	1495	t.o.	t.o.	t.o.	t.o.	2	4	7	92	14	238
S_{15}	1	6	216	t.o.	t.o.	t.o.	2	3	5	9	14	6.6
S_{16}	1	11	1690	t.o.	t.o.	t.o.	2	3	5	9	14	6.6
S_{17}	1	88	t.o.	t.o.	t.o.	t.o.	2	3	6	14	12	7.4
S_{18}	1	37	t.o.	t.o.	t.o.	t.o.	2	12	107	t.o.	14	t.o.
S_{19}	1	99	t.o.	t.o.	t.o.	t.o.	2	3	5	8	12	6
S_{20}	1	42	165878	t.o.	t.o.	t.o.	2	4	5	16	13	8
S_{21}	1	147	t.o.	t.o.	t.o.	t.o.	2	4	7	93	13	23.8
S_{22}	1	37	153679	t.o.	t.o.	t.o.	2	8	64	t.o.	15	t.o.
S_{23}	1	102	t.o.	t.o.	t.o.	t.o.	2	3	6	8	13	6.4
S_{24}	1	79	t.o.	t.o.	t.o.	t.o.	2	3	5	10	17	7.4
PSO	2745	15986	565155	t.o.	t.o.	t.o.	3323	6647	12716	20519	31500	14941
GA	735	1162	1087	t.o.	t.o.	t.o.	695	692	725	777	752	728.2
BAT	545	902	775	1038	1189	889.8	333	387	438	496	560	442.8

Table 5. Knight's Tour, Quasigroup and Langford results, in terms of Solving Time

S_j	Knight's Tour		$\bar{X}_{KT}$	Quasigroup					$\bar{X}_{QG}$	Langford $L_{m=2}$					$\bar{X}_L$
	5	6		1	3	5	6	7		8	12	16	20	23	
S_1	1825	90755	46290	1	1	t.o.	45	256	t.o.	3	20	70	191	79	72.6
S_2	t.o	t.o	t.o	1	1	t.o.	8020	t.o.	t.o.	10	242	70526	t.o.	t.o.	t.o.
S_3	2499	111200	56850	1	1	7510	15	10	1507.4	3	4	231	546	286	214
S_4	t.o	89854	t.o	1	1	t.o.	45	307	t.o.	4	29	115	318	140	121.2
S_5	t.o	t.o	t.o	1	1	t.o.	t.o.	943	t.o.	3	43	1217	61944	68254	26292
S_6	t.o	t.o	t.o	1	1	t.o.	3605	16896	t.o.	3	32	489	11	19	110.8
S_7	t.o	39728	t.o	1	1	9465	15	10	1898.4	3	4	237	553	285	216.4
S_8	t.o	t.o	t.o	2	1	t.o.	t.o.	16	t.o.	2	22	7	240	19	58
S_9	1908	93762	47835	2	1	t.o.	40	240	t.o.	3	20	69	185	79	71.2
S_{10}	t.o	t.o	t.o	1	1	t.o.	t.o.	13481	t.o.	10	270	55291	t.o.	t.o.	t.o.
S_{11}	2625	102387	52506	1	1	9219	15	10	1849.2	4	4	250	538	285	216.2
S_{12}	t.o	109157	t.o	1	1	t.o.	45	348	t.o.	4	29	118	312	140	120.6
S_{13}	t.o	t.o	t.o	1	1	t.o.	t.o.	1097	t.o.	3	44	1273	61345	71209	26774
S_{14}	t.o	t.o	t.o	1	1	t.o.	3565	18205	t.o.	3	32	530	11	19	119
S_{15}	t.o	46673	t.o	1	1	10010	15	11	2007.6	3	5	235	541	278	212.4
S_{16}	t.o	t.o	t.o	2	1	t.o.	t.o.	15	t.o.	2	21	8	237	19	57.4
S_{17}	1827	96666	49247	1	1	9743	7075	9	3368.8	3	18	66	170	75	66.4
S_{18}	t.o	t.o	t.o	1	1	t.o.	t.o.	1878	t.o.	11	242	55687	t.o.	t.o.	t.o.
S_{19}	2620	97388	50004	1	1	20	125	12	31.8	3	4	245	562	272	217.2
S_{20}	t.o	90938	t.o	1	1	10507	6945	9	3492.6	4	29	107	294	126	112
S_{21}	t.o	t.o	t.o	1	1	t.o.	t.o.	1705	t.o.	3	33	510	11	20	115.4
S_{22}	t.o	t.o	t.o	1	1	t.o.	t.o.	9	t.o.	3	43	1297	58732	73168	26648
S_{23}	t.o	40997	t.o	1	1	21	130	12	33	3	5	240	569	276	218.6
S_{24}	t.o	t.o	t.o	1	1	t.o.	t.o.	14	t.o.	5	13	584	15437	10	3209.8
PSO	21089	170325	95707	2160	1250	59158	44565	28612	27149	5000	10430	20548	28466	30468	18982
GA	456375	203020	329698	675	665	11862	947	795	2989	762	1212	1502	1409	1287	1234.4
BAT	3194	4570	3882	262	214	645	626	516	452.6	415	506	623	742	810	619.2

instances ($n = 16, 20, 23$) of the Langford problem. Finally, considering the average number of backtracks for solving the complete set of problems, BAT takes the first place.

Runtimes are depicted in tables 4 and 5 where the proposed approach is also quite competitive. For small instances of the magic square and latin square, BAT is slower than enumeration strategies, but for hard instances ($n = 5, 6, 7$, and 8 for only latin square) BAT begins to exhibit excellent performance. Regarding knight's tour problems, BAT is again the fastest one, and more than 40 times faster particularly for $n = 6$. For small instances of the quasigroup ($n = 1, 3$) and langford problems ($n = 8, 12$), BAT is again slower that enumeration strategies, but for hard instances of quasigroup ($n = 5, 6, 7$) and Langford ($n = 16, 20, 23$), BAT evidence a better performance than PSO, GA and the use of single strategies. Finally, considering the average runtime, BAT keeps its first place.

5   Conclusions and Future Work

In this paper, we have presented a new and more efficient approach for online control for solving CSPs based on bat optimization. The idea is to interleave a portfolio of enumeration strategies in order to use the most promising one at each step of the process. The online control is carried out by means of a weighted sum model, which is finely tuned by a bat algorithm. The promising results obtained by the proposed approach can be explained by two interesting capabilities of bat algorithms: autozooming and parameter control. The first one enables the algorithm to automatically zooming within areas of the search space where promising solutions have been encountered, resulting in more rapid convergence rates [18]. The second one allows the variation of loudness and pulse emission rate (as shown in Eq. 5 and 6) during solving time. This allows one to automatically swap from exploration to exploitation when the optimal solution is approaching. The benefits of both features can particularly be observed when solving harder instances of CSPs, where BAT runtimes are noticeably better compared to GA and PSO.

The results obtained from the experiments open up opportunities for further research. In the medium-term, we plan to incorporate new and more sophisticated enumeration strategies to the portfolio, as well as to experiment with new modern metaheuristics for the optimization of the WSM, such as Cuckoo Search [16], Artificial Bee Colony [11], and Gravitational Search [13]. Another interesting idea is about the design of a similar adaptive framework for interleaving different domain filtering techniques.

References

1. Crawford, B., Soto, R., Montecinos, M., Castro, C., Monfroy, E.: A Framework for Autonomous Search in the Ecl^{i}ps^e Solver. In: Mehrotra, K.G., Mohan, C.K., Oh, J.C., Varshney, P.K., Ali, M. (eds.) IEA/AIE 2011, Part I. LNCS, vol. 6703, pp. 79–84. Springer, Heidelberg (2011)
2. Barták, R., Rudová, H.: Limited assignments: A new cutoff strategy for incomplete depth-first search. In: Proceedings of the 20th ACM Symposium on Applied Computing (SAC), pp. 388–392 (2005)
3. Boussemart, F., Hemery, F., Lecoutre, C., Sais, L.: Boosting systematic search by weighting constraints. In: Proceedings of the 16th Eureopean Conference on Artificial Intelligence (ECAI), pp. 146–150. IOS Press (2004)
4. Crawford, B., Castro, C., Monfroy, E., Soto, R., Palma, W., Paredes, F.: Dynamic Selection of Enumeration Strategies for Solving Constraint Satisfaction Problems. Rom. J. Inf. Sci. Tech. (2012) (to appear)

5. Crawford, B., Soto, R., Castro, C., Monfroy, E., Paredes, F.: An Extensible Autonomous Search Framework for Constraint Programming. Int. J. Phys. Sci. 6(14), 3369–3376 (2011)
6. Crawford, B., Soto, R., Monfroy, E., Palma, W., Castro, C., Paredes, F.: Parameter tuning of a choice-function based hyperheuristic using particle swarm optimization. Expert Syst. Appl. 40(5), 1690–1695 (2013)
7. Epstein, S., Petrovic, S.: Learning to solve constraint problems. In: Proceedings of the Workshop on Planning and Learning (ICAPS) (2007)
8. Epstein, S.L., Freuder, E.C., Wallace, R.J., Morozov, A., Samuels, B.: The adaptive constraint engine. In: Van Hentenryck, P. (ed.) CP 2002. LNCS, vol. 2470, pp. 525–542. Springer, Heidelberg (2002)
9. Grimes, D., Wallace, R.J.: Learning to identify global bottlenecks in constraint satisfaction search. In: Proceedings of the Twentieth International Florida Artificial Intelligence Research Society (FLAIRS) Conference, pp. 592–597. AAAI Press (2007)
10. Hamadi, Y., Monfroy, E., Saubion, F.: Autonomous Search. Springer (2012)
11. Karaboga, D., Basturk, B.: On the performance of artificial bee colony (abc) algorithm. Appl. Soft Comput. 8(1), 687–697 (2008)
12. Maturana, J., Saubion, F.: A compass to guide genetic algorithms. In: Rudolph, G., Jansen, T., Lucas, S., Poloni, C., Beume, N. (eds.) PPSN 2008. LNCS, vol. 5199, pp. 256–265. Springer, Heidelberg (2008)
13. Rashedi, E., Nezamabadi-pour, H., Saryazdi, S.: Gsa: A gravitational search algorithm. Inf. Sci. 179(13), 2232–2248 (2009)
14. Wallace, R.J., Grimes, D.: Experimental studies of variable selection strategies based on constraint weights. J. Algorithms 63(1-3), 114–129 (2008)
15. Xu, Y., Stern, D., Samulowitz, H.: Learning adaptation to solve constraint satisfaction problems. In: Proceedings of the 3rd International Conference on Learning and Intelligent Optimization (LION), pp. 507–523 (2009)
16. Yang, X.-S., Deb, S.: Cuckoo search via lévy flights. In: Proceedings of World Congress on Nature & Biologically Inspired Computing (NaBIC), pp. 210–214. IEEE (2009)
17. Yang, X.-S., Deb, S., Loomes, M., Karamanoglu, M.: A framework for self-tuning optimization algorithm. Neural Computing and Applications 23(7-8), 2051–2057 (2013)
18. Yang, X.-S., He, X.: Bat algorithm: literature review and applications. IJBIC 5(3), 141–149 (2013)
19. Yang, X.-S.: A new metaheuristic bat-inspired algorithm. In: González, J.R., Pelta, D.A., Cruz, C., Terrazas, G., Krasnogor, N. (eds.) NICSO 2010. Studies in Computational Intelligence, vol. 284, pp. 65–74. Springer, Heidelberg (2010)
20. Yang, X.-S.: Bat algorithm for multi-objective optimisation. IJBIC 3(5), 267–274 (2011)

LEXMATH - A Tool for the Study of Available Lexicon in Mathematics

Pedro Salcedo Lagos[1], María del Valle[1], Ricardo Contreras Arriagada[2(✉)],
and M. Angélica Pinninghoff[2]

[1] Department of Research and Educational Informatics,
University of Concepción, Chile
[2] Department of Computer Science
University of Concepción, Chile
{psalcedo,mdelvall,rcontrer,mpinning}@udec.cl

Abstract. To have a good command of the language, not only allows us to write effectively a message, but it also helps us to understand what the other person wants to communicate. Students are in the process of learning to master the language. In particular, as they increase their level of education, they find a series of new terms (lexicon), which need to be understood using appropriate strategies, not only in the subjects of humanistic education, but also in scientific training and especially in mathematics. In the study of the lexicon, from their different points of view, the use lexical statistics gains in importance, which gives us the possibility of quantifying the lexical units of a language using different statistical indicators. Through this work the LexMath tool will be described. This software was developed for the purpose of having an online software on the Internet, for measuring the vocabulary of a particular group in a specific area and present easily the main indicators for the study of the latent lexicon of a given population, and the graphs which enable the study of semantic relations formed between them.

Keywords: Lexicon · Hypermedia · Mathematics · Graphs · Language

1 Introduction

The lexicon is the vocabulary of a particular language, field, social class, person or specific domain of knowledge. The available lexicon for a person is the set of words he knows and uses. Lexical availability refers to the field of research that consists in the recollection and analysis of the available lexicon in a specific community when dealing with a specific topic [12], [13].

Lexical availability studies began in the 50s associated to the French linguistic with the research of R. Michea [20]. The objective of the research was to make easy, for immigrants in France, the learning of the French language. To collect the available lexicon in the community, some points of interest were considered, and around these points, individuals considered in the sample should generate a list of lexical units. Subsequently, in different countries different researchers

© Springer International Publishing Switzerland 2015
J.M. Ferrández Vicente et al. (Eds.): IWINAC 2015, Part II, LNCS 9108, pp. 11–19, 2015.
DOI: 10.1007/978-3-319-18833-1_2

have developed analogous efforts, [21] in Canada, [11] in England, [19], [22] in Puerto Rico, [5] in Spain. In Chile, specifically at the University of Concepción, it is possible to highlight the works of Mena and Echeverría [12], [13].

The central question is, why it is important to study lexical availability in people? The answer is that different studies indicate that through the analysis of lexical availability it is possible to measure the level of knowledge for a specific topic. In other words, the level of knowledge is related to the lexicon a person can handle. The available lexicon let us know the vocabulary a particular society uses for referring to a determined field, allowing researchers to establish the way in which a community understand a concept, to establish weaknesses, and other attributes related to the way in which people generate relationships between words and concepts.

To establish, through the lexicon, the level of knowledge that a person has about a particular topic, it is necessary to measure the lexicon. The way in which it is accomplished is by asking to a group that belong to a community (a class in a school, for instance), to write all the words associated to a topic they can remember, during a short time interval, let us say two minutes. Then, some statistic analysis is carried out for getting the number of different words, the frequency of appearance of words, distribution or position in which words appear, and so on.

Because of the huge volume of information, at the lexicon level and from the point of view of statistics, it is reasonable to develop a computational platform for supporting research in the field, for both, information gathering and processing of the collected information. The core idea of this work is to introduce the platform operation description. The platform, specifically developed for this purpose is called LexMath.

This article is structured as follows; the first section is made up of the present introduction; the second section describes the problem we are dealing with, the third section describe the LexMath platform; the fourth section describes how the platform was used, and the final section shows the conclusions of the work.

2   The Problem

Talking about the process of concepts learning, Baddeley in his book "Memoria" [6] says that the concepts are learned in similar contexts, that most of material that a lecturer teach is not transferred to students, due to a lack of general knowledge, to the misunderstanding on the applicability of the particular object under consideration. In other words, depending on the amount of known terms or words in a close relationship with the concept the lecturer is teaching, the more easy is to acquire it.

Besides that, there are a set of terms related to the concept of lexicon, that are used by researchers:

– Usual or basic lexicon. Subset of a vocabulary used in everyday life, that in characterized for a high degree of use, appearing frequently in every lexical interaction, and that is independent of the topic under consideration.

- <u>Mental lexicon</u>. Is the capability of understanding and using lexical units that allow people to realize an interpretation or to generate units, that were not previously detected, and combine these units with other available units. This mental lexicon represents the individual knowledge as part of the vocabulary a community holds.
- <u>Lexical availability index (LAI)</u>. It is a value that indicates the availability degree of a word in an interest center. An interest center is a particular topic that manage a specific vocabulary. The LAI considers the frequency that a word is used in every position in which it is used (named) [12].

To obtain a basic lexicon it is usual to work with texts and to use frequency and dispersion as parameters; the available lexicon is obtained through surveys, by using a stimulus encouraging the individual to update his/her mental lexicon, that in words of Emmorey and Fromkin [14] is the "grammatical component that contains information about the words that are necessary to the speaker". The information of the words is obtained through phonological, morphological, syntactic and semantic information (meaning or conceptual structure [18]). According to Hall [17] words in the mental lexicon are acquired and memorized through speaking, orthography, syntactic frame and concept or word meaning.

To obtain the available lexicon in a community, different interests center are prepared (food, games, professions, etc.). Around these centers of interest centers individuals create lists of lexical units in a limited time; only two minutes for each center of interest. These kind of experiences are found in [24], that described a dictionary containing the available lexicon in Navarra (Spain), or in [1] that exhibit the available lexicon in Jaén (Spain). Similar experiences can be found in Aragón [3], Ceuta [4], Las Palmas de Gran Canaria [8], Río Piedras in Puerto Rico [9], and in the community of Valencia [10].

In mathematics, the closest reference is the work of Echeverría et al. [13] and [15]; the first apply a survey of lexical availability to students and lecturers in Mathematical Engineering career (University of Concepción, Chile). The aim of this survey was to identify which is the available lexicon for these students in interests centers associated with topics that students are expected to know (calculus, algebra, statistics, physics and geometry). In [15] the lexicon available in a set of 1557 high school students is quantified and described, showing that there is an increment in the available lexicon as student increase their formative level.

It is important to note that the mechanism employed to obtain different lexicons by using conventional approaches, writing words in a paper in less that two minutes, requires important resources: surveyors, paper, time, among others. In a research project Fondecyt [16] this method has been applied to a sample of 1557 students on seven centers of interest, taking a set of 60000 words for analysis, after correcting some orthographical errors, correcting some grammatical errors, and generating rules for transforming similar words, like diminutives [23]. By considering all the previous difficulties, it was taken the decision of developing a tool for automating this work.

3   LexMath

LexMath is created to establish an innovation in the way in which the available lexicon for a group of students is collected (see Figure 1 and 2). LexMath proposes methods for increasing the lexicon through an adaptive hypermedia; offering adequate tools for a teacher in such a way that it can be considered as a useful mechanism supporting the teaching strategies, in particular when working with Numbers, Geometry, Algebra, Data and Randomness, which are considered in the subtopic of mathematica during the normal cycle of formal learning.

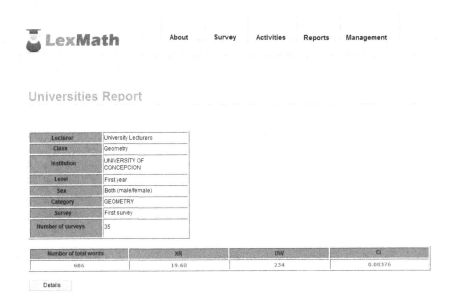

Fig. 1. Available lexical report

Figure 1 shows a screen with results from LexMath. In the top border different buttons make visible the platform functionalities. *About* describes LexMath; *Survey* allows for the students to write the words that they are thinking, related to an interest center, in a time interval of two minutes; *Activities* proposes the activities that are more suitable to the lexicon the student holds (in adaptive way); *Reports* permits to display a table for showing the lexical availability (list of words ordered according to the LAI index), related to a specific question. *Management* (a restricted option) allows to add, modify or delete information in the platform. The center of the screen displays information about a specific center of interest (in this case Geometry) and some characteristics about the group being tested. The bottom part of the screen shows results: total of words, that is the number of collected words; XR is the average of words, i.e., the number of total registered words divided into the number of individuals in the sample,

DW is the number of different words collected; and CI is the Cohesion index, that represents the similarity of words appearing in the survey.

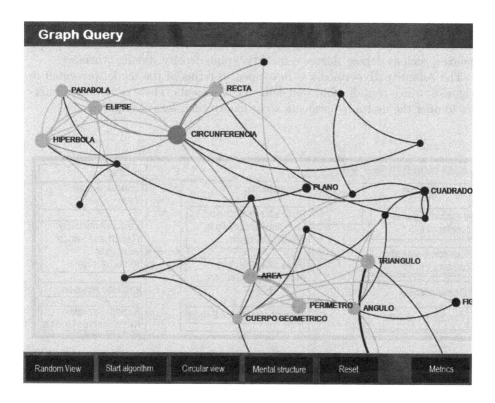

Fig. 2. Semantic structure as a graph

LexMath allows to visualize semantic relationships, taking into account the frequency of different sequences, when testing lexical availability (see Figure 2). Nodes in the graph are the words, the size of each node increases as increases the number of times that a word appears in the answers of a group. Edges indicate the sequence in which words are written. The edge's thickness increases as the words connected by these edges are written in the same order. It is possible that in a highly connected network visualization is not an easy task. Due to this issue, LexMath allows to delete relationships that are not relevant (i.e., relationships that appear with a low frequency). It is useful for researchers if they are interested in some specific relationships; in other words, they can focus only on strong relationships.

Figure 2 is a screen with a graph that shows semantic relationships related to Geometry. Additionally, the platform allows to show the graph in different schemas, as indicated through the buttons appearing on the bottom part of the image. *Random View* generates a view in which nodes are randomly distributed;

Start algorithm allows nodes distribution through an algorithm for obtaining an improved view of the graph; *Circular view* display nodes in a circumference and distributes edges inside the circumference for having a different graph visualization; *Mental structure* allows to hide less frequent nodes (lower weight nodes) displaying only the most significant clusters; *Reset* recover the initial display of the graph; and finally *Metrics* is used to show the values of the main graph indexes, such as degree, degree centrality, graph density, among others.

The Adaptive Hypermedia is developed in terms of the model presented in Figure 3, which contains four well defined components. These components interact to offer the user a hypermedia according to their lexical requirements.

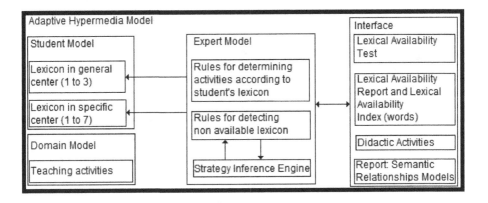

Fig. 3. Semantic relationships in LexMath

- Student model component: contains the databases that hold the lexical availability for each student, in general and specific centers of interest.
- Domain model component: contains the databases associated to different media and activities, according to a didactic proposal, for increasing the available lexicon in different centers of interest, and the ideal lexicon to be obtained by applying the same survey to the teachers.
- Expert model component: contains all the rules that are necessary to establish the general lexicon for a student and the non available lexicon. Additionally it contains the inference engine, which is in charge of determining the adequate activities in terms of the student model and the domain model.
- Interface: This is the component that allows to collect the lexicon, through personalized activities and to generate different reports, including statistical reports.

For generating the graph shown in Figure 2, we use the Script Gexfjs.js (http://brandonaaron.net), open source for academic use, supported by the Gephi technology [7].

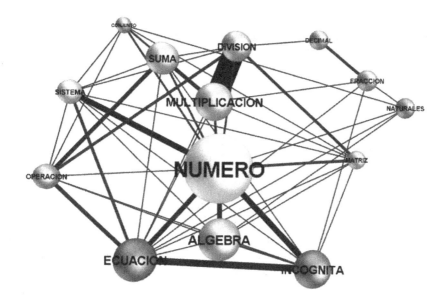

Fig. 4. Semantic relationships for the center of interest Algebra

4   LexMath: Application

During the development of LexMath [23], the use of resources was an important issue, and it was established a strategy for applying the tests in exactly 2 minutes through Internet, finding orthographical errors and grammatical errors, storing all the information in a database, including psychosocial characteristics for every student considered in the survey. Additionally, the system allows to access some tools that automatically generate statistics views, and display semantic networks for visualizing the most relevant semantic structures (lexical profiles or mental models).

In Concepción (Chile), it was conducted a research about lexical analysis supported by LexMath. The hypothesis is that exist differences in the vocabulary associated to the field of mathematics, according to socio-economical level different institutions present. The goal of this research was to analyze the lexicon and semantic networks on a sample of high school students.

Surveys were applied to 14 educational institutions, representing a wide spectrum in socio-economical level. This sample considered a total population of 1557 students, and the aim of the surveys was to collect the lexicon students presented around seven centers of interest.

In Figure 4 is shown an example of a graph generated by using LexMath, it is possible to identify the semantic relationships established when cosidering a particular center of interest: Algebra. In the set of 14 nodes that appears on the graph, it seems that the most important words, in terms of number of relationships and number of times that they were mentioned by students, are NUMERO (number), ECUACION (equation) and INCOGNITA (unknown). The first two words hold eleven of the thirteen possible links with the rest of the nodes, while the third (INCOGNITA) holds eight of these links.

5 Conclusions

The main observed result is that LexMath allows to carry, on line, most of necessary tasks for realizing the lexical analysis for a specific group, faster than the traditional procedure that uses paper, displaying in a few seconds statistics and graphics. The mentioned research hypothesis was confirmed, in the sense that exist differences in semantic networks for identical center of interest, when considering different social-economical characteristics in students. It is important to note that students with a lower social-economical level show less nodes and clusters, i.e., present visible weaknesses when dealing with specific concepts. No doubt, it is a very important issue that should be addressed in future initiatives concerned with quality and equality opportunities.

Acknowledgement. This work is partially supported by the Chilean National Fund for Scientific and Technological Development, FONDECYT, through research project number 1140457.

References

1. Lara, A.:: El léxico disponible de los estudiantes de Jaén, Jaén University (2006)
2. Aguilar, M., Navarro, J., López, J., Alcalde, C.: Pensamiento formal y resolución de problemas matemáticos. Psicothema 14(2), 382–386 (2002), Revised in psicothema.com/pdf/736.pdf
3. Arnal, M. L.(coord.): Léxico disponible de Aragón, Zaragoza: Libros Pórtico (2004)
4. Ayora, C.: Disponibilidad léxica en Ceuta: aspectos sociolinguísticos, Cádiz University (2004)
5. Azurmendi, M.J.: Elaboración de un modelo para la descripción sociolinguística del bilinguismo y su aplicación parcial a la comarca de San Sebastián. San Sebastián: Caja de Ahorros Provincial de Guipúzcoa (1983)
6. Baddeley, A., Eysenck, M., Anderson, M.: Memoria. Alianza Editorial, Madrid (2010)
7. Bastian, M., Heymann, S., Jacomy, M.: Gephi: an open source software for exploring and manipulating networks. In: International AAAI Conference on Weblogs and Social Media (2009)
8. Bellón: Léxico disponible de la provincia de Córdoba, Las Palmas de Gran Canaria, Doctoral Thesis (2003)
9. Butrón, G.: El léxico disponible: índices de disponibilidad, Río Piedras: Universidad de Puerto Rico (Doctoral Thesis, unpublished) (1987)
10. Casanova, M.: La disponibilidad léxica en la Comunidad Valenciana. In: Blas, J.L., Casanovay, M., Velando, M.(coords.): Discurso y sociedad. Contribuciones al Estudio de la lengua en Contexto Social, Castellón de la Plana: Universidad Jaume I, 737–752 (2006)
11. Dimitrijevic, N.: Lexical Availability. Groos, Heidelberg (1969)

12. Echeverría, M.: Crecimiento de la disponibilidad léxica en estudiantes chilenos de nivel básico y medio. En: Morales, H.L.(ed.) La enseñanza del español como lengua materna, pp. 61–78. Universidad de Puerto Rico, Río Piedras (1991)

13. Echeverría, M., Urzua, P., Sáez, K.: Disponibilidad Léxica Matemática. Análisis cuantitativo y cualitativo. In: Revista de Lingüística Teórica y Aplicada Concepción (Chile), II Sem. 2006, vol. 44 (2), pp. 59-76. Universidad de Concepción, Chile (2006)

14. Emmorey, K., Fromkin, V.: The Mental Lexicon. In: Newmeyer, F. (ed.) Linguistics: The Cambridge Survey, vol. III. CUP, Cambridge (1988)

15. Ferreira, A., Salcedo, P., del Valle: Estudio de disponibilidad léxica en el ámbito de las matemáticas. Estudios Filológicos, Universidad Austral de Chile, Valdivia 54, 69–84 (2014)

16. Pedro Salcedo L., Maria del Valle L., Anita Ferreira C., Nail, O.: Project: Fondecyt 1120911. Disponibilidad Léxica Matemática en Estudiantes de Enseñanza Media y su Aplicación en Hipermedios Adaptativos, Investigadores (2012, 2013), http://www.lexmath.com

17. Hall, J.H.: Morphology and Mind. Routledge, Londres (1992)

18. Jackendoff, R.: Semantic structures. MIT Press, Cambridge (1990)

19. López Morales, H.: Léxico disponible de Puerto Rico. Arco/Libros, Madrid (1999)

20. Michea, R.: Mots fréquents et mots disponibles. Un aspect nouveau de la statistique du langage. en Les Langues Modernes 47, 338–344 (1953)

21. Mackey, W.C.: Le vocabulaire disponible du franaise, 2 vols. Paris-Bruxelles-Montreal (1971)

22. Román, B.: Disponibilidad léxica de los escolares de primero, tercero y quinto grados del distrito escolar de Dorado (Master Thesis). Universidad de Puerto Rico, Recinto de Río Piedras (1985)

23. Salcedo, P., Nail, O., Arzola, C.: "Análisis de Relaciones Semánticas del Léxico Disponible en Matemáticas en un Hipermedio Adaptativo". Nuevas Ideas en Informática Educativa. In: Actas del Congreso Internacional de Informática Educativa. Santiago de Chile, vol.8, pp. 154–158 (Diciembre2012)

24. Saralegui, C., Tabernero, C. (2008). Aportaciones al proyecto Panhispánico de Léxico Disponible: Navarra. In: XXXVII Simposio Internacional de la Sociedad Española de Lingüística (SEL), Navarra (2008) ISBN: 84-8081-053-X

25. Urzúa, P., Sáez, K., Echeverría, M.: Disponibilidad léxica matemática, análisis cuantitativo y cualitativo. Revista de Lingüística Teórica y Aplicada (RLA) 44(2) (2006)

Low-Power Occupancy Sensor
for an Ambient Assisted Living System

Francisco Fernandez-Luque, Juan Zapata[✉], and Ramón Ruiz

Department of Electrónica, Tecnología de Computadoras y Proyectos
ETSIT- Escuela Técnica Superior de Ingeniería de Telecomunicación
Universidad Politécnica de Cartagena
Antiguo Cuartel de Antigones. Plaza del Hospital 1, 30202 Cartagena, Spain
{ff.luque,juan.zapata,ramon.ruiz}@upct.es
http://www.detcp.upct.es

Abstract. In this work, we introduce an Ambient Assisted Living (AAL) system that allows to infer a potential dangerous action of an elderly person living alone at home. This inference is obtained by a specific sensorisation with sensor nodes and a reasoning layer embedded in a personal computer that learns of the users behavior patterns and advices when actual one differs significantly in the normal patterns. In this type of systems, energy is a limited resource therefore sensor devices need to be properly managed to conserve energy. In this paper, a force-capacitive transducer based sensor has been proposed, implemented and tested. This sensor is based on Electro-Mechanical Films (EMFi^{TM}) transducer which is able to detect force variations in a quasi-passive way. The transducer is a capacitor with variable capacity depending on the force exerted on its surface. The characterization of the transducer conducted by us in this way is not present in the literature. This detection of force is used to trigger an active mechanism to measure the weight by means of the transducer capacity, now modelled by us. A low-power wireless sensor node prototype including this new sensor has been assembled and validated with a wide range of weights. The occupancy detection was successful and the power consumption of the node was increased at less than a 15%, which is acceptable for implementation.

Keywords: Low power sensors · EMFI · Signal transforms · Ambient Assisted Living · Ubiquitous monitoring · WSN

1 Introduction

Increasing health care costs and an ageing population are placing significant strains upon the health care system. Small pilot studies have shown that meeting seniors' needs for independence and autonomy, coupled with expanded use of home health technologies, have provided outcomes that improved care. Difficulty with reimbursement policies, governmental approval processes, and absence of efficient deployment strategies has hampered adopting non-obtrusive intelligent monitoring technologies.

© Springer International Publishing Switzerland 2015
J.M. Ferrández Vicente et al. (Eds.): IWINAC 2015, Part II, LNCS 9108, pp. 20–29, 2015.
DOI: 10.1007/978-3-319-18833-1_3

Our research project aims to implement devices which detect behaviour patterns from their users (elderly person living alone at home) and use them to take alert actions when significant variations happened [1].

The occupancy detection in bed or seat is very important in AAL systems [3]. The main aim of this kind of sensor system is to monitor the bedtime use of an old person or a patient who is living all alone in a house. In this kind of situations, the bed-monitoring sensor can act as a lifesaver. By using the historical data points from its database, the system will decide whether to trigger an alarm signal, which in turn alerts the concerned person, who can be a close relative or a caregiver, for an immediate medical help. The developed AAL system incorporated an implementation of these detectors based on the contact mats deployed under the mattress, in the case of bed, or under a cushion, in the case of the chair or couch [1]. This system has presented difficult problems to solve without a change in technology, since its functionality is dependent on various environmental factors such as the type of mattress, mattress position or weight of the user. On the other hand, in ambient intelligence systems based on wireless sensor network like ours, one of the most precious commodities is power. Sensor nodes can only operate as long as its battery maintains power. This trend makes imperative that a power-efficient occupancy sensor node needs to be designed.

In this paper, we propose to solve both problems (power consumption and functionality) by using MEMS (Micro-Electro-Mechanical Systems) in bed/chair occupancy sensor nodes. In particular, Electro-Mechanical Film [4] (EMFi^{TM}) based transducers can be used to implement foil capacitors, whose capacity is dependent on the pressure supported by the surface.

The related work to detect bed or seat occupancy and EMFi technology-based sensors are presented in Section 2. The application scenario and, specifically, the occupancy detection issue are approached in Section 3. The design and implementation, on a prototype, of the new sensor is described in Section 4. Results obtained from the different modules and the entire device are shown in Section 5. Some conclusions and possible future work lines are referenced in Section 6. Finally, acknowledgments are collected in Section 6.

2   Related Work

Pressure mats have been appropriate as a first approach to detect the bed or seat occupancy. However, their functionality is dependent on various environmental factors such as the type of mattress and bed base or weight of the user. Its operation is simple. It consists of two foils separated by a layer of foam with cylindrical holes. Both sensor foils conform a normally-open (NO) contact that closes under the weight of the user. The main disadvantage of this device is that, once made, its sensitivity is determined by the diameter of the holes. The sensitivity required is dependent on the type of mattress, bed base and weight of the user. The use of mats with different sensitivity for each individual is not feasible on a large scale. For these reasons finding adaptive systems to maintain consumption and bounded cost are desirable. Other recent work in the same field have opted for the use of sensors based on resistive transducers [5,6]. An example of this type of sensor is based on the FlexiForce® transducer from the manufacturer Tekscan [7]. This consists of two layers of substrate composed of layers of polyester. The transducer is modeled by a variable resistance depending on pressure on the sheet. Gaddam et al. from Massey University in New Zealand have developed a bed occupancy sensor that generates a digital signal indicative of such presence by implementing a configurable threshold [5,6]. Another device developed by the researchers is a smart occupancy/weight sensor which lets you know the weight distribution on the surface of the bed [6]. This type of sensor uses a FlexiForce in each of the legs of the bed and generates a weight vector which is subsequently processed.

These devices may be suitable for supporting AAL systems, except for their high energy consumption. In this paper a third alternative is proposed for the use of occupancy/weight sensor based on MEMS. Specifically, transducers based on EMFi (*Electro-Mechanical Film* [4]) can be used to implement flat capacitors whose capacity is dependent of the force applied to the surface. For some time, this type of piezoelectric film sensors have been used in various sensing tasks. EMFi sensor has been used to detect heart rate and breathing by Technical Research Centre of Finland (VTT) and some commercial companies. Tazawa et al. from Muroran Institute of Technology, Japan, researched avian embryos, and used various technologies to measure the Ballistocardiograms (BDG). They also used piezoelectric film in one of their publications [8]. Recently, Choi and Jiang, from Yamaguchi University describe a new wearable sensor probe with a couple of conductive fabric sheets material and a PVDF film material is developed [9].

Our proposal is as follows. EMFi-based devices, properly covered, may be placed under one leg of the bed to allow to obtain an approximate measure of the applied force. Periodic signals provided by the sensor node would be filtered by a specific circuit tuned by the EMFi transducer. Thus the capacity of the transducer can be measured and hence the applied force. The sensor node would be calibrated on place, before their operation. This solves the problem of lack of adaptability. The calibration routines can be done using information extracted from the context. The measure of weight would be an active process and therefore involve a certain consumption. However, a circuit proposed by the EMFi manufacturers [10] can be implemented using a passive detector of weight change. The output of the detector would be used as an interrupt to the microcontroller of the sensor node and active measurements would be made only after detection of a variation.

3   Approaching the Occupancy Detection Issue

3.1   Application Scenario

A first prototype scenario has been developed in which a user will have a home assistance system that is able to monitor his or her activity in order to detect incidents and uncommon activities. The prototype house or scenario has a bedroom, a hall, a corridor, a toilet, a kitchen, and a living room.

Current version of the developed intelligent warning device is a minicomputer-based system connected to a network of sensors deployed on walls, doors and furniture in the house of the user [11]. The sensorial abstraction layer conforms the lower level of the monitoring system. It consists of a wireless sensor network and a driver that acts as an adapter between the intelligent alert application and the network. The driver is programmed into the high-level Java language, as well as the user application, and both run on a PC running Linux. The driver implements an Application Programming Interface (API) defined for this purpose whereby the sensory abstraction layer provides service to the user application.

3.2   EMFi Transducer: Properties and Features

Electromechanical Film (EMFi) is a thin, cellular, biaxially oriented polypropylene film [12,13]. High sensitivity, light weight and relatively low cost are the main advantages on EMFi. The film has a permanent charge that changes when pressure is applied to the film. The applied pressure compresses the air voids of the thicker middle layer, which causes the charge change picked up by the surface electrode layers. This transducer can be modelled as a charge source dependant of dynamical forces applied on its surface as follows [14,4]: $\Delta Q = k_s \Delta F$, where k_s is the transducer sensibility. In Figure 1, EMFi sensors theory of operation is showed. Main properties of the commercial model from manufacturer Emfit Ltd. [10] are shown at Table 1.

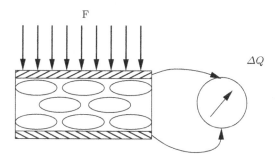

Fig. 1. EMFi sensors theory of operation

On the other hand, the transducer is a capacitor with variable capacity depending on the force exerted on its surface. The characterization of the transducer in this way is not present in the literature. Some basic tests have been conducted using a small structure that allows to apply a controlled force to the transducer. A LCR Meter module (Hameg MH8018) has been used in order to obtain a first approach. The tests were conducted on samples of a set of EMFi transducers from the S-Series provided by the EMFIT Ltd. manufacturer in its circular format of 1 cm radius.

This procedure has been applied to 4 different samples and repeated 2 to 5 times on each. The capacitive value has been registered at every step, not only

Table 1. EMFIT transducer properties

Property	Symbol	Value	Unit	Tolerance
Storage Temp.	Ts	-40 to 50	$^{\circ}C$	
Operation Temp.	Tr	-20 to 50	$^{\circ}C$	
Thickness	D	70	μm	
Sensitivity	Sq	25-250	pC/N	$\pm 5\%$
Young Module	TD	0.5	Mpa	$\pm 50\%$
Op. Range	P	N/cm^2	> 100	

just after the force incrementation but also 30 s after that. That is because a significant transitory has been detected and it must be taken into account. Two tests have been performed registering the capacitance at 30 s, 60 s, 90 s and 120 s after the force change, in order to characterize the transients.

After collecting data, polynomial fit has been used to find the coefficients of a polynomial $P(X)$ of degree $N = 2$ that best fit the data in a least-squares sense. Result is a row vector of length 3 containing the polynomial coefficients in descending powers, $P(1)X^N + P(2)X^{(N-1)} + ... + P(N)X + P(N+1)$. The error of estimations using this polynomial has been determined by means of the standard deviation from every data over the fitted curve. Results of fitting show that error is low enough to use the capacity as estimation of weight.

4 Design and Implementation of EMFi-Based Occupancy Sensor

The low-power sensor device consists of two functional blocks: a quasi-passive force change detector and an active occupancy sensor, by means of weight measuring.

The force change detector design is based on the schematics proposed by the EMFIT manufacturer for dynamic forces measuring [15]. This schematics use the sensory model in which the transducer behaves as a charge source which is dependent on the force applied between it's sheets. The circuit shown upper right in Figure 2 is a charge amplifier which transforms the signal from the transducer into a voltage signal. The output of this circuit is proportional to the variation of the pressure supported by the transducer. This signal is then threshold filtered to generate interrupts to the microcontroller, thereby acting as a detector of force changes, with low power consumption.

The second block, the occupancy sensor, employs the EMFi transducer with its model as a force dependant capacitor. The capacity of the transducer is measured by means of a capacitive voltage divider. As shown in Figure 2, when a capacitance measure is needed then the voltage divider is feed with an AC signal. The resulting output is a signal with the same frequency but amplitude-modulated by the force on the transducer. After its impedance conditioning, this signal can be directly sampled and processed by a microcontroller to get a capacitance estimation; which is also an estimation of the weight and the occupancy

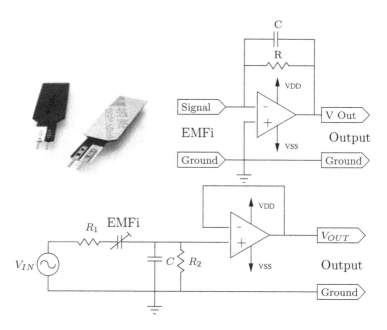

Fig. 2. On the left, EMFIT S-Series. On the right, charge amplifier for EMFIT. Below, voltage divider to get the EMFi transducer capacity.

probability. On the other hand, the signal can be rectified and conditioned by hardware in order to obtain an analog signal indicative of weight supported by the transducer. A variant of the schematics is also proposed in which, capacitor C and EMFi transducer swap positions.

The application programmed on the microcontroller (firmware) manages the external interrupts from the force change detector, feeds the capacitive voltage divider with a square signal of 100 kHz and, finally, samples and processes the output. The sequence is as follows: (1) When a force change occurs, the device triggers a weight measuring over the voltage divider; (2) it compares the result to the previously calculated threshold and fires an event to communicate the bed/seat occupancy; (3) the event is transmitted via radio to the PC through the base station. Additionally, a Light-Emitting Diode (LED) is lighted when occupancy detection is positive.

The output signal of the transducer is not static so the measurement is calculated from 150 samples from the ADC. Several algorithms have been tested to get an stable and significant value. Among them, the following one has been selected: First, ten values are used to get a starting average value. After that, every single sample (m_n) is added to the current average value, M_n, where

$$M_n = M_{n-1}(1 - \frac{c}{n}) + m_n\frac{c}{n}$$

and n is the sample number and c is a weight coefficient which is established by comparing the current sample with the previous average as follows:

$$c = \begin{cases} 1 & \text{if } \dfrac{m_n}{M_{n-1}} \in [0.9, 1.5], \\ \dfrac{1}{3} & \text{if } \dfrac{m_n}{M_{n-1}} \in [0.5, 0.9) \cup (1.5, 2], \\ 0 & \text{otherwise} \end{cases}$$

The output of the voltage divider is split into three regions or states by thresholding as shown in Figure 3. The first region contains the values associated with the idle state, the second region contains the values associated with the busy state and the third region, located between the first two, contains the values too close to the average which are not determinative and therefore considered invalid. The regions are separated by two thresholds: unoccupancy and occupancy threshold. An array with five values for each of the two states, occupied and unoccupied, is collected to determine the thresholds. Then the mean of each vector is obtained. Thresholds are defined as the points between the two averages which are far from them 1/3 of the distance between them. When the thresholds for the regions have been obtained, it is possible to verify the occupancy of the seat. This occupancy verification process begins by detecting a change in weight. This variation in weight may be due to a change of occupation or movements of the person occupying the seat. For occupancy verification an array of five values of the voltage divider is acquired. Each of these values are obtained from the processing of 150 samples of the ADC.

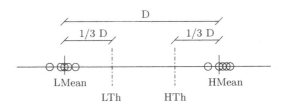

Fig. 3. Output values of weight are divided in three regions by thresholding

5   Results

The EMFi transducer had previously been modeled [16]. Mainly, EMFi has been modeled as a charge source dependent on variations in the force applied on the surface. There is no model that allows to know how the capacity varies with the application of different static forces. Several studies have been performed on 4 samples of a set of EMFi transducers from the S-Series provided by the EMFIT Ltd. manufacturer. At the same time, capacity was measured using a digital LCR (Inductance (L), Capacitance (C), and Resistance (R)) meter.

Next, the collected data have been used in order to obtain a polynomial by least squares adjustment. From results we can conclude that the second-order polynomial fit by least squares approximation is accurate enough to justify its use as a model to get an estimation of the weight from the measurement of capacity. The design of the new occupancy sensor consists of two different modules, the first one detects force changes while the second one effectuates weight readings. The prototype of force change detector has been assembled according to the schematics shown in Figure 2 with the following values: R=1 MΩ and C=220 pF. This value for R is recommended by the manufacturer of the transducer, while the value of C is required for a time constant of $\tau = RC = 220 \mu s$. This time constant is sufficient for the microcontroller detects the signal.

When force is decreased (standing up action), signal goes to V_{cc} and comes back to rest in the same way. This signal is digitalized by thresholds to get two signals that actuate as external interrupts to the microcontroller, as shown in Figure 4. Complex movements on the bed or couch, with different intensities, could lead to partial and unreliable detection, as the oscillations observed in Figure 4. This justifies the need of weight measurement on the system at rest to ensure the occupancy. The analog signal is digitized and conditioned to be a source of interrupt to the microcontroller using a CMOS circuit. In Figure 4 is shown the digital output of the detector. Tests have been effectuated on the occupancy sensor prototype for several values for the parameters pointed at Section 4. Resistor R_1 has been adjusted to suppress the over-shot (OS) keeping the rising time (T_r) as low as possible. These values were adjusted to: $R_1 = 150 \Omega$, $OS = 0\%$ y $T_r = 70$ ns.

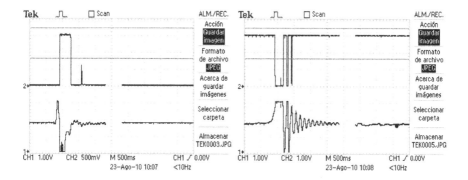

Fig. 4. Output signal of force change detector (digital stage). On the left, force increment detection (action of sitting); on the right, force decrement detection (action of standing up).

To estimate the average consumption of the device it is necessary to estimate an approximate read rate: (1) In a vacant bed or chair, it is considered that changes in weight were not detected and therefore do not produce readings. (2) In a conservative way, it is considered that an occupied bed or chair presents a mean period between detections of variation of $10\,s$. It is possible to implement a state machine to establish a maximum rate of measurement, assuming the consequent delay in the generation of events. (3) A person may occupy a bed or chair for an average of 8 hours per day. With these three assumptions, the duty cycle for the occupancy sensor is $DC = \dfrac{8\,h \cdot 370\,ms}{24\,h \cdot 10\,s} = 0.012\%$ and the average consumption generated by it would be $I = DC \times I_{Check} = 120\,\mu A$, around 5% of the average consumption of the node.

6 Discussion and Conclusions

The goal of this work was to obtain an occupancy sensor for bed or couch that could improve the contact mats previously employed in the sensor network for a set of Ambient Assisted Living (AAL). This interest is due to the lack of adaptability that mats presented to various factors such as hardness and type of mattress and bed base, in the case of the bed, and the hardness of the structure in the case of the couch, as well as the weight of the person in the one case and the other. Other sensors based on resistive transducers do not meet the low power constraints. The proposed sensor, based on the EMFi transducer, is independent of these factors except the user's weight, however, the possibility of calibration makes this aspect does not become a problem. A new model has been determined for the EMFi transducer to use it to measure static weight, while the previous model allowed to measure force variations. This new model allowed to design a reliable occupancy sensor for AAL applications. Finally, future direction would point as redesigning the sensor so that it can be a stand-alone device that does not require a microcontroller to operate. An implementation of this could even be transferred to the field of microelectronic design for manufacturing as an integrated circuit, which would be translated into an especially significant reduction in cost and energy consumption.

Acknowledgement. This work was supported by the Spanish Ministry of Ciencia e Innovación (MICINN) under grant TIN2009-14372-C0302 and for the Fundación Séneca under grant 15303/PI/10.

References

1. Fernández-Luque, F., Zapata, J., Ruiz, R., Iborra, E.: A wireless sensor network for assisted living at home of elderly people. In: Mira, J., Ferrández, J.M., Álvarez, J.R., de la Paz, F., Toledo, F.J. (eds.) IWINAC 2009, Part II. LNCS, vol. 5602, pp. 65–74. Springer, Heidelberg (2009)
2. Blaya, J.A.B., Palma, J., Villa, A., Perez, D., Iborra, E.: Ontology based approach to the detection of domestic problems for independent senior people. In: Mira, J., Ferrández, J.M., Álvarez, J.R., de la Paz, F., Toledo, F.J. (eds.) IWINAC 2009, Part II. LNCS, vol. 5602, pp. 55–64. Springer, Heidelberg (2009)
3. Gaddam, A., Mukhopadhyay, S., Gupta, G.: Necessity of a bed-sensor in a smart digital home to care for elder-people. In: 2008 IEEE Sensors, pp. 1340–1343 (2008), doi:10.1109/ICSENS.2008.4716693

4. Lekkala, J., Paajanen, M.: Emfi-new electret material for sensors and actuators. In: 1999 Proceedings of the 10th International Symposium on Electrets, ISE 10, pp. 743–746 (1999), doi:10.1109/ISE.1999.832151

5. Gaddam, A., Mukhopadhyay, S., Gupta, G.: Development of a bed sensor for an integrated digital home monitoring system. In: IEEE International Workshop on Medical Measurements and Applications, MeMeA 2008, pp. 33–38 (2008), doi:10.1109/MEMEA.2008.4542993

6. Gaddam, A., Mukhopadhyay, S., Gupta, G.: Intelligent bed sensor system: Design, expermentation and results. In: 2010 IEEE Sensors Applications Symposium (SAS), pp. 220–225 (2010), doi:10.1109/SAS.2010.5439390

7. Flexiforce® Sensors - Specifications and Features, Tech. rep., Tescan, Inc.

8. Tazawa, H., Hashimoto, Y., Takami, M., Yufu, Y., Whittow, G.C.: Simple, noninvasive system for measuring the heart rate of avian embryos and hatchlings by means of a piezoelectric film. Med. Biol. Eng. Comput. 31(2), 129–134 (1993), http://www.biomedsearch.com/nih/Simple-noninvasive-system-measuring-heart/8331992.html

9. Choi, S., Jiang, Z.: A wearable cardiorespiratory sensor system for analyzing the sleep condition. Expert Syst. Appl. 35, 317–329 (2008), http://portal.acm.org/citation.cfm?id=1379458.1379570, doi:10.1016/j.eswa.2007.06.014

10. Thin film ferro-electret sensors, http://www.emfit.com/en/sensors/products_sensors/

11. Ruiz, R., Fernández-Luque, F.J., Zapata, J.: Wireless sensor network for ambient assisted living. In: Merrett, G.V., Tan, Y.K. (eds.) Wireless Sensor Networks: Application-Centric Design, InTech, pp. 127–146 (2010)

12. Paajanen, M.: Electromechanical film (emfi) — a new multipurpose electret material. Sensors and Actuators A: Physical 84(1-2), 95–102 (2000), http://linkinghub.elsevier.com/retrieve/pii/S0924424799002691

13. Paajanen, M.: Modelling the electromechanical film (emfi). Journal of Electrostatics 48(3-4), 193–204 (2000), http://linkinghub.elsevier.com/retrieve/pii/S0304388699000650

14. Paajanen, M., Lekkala, J., Valimaki, H.: Electromechanical modeling and properties of the electret film emfi. IEEE Transactions on Dielectrics and Electrical Insulation 8(4), 629–636 (2001), doi:10.1109/94.946715

15. Preamplifiers for EMFIT sensors, Tech. rep., EMFIT, Ltd. (2004)

16. Tuncer, E., Wegener, M., Gerhard-Multhaupt, R.: Modeling electro-mechanical properties of layered electrets: application of the finite-element method. Journal of Electrostatics 63(1), 21–35 (2005), http://www.sciencedirect.com/science/article/pii/S0304388604001329, doi:10.1016/j.elstat.2004.06.002

17. Sixsmith, A., Johnson, N.: A smart sensor to detect the falls of the elderly. IEEE Pervasive Comput. 3(2), 42–47 (2004), doi:10.1109/MPRV.2004.1316817

Evolution of Synaptic Delay Based Neural Controllers for Implementing Central Pattern Generators in Hexapod Robotic Structures

José Santos[✉] and Pablo Fernández

Computer Science Department, University of A Coruña, Spain
jose.santos@udc.es

Abstract. We used synaptic delay based neural networks for implementing Central Pattern Generators (CPGs) for locomotion behaviors in hexapod robotic structures. These networks incorporate synaptic delays in their connections which allow greater time reasoning capabilities in the neural controllers, and additionally we incorporated the concept of the center-crossing condition in such networks to facilitate obtaining oscillation patterns for the robotic control. We compared the results against continuous time recurrent neural networks, one of the neural models most used as CPG, when proprioceptive information is used to provide fault tolerance for the required behavior.

1 Introduction

Central Pattern Generators (CPGs) are neural circuits in peripheral nervous systems like the spine that can be activated to produce rhythmic motor patterns such as walking, breathing, flying or swimming. CPGs are able to operate in the absence of sensory or descending inputs that carry specific timing information. Our aim is focused on the control of legged robotic structures by CPGs. Legged locomotion is characterized by cyclic activity of the limbs, and the defining feature of CPGs is a high degree of recurrence, which greatly biases the dynamics of the system toward cyclic activation patterns [10].

CPGs were implemented using models such as vector maps, systems of coupled oscillators and connectionist models [6]. In the neural computing community, Beer [1] introduced the model of Continuous Time Recurrent Neural Networks (CTRNN), one of the most used to represent CPGs. Later, Mathayomchan and Beer introduced the concept of center-crossing in CTRNNs [8] to easily obtain neural oscillators. The basic idea behind center-crossing CTRNNs is to force the neurons to work, most of the time, in the most sensitive area of their transfer functions, which facilitates that the NN oscillates for producing the required rhythmic behavior.

In previous works [2][12] we used center-crossing CTRNNs as neural controllers in biped structures, and additionally we defined "adaptive center-crossing CTRNNs" to improve the ability of these recurrent neural networks to produce rhythmic activation behaviors. The adaptive process, which modifies the

© Springer International Publishing Switzerland 2015
J.M. Ferrández Vicente et al. (Eds.): IWINAC 2015, Part II, LNCS 9108, pp. 30–40, 2015.
DOI: 10.1007/978-3-319-18833-1_4

CTRNN parameters in run-time to reach the center-crossing condition, facilitates obtaining the networks that act as CPGs to control biped structures. Additionally, we combined the center-crossing condition in ANNs that incorporate synaptic delays in their connections [3][13], testing the integration of the center-crossing condition with the time reasoning capabilities of such networks, with the aim of obtaining the required neural oscillators by means of simulated evolution [11]. The results showed that these center-crossing recurrent synaptic delay based neural networks outperformed the results of CTRNNs in biped structures [11].

In the present work we use Differential Evolution as evolutionary method to obtain the neural controllers in hexapod structures. We used CTRNNs and synaptic delay based NNs (SDBNNs), with the integration of the center-crossing condition in both models, testing the fault tolerance of these types of neural controllers in different situations that legged robotic structures should tackle.

2   Methods

2.1   Robotic Structure

We used a hexapod robotic structure implemented with the physics simulator Open Dynamics Engine (ODE) [14]. The articulated structure consists of rigid bodies connected through joints.

The hexapod has a central body part, which links with each leg by means of a joint with two degrees of freedom (in x and z axis). These joints were simu-

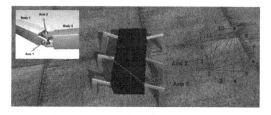

Fig. 1. Control of the hexapod structure by a CTRNN. The values of the NN nodes control the different joints of the simulated hexapod in the ODE environment. The inset shows the hinge joint used at each of the contacts between the central body and the legs (2 DOF).

lated as torsional springs. On the contrary, the joint connecting the two parts of each leg is fixed. The outputs of the neural network models were scaled to provide a velocity that can reach the angle limits of the joints. Figure 1 indicates how the joints of the implemented hexapod structure are controlled by a neural controller, a fully interconnected CTRNN in the Figure.

The mass of each body part was proportional to its volume, the gravity was fixed to $-9.81 m/s^2$, a time step of $0.01s$ was used in the ODE simulation for each iteration in the environment and the maximum ground friction was used in the simulator to avoid the possibility of the legs sleeping.

2.2   Neural Controller Models

Continuous Time Recurrent Neural Networks. Beer [1] introduced the model of Continuous Time Recurrent Neural Networks (CTRNN), one of the most used to represent CPGs. In a CTRNN, the state of a single neuron i is computed by the following equation:

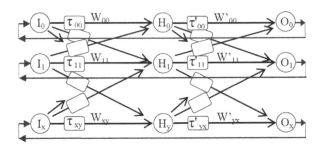

Fig. 2. Synaptic delay based neural network with feedback connections

$$\dot{y}_i = f(y_i, ..., y_n) = \frac{1}{\tau_{ci}}(-y_i + \sum_{j=1}^{n} w_{ji}\sigma(y_j + \theta_j) + I_i) \qquad i = 1, 2, ..., N \quad (1)$$

where y_i is the state of neuron i, τ_{ci} is a time constant which defines the rate of decay of the state, w_{ji} is the weight of the incoming connection from neuron j, σ is the sigmoid activation function, θ_i is the bias term or firing threshold of the node and I_i is an external input. Totally connected CTRNNs will be used, where each node follows this equation.

Center-crossing Condition. Mathayomchan and Beer introduced the concept of center-crossing in CTRNNs [8] for easily obtaining neural oscillators. In a center-crossing CTRNN "the null surfaces of all neurons intersect at their exact centers of symmetry" [8]. The null surface of a neuron is where the sum of the neuron bias and all the synaptic inputs is 0. This ensures that each neuron's activation function is centered over the range of net inputs that it receives.

Using a sigmoid activation function in the NN nodes, a neuron has a firing frequency of 0.5 at its null surface ($\sigma(0) = 0.5$), that is, center-crossing networks have neurons that on average have firing frequencies around this value. Hence, the center-crossing condition occurs when the neuron biases of all neurons are set to the negative of the sum of the input weights divided by 2:

$$\theta_i = \frac{-\sum_{j=1}^{N} w_{ji}}{2} \qquad (2)$$

This means that the bias exactly counteracts the sum of all the synaptic inputs when the connected neurons have a firing frequency of 0.5. Thus, the nodes of such type of networks should have an average firing value of 0.5, which implies that the neurons are in states of maximum sensitivity most of the time.

The importance of the network process is that now small changes in synaptic inputs around the null surface can lead to a very different neuron firing frequency. Outside this range, the change of a net input has only sparse effect on the firing frequency. Bearing in mind that the objective is to obtain neural oscillators that act as CPGs, the richest dynamics should be found in the neighborhood of the center-crossing networks in the search space of parameters. Hence, as Mathayomchan and Beer indicate [8], one would expect that an evolutionary algorithm would benefit from focusing its search there.

Synaptic Delay Based Neural Networks. In a Synaptic Delay Based Neural Network (SDBNN), in addition to the weights of the synaptic connections, their length is modeled through a parameter that indicates the delay a discrete event suffers when going from the origin neuron to the target neuron.

The network consists of several layers of neurons connected as a multiple layer perceptron (MLP). The only difference with a traditional MLP is that the synapses contain a delay term in addition to the classical weight term. That is, now the synaptic connections between neurons are described by a pair of values, w and τ, where w is the weight describing the ability of the synapses to transmit information and τ is a delay, which in a certain sense provides an indication of the length of the synapse. Figure 1 illustrates the neural structure, adding recurrences from the outputs to the inputs, in order to use the structure as Central Pattern Generator.

Each neuron in a given layer can choose which of the previous outputs of the neurons in the previous layer it wishes to input in a given instant of time. The process is summarized in the following equation, which defines the output of a neuron k in an instant of time t:

$$O_{kt} = \sigma\left(\sum_{i=0}^{N}\sum_{j=0}^{t} \delta_{j(t-\tau_{ik})} w_{ik} h_{ij}\right) \tag{3}$$

where σ is the transfer function of the neuron (sigmoid function), h_{ij} is the output of neuron i of the previous layer in instant j and w_{ik} is the weight of the synaptic connection between neuron i and neuron k. δ represents the *Kronecker Delta*. The first summation is over all the neurons that reach neuron k (those of the previous layer) and the second one is over all the instants of time considered. The result of this function is the summation of the neuron outputs of the previous layer in times $t - \tau_{ik}$ (where τ_{ik} is the delay in the corresponding neural connection) weighed by the corresponding weight values.

The neural structure was used in temporal pattern recognition, prediction of temporal series or as a model of dynamic reconstruction of a time series [3][4][13]. Here the network structure is used as neural controller, and the parameters (weights and synaptic connections delays) are evolved. To obtain neural oscillators that can act as CPGs (without the use of proprioceptive information), a recurrent structure was used, with the feedbacks shown in Figure 2 from outputs to inputs.

When the center-crossing condition is introduced in these synaptic delay based NNs, each node of the NN incorporates a bias term defined by Equation 2, facilitating the oscillation behavior in the activation of the nodes. Moreover, the output nodes will incorporate a time constant τ_c that defines a decay of the activation of the nodes through time (Equation 1) and which facilitates obtaining different activation patterns.

2.3   Differential Evolution

We used Differential Evolution (DE) for automatically obtaining the neural controllers that act as CPGs. DE [9] is a population-based search method. DE creates new candidate solutions by combining existing ones according to a simple formula of vector crossover and mutation. Then the algorithm keeps whichever candidate solution has the best score or fitness on the optimization problem at hand. The central idea of the algorithm is the use of difference vectors for generating perturbations in a population of vectors. This algorithm is specially suited for optimization problems where possible solutions are defined by a real-valued vector. The basic DE algorithm is summarized in the pseudo-code of Algorithm 2.1 (which tries to minimize the fitness).

Algorithm 2.1: DIFFERENTIAL EVOLUTION($Population$)

```
for each Individual ∈ Population
  do {Individual ← INITIALIZERANDOMPOSITIONS()
repeat
  for each Individual  x ∈ Population
      ⎧ x₁, x₂, x₃ ← GETRANDOMINDIVIDUAL(Population)
      ⎪ // must be distinct from each other and x
      ⎪ R ← GETRANDOM(1, n)  // the highest possible
      ⎪ // value n is the dimensionality of the problem to be optimized
      ⎪ for each i ∈ 1 : n
  do ⎨ // Compute individual's potentially new position y = [y₁, ..., yₙ]
      ⎪        ⎧ rᵢ ← GETRANDOM(0, 1)   // uniformly in open range (0,1)
      ⎪   do   ⎨ if ((i = R) || (rᵢ < CR))
      ⎪        ⎪      yᵢ = x₁ᵢ + F(x₂ᵢ − x₃ᵢ)
      ⎪        ⎩   else yᵢ = xᵢ
      ⎩ if (f(y) ≤ f(x))   x = y  // replace x with y in Population
until TERMINATIONCRITERION()
return (GETLOWESTFITNESS(Population))  // return candidate solution
```

One of the reasons why Differential Evolution is an interesting method in many optimization or search problems is the reduced number of parameters that are needed to define its implementation. The parameters are F or differential weight and CR or crossover probability. The weight factor F (usually in $[0, 2]$) is applied over the vector resulting from the difference between pairs of vectors (x_2 and x_3). CR is the probability of crossing over a given vector of the population (x) and a mutant vector created from the weighted difference of two vectors ($x_1 + F(x_2 − x_3)$). The index R guarantees that at least one of the parameters will be changed in such generation of the candidate or trial solution (y). Finally, the fitness of the trial vector ($f(y)$) and the target vector ($f(x)$) are compared to determine which one survives for the next generation, so the fitness of the best solution of the population is improved or remains the same through generations.

The usual variants of DE choose the base vector x_1 randomly (labeled as variant $DE/rand/1/bin$, bin represents the "binomial" crossover described in

Algorithm 2.1) or as the individual with the best fitness found up to the moment (x_{best}) (variant $DE/best/1/bin$). To avoid the high selective pressure of the latter, we used a tournament to pick the vector x_1, which also allows us to easily establish the selective pressure by means of the tournament size.

As Feoktistov [5] indicates, the fundamental idea of the algorithm is to adapt the step length ($F(x_2 - x_3)$) intrinsically along the evolutionary process. At the beginning of generations the step length is large, because individuals are far away from each other. As the evolution goes on, the population converges and the step length becomes smaller and smaller, providing this way an automatic balance in the search.

The individuals of the genetic population encode the parameters that define the corresponding neural controller (connection weights, temporal delays, time constants) and the fitness is defined as the distance traveled by the robot in a straight line, as detailed next.

3   Results

3.1   Experimental Setup

A population of 100 individuals was used in the different evolutionary runs. We used DE parameters $F = 0.35$ and $CR = 0.9$ (in the intervals suggested in [9]), while we fixed a tournament size of 8% of the population to select the base individual x_1 (Algorithm 2.1), which imposes a low selective pressure. The fitness was defined, as in most of previous works [6][8][10], as the distance traveled in a straight line by the legged structure in a given time (8 seconds in the experiments), penalizing it when the robot does not follow a straight trajectory.

In the case of fully interconnected center-crossing CTRNNs, all the neurons were motor neurons to control each of the joint angles in the hexapod structure (Fig. 1). We used a neuron to control each of the two degrees of freedom in the 6 joints, so the genotypes have 156 parameters to evolve (144 weights and 12 time constants in the 12 NN nodes). Using the second possibility of center-crossing recurrent synaptic delay based NNs, with a typical configuration using an input and an output layer of 12 nodes, and two hidden layers with 4 nodes, the number of encoded parameters is quite larger: 236 (112 weights, 112 synaptic delays, 12 time constants of the 12 output nodes). In all cases the bias terms were defined by the center-crossing condition (Equation 2).

In all NN models, the connection weights were constrained to lie in the range [-16, 16] (as in [8] and [10]) and the time constants (τ_c) were constrained in the range [0.5, 5], as in [10]. Nevertheless, this interval depends on the required behavior since the time constants greatly determine the oscillation period. For the synaptic delay based NNs, the synaptic delays (τ) were constrained in the range [0, 25]. This means that each node can use, for each connection with a previous node of the previous layer, the current information of the input node (a synaptic delay with a value of 0) or the information of the node but 25 time instants in the past. As there are two hidden layers (three connection sets), an output node can "reason" with the 75 past values of the (processed) input

information. In all cases, all the defining parameters were encoded in the range
[-1, 1] and decoded in such intervals. Finally, to avoid the transient perturbations
at the beginning of the temporal evolution of the recurrent neural networks, each
network controller was iterated a given number of steps (250) before taking the
control of the hexapod joints.

3.2   Neural Controllers Comparison

In a first test the
neural controllers do
not use proprioceptive
information to imple-
ment the GPGs. We
used the fitness pro-
vided by a CTRNN as
baseline in the com-
parison, since it is one
of the neural models
most used to imple-
ment CPGs. Figure 3
shows a comparison of
fitness evolution when
DE is used for au-
tomatically obtaining
a CTRNN controller
and a SDBNN con-
troller. The hexapod

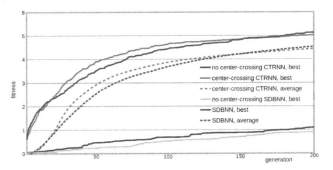

Fig. 3. Fitness evolution through generations when Differential Evolution is used to optimize the parameters of a
CTRNN controller and a SDBNN controller for the hexapod structure. The hexapod is tested on a stairs surface. All
graphs are an average of 15 different runs of the evolutionary
algorithm.

has to climb up stairs, a more difficult task for both neural controllers than
learning a rhythmic pattern for a flat surface. In both neural models we used a
short interval [0.5, 2.5] for the time constants (τ_c) in order to obtain temporal
patterns with short periods which provide shorter and faster movements. Both
neural controllers produce the required temporal pattern so that the robot can
climb up the stairs (and in simpler cases such as the robot on a flat surface).
The reason is in the stable behavior provided by the hexapod structure, so fitness evolutions are quite similar. On the contrary, as tested in previous works
[11], the SBDNN clearly outperforms the CTRNN in biped structures which
present a more unstable behavior. Figure 3 also shows the advantage of using
the center-crossing condition, since without its use the neural controllers cannot
obtain stable oscillation patterns to perform the required behavior. In this last
case, the bias terms are evolved as the other NN parameters.

Table 1 summarizes the basic statistical information regarding the evolution
runs (15 different runs of DE using the two neural controllers). We omitted the
cases when the center-crossing condition is not used. It is clear that the SDBNN
controllers provide more stable behaviors, since the standard deviations in the
best individuals and in the average quality in the different runs is lower than the
half of the CTRNN case.

Table 1. Final fitness values corresponding to evolutions in Figure 3

			Final value	Standard deviation
Hexapod on stairs (Fig. 3)	CTRNN	Best fitness	5.0222	0.8259
		Average	4.4383	0.7923
	SDBNN	Best fitness	5.1301	0.3481
		Average	4.5313	0.3273

3.3 Fault Tolerance of Synaptic Delay Based Neural Controllers

Our main aim is to test the fault tolerance capability of the neural controllers to face malfunctioning in the actuators. Figure 4 shows an example, with the trajectory followed by the hexapod (best evolved controller) when one of the front-most legs has a malfunctioning after $t = 8$ seconds from the beginning. After that time instant the leg ignores the velocity imposed by the corresponding neural controller. The fitness function was the distance traveled in a straight line (in 35 seconds), penalizing it when the robot does not follow a straight trajectory.

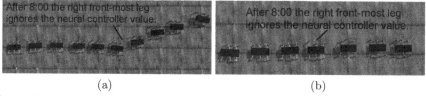

(a)   (b)

Fig. 4. Trajectory of the hexapod when, at $t = 8$, the right front leg ignores the corresponding neural value. (a) Trajectory using the best evolved CTRNN controller; (b) Trajectory using the best evolved synaptic delay based neural network as controller.

In this example, the neural controllers receive as (proprioceptive) inputs the 12 angles of the articulations of the legs. In the case of the CTRNN, as Equation 1 indicates. In the case of the SDBNN, as additional inputs in the input layer (Figure 2, without extra feedback connections to these new inputs). Figure 4 (left part) shows that the best evolved CTRNN was not able to use the additional information provided by the joint angles to correct the trajectory. Figure 5 (upper part) shows the oscillation pattern provided by one of the CTRNN nodes, a periodic pattern that cannot face the new situation and which generates the curved trajectory. The videos of the trajectories in the examples can be viewed in [7].

Note that the angle values of the malfunctioning leg introduce noise to the NN controller, as seen in the bottom part of Figure 5. Nevertheless, in the case of the best SDBNN controller, after $t = 8$, since the amplitude of the values in the angles of the malfunctioning leg are different, the neural controller can reason with such temporal values to adapt or correct the movements in the other legs so that the robot follows the straight trajectory. This can be seen in the right part of Figure 4 as well as in the central part of Figure 5, the latter displaying the temporal pattern of activation in one of the SDBNN output nodes, which shows how the period is slightly changed to face the new situation detected in the temporal inputs.

In a second example we impose that one of the central legs is fixed and without touching the ground. Thus, there is again an imbalance in the number of supporting legs in both sides of the hexapod, so the neural controller must automatically reconfigure the activation pattern so that the robot travels using five legs. Figure 6 (left part) shows the trajectory followed by the hexapod when the right central leg is fixed and raised, between $t = 5$ and $t = 9$ seconds from the beginning. The fitness is defined as in the previous example. The right part shows the oscillation pattern provided by the best evolved SDBNN, which corresponds to the one of the left central leg. The evolved SDBNN controller can correct again the situation to obtain a straight trajectory in the robot (The CTRNN is not able to correct the situation, as in previous example). The oscillation pattern shows how the activation in the nodes changes in such interval to obtain another

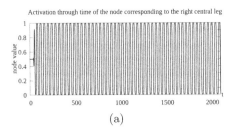

(a)

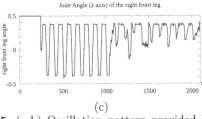

(b)

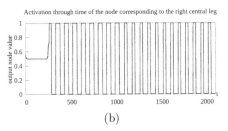

(c)

Fig. 5. (a-b) Oscillation pattern provided by the best evolved neural controllers when, at $t = 8$, the right front leg ignores the corresponding neural value. The pattern corresponds to the NN node that controls a central leg. (a) Oscillation pattern provided by the CTRNN, (b) Oscillation pattern provided by the synaptic delay based neural network, which shows the change of the temporal activation pattern at $t = 8$ (iteration 1050). (c) Joint angle (in z axis) corresponding to the malfunctioning front leg.

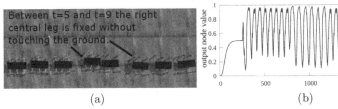

(a)

(b)

Fig. 6. Between $t = 5$ and $t = 9$ the right central leg is fixed and raised. (a) Trajectory of the hexapod (best evolved SDBNN), (b) Oscillation pattern provided by the SDBNN in the node that controls the left central leg, which shows the change of the temporal activation pattern at $t = 5$ (iteration 750) and $t = 9$ (iteration 1150).

rhythmic pattern in the interval. Once the leg recovers its normal functioning ($t = 9$), the network recuperates again the first oscillation pattern in few iterations.

4   Conclusions

We tested the use of synaptic delay based neural networks for implementing Central Pattern Generators for locomotion behaviors in hexapod robotic structures. The importance of the neural structure is that now the incorporation of synaptic delays allows the network to obtain a model of the temporal processes required. In this sense, it is used as a model for defining the required time series (oscillation behavior). Moreover, the incorporation of the center-crossing condition in such networks allows to easily obtain the necessary oscillation temporal patterns. The results show that SDBNNs, on the contrary to CTRNNs, can generate adaptive patterns of temporal activation when proprioceptive information is used to correct the oscillation patterns of the CPG, which endows the SDBNN controllers with fault tolerance for the locomotion behaviors.

Acknowledgements. This work was funded by the Ministry of Economy and Competitiveness of Spain (project TIN2013-40981-R).

References

1. Beer, R.: On the dynamics of small continuous time recurrent neural networks. Adaptive Behaviour 3(4), 469–509 (1995)
2. Campo, A., Santos, J.: Evolution of adaptive center-crossing continuous time recurrent neural networks for biped robot control. In: Proceedings European Symposium on Artificial Neural Networks (ESANN 2010), pp. 535–540 (2010)
3. Duro, R.J., Santos, J.: Discrete time backpropagation for training synaptic delay based artificial neural networks. IEEE Transactions on Neural Networks 10(4), 779–789 (1999)
4. Duro, R.J., Santos, J.: Modeling temporal series through synaptic delay based neural networks. Neural Computing and Applications 11, 224–237 (2003)
5. Feoktistov, V.: Differential Evolution: In Search of Solutions. Springer, NY (2006)
6. Ijspeert, A.J.: Central pattern generators for locomotion control in animals and robots: A review. Neural Networks 21(4), 642–653 (2008)
7. Hexapod videos: http://www.dc.fi.udc.es/ai/~santos/Iwinac2015/Iwinac2015.html
8. Mathayomchan, B., Beer, R.: Center-crossing recurrent neural networks for the evolution of rhythmic behavior. Neural Computation 14, 2043–2051 (2002)
9. Price, K.V., Storn, R.M., Lampinen, J.A.: Differential Evolution. A Practical Approach to Global Optimization. Natural Comp. Series. Springer (2005)
10. Reil, T., Husbands, P.: Evolution of central pattern generators for bipedal walking in a real-time physics environment. IEEE Transactions on Evolutionary Computation 6(2), 159–168 (2002)
11. Santos, J.: Evolved center-crossing recurrent synaptic delay based neural networks for biped locomotion control. In: Proceedings IEEE Congress on Evolutionary Computation (IEEE-CEC 2013), pp. 142–148 (2013)

12. Santos, J., Campo, A.: Biped locomotion control with evolved adaptive center-crossing continuous time recurrent neural networks. Neurocomputing 86, 86–96 (2012)
13. Santos, J., Duro, R.J.: Influence of noise on discrete time backpropagation trained networks. Neurocomputing 41(1-4), 67–89 (2001)
14. Smith, R.: Open Dynamics Engine (2003), http://opende.sourceforge.net

Genetic Algorithm for the Job-Shop Scheduling with Skilled Operators

Raúl Mencía, María R. Sierra, and Ramiro Varela[✉]

Department of Computer Science,
University of Oviedo, Campus of Gijón, 33204 Gijón, Spain
ramiro@uniovi.es
http://www.di.uniovi.es/iscop

Abstract. In this paper, we tackle the job shop scheduling problem (JSP) with skilled operators (JSPSO). This is an extension of the classic JSP in which the processing of a task in a machine has to be assisted by one operator skilled for the task. The JSPSO is a challenging problem because of its high complexity and because it models many real-life situations in production environments. To solve the JSPSO, we propose a genetic algorithm that incorporates a new coding schema as well as genetic operators tailored to dealing with skilled operators. This algorithm is analyzed and evaluated over a benchmark set designed from conventional JSP instances. The results of the experimental study show that the proposed algorithm performs well and at the same time they allowed us to gain insight into the problem characteristics and to draw ideas for further improvements.

1 Introduction

The job shop scheduling problem with skilled operators (JSPSO) is a generalization of the job shop scheduling problem with operators (JSO) defined in [1], which in turn generalizes the classic job shop scheduling problem (JSP). In the JSP the tasks are distributed into jobs, and each job defines a sequential ordering for the processing of its tasks. The JSO extends the JSP in such a way that each task must be assisted by one of the available human operators, all of them with the same skills. In the JSPSO, only a subset of the operators are skilled to assist each task, what is more realistic in many production environments. Therefore, finding good solutions for the JSPSO is an issue of major interest for human resources management as it would allow the company to make the best use of its employees and to plan their training for future projects.

The JSPSO is also a special case of the scheduling problem with arbitrary precedence relations and skilled operators defined in [2] and denoted JSSO. Therefore, any solution method devised for the JSSO could be applied to solve the JSPSO. However, the particular structure of the precedence relations in the JSPSO allows for devising efficient heuristics that cannot be applied to the JSSO.

To solve the JSPSO we use a genetic algorithm and propose a new coding schema that exploits the job structure of the JSSO and the fact that a task

J.M. Ferrández Vicente et al. (Eds.): IWINAC 2015, Part II, LNCS 9108, pp. 41–50, 2015.
DOI: 10.1007/978-3-319-18833-1_5

may be assisted by different operators. This schema is designed from the well-known model of permutations with repetition proposed in [3]. A permutation with repetition expresses a processing ordering for the tasks that is compatible with the sequential order of the tasks within the jobs. In the new encoding, a permutation with repetition is combined with a second sequence of symbols that expresses operator preferences for each one of the tasks. We also proposed genetic operators tailored for this encoding. The resulting genetic algorithm is evaluated over a benchmark set designed from conventional instances of the JSP. The results of this evaluation shown that the proposed algorithm is a good approach to solve the JSPSO and at the same time they allow us to draw some proposals for further improvement.

The remaining of the paper is organized as follows. In section 2, we give a formal definition of the JSPSO as a constraint satisfaction problem with optimization. In section 3 we introduce the genetic algorithm devised to solve the JSPSO; in particular we describe the new encoding and genetic operators. Then, in section 4 we summarize the results of the experimental evaluation of the proposed GA. The paper finishes with some conclusions and ideas for future work in Section 5.

2   Description of the Problem

Formally the job-shop scheduling problem with skilled operators can be defined as follows. We are given a set of n jobs $\mathcal{J} = \{J_1, \ldots, J_n\}$, a set of m resources or machines $\mathcal{M} = \{M_1, \ldots, M_m\}$ and a set of p operators $\mathcal{O} = \{O_1, \ldots, O_p\}$. Each job J_i consists of a sequence of m tasks $(\theta_{i1}, \ldots, \theta_{im})$. $\mathcal{T}$ denotes the set of all tasks in the problem.

Each task $u \in \mathcal{T}$ requires two resources during its processing time p_u: a particular machine $m_u \in \mathcal{M}$ and one of the operators $o \in \mathcal{O}_u \subseteq \mathcal{O}$, where $\mathcal{O}_u$ denotes the subset of operators skilled to assist the task u. We assume that the processing time p_u is independent of the assisting operator.

A feasible schedule is a complete assignment of the starting times, st_u, and operators, o_u, to all tasks $u \in \mathcal{T}$ subject to the following constraints:

  i. The operations of each job are sequentially scheduled.
 ii. Each machine can process at most one operation at any time.
iii. The processing of a task on the machine cannot be interrupted.
 iv. Each task must be assisted by one skilled operator during its whole processing time.
  v. The assisting operator is the same over the whole processing time of a task.
 vi. An operator cannot assist more than one task at a time.

The objective is finding a feasible schedule that minimizes the completion time of all the operations, namely the makespan.

Obviously, if the number of operators is sufficiently large, the JSPSO degenerates into the classic JSP. So, the interesting cases of the JSPSO arise when the operators are scarce. For example when $p < min(n, m)$ or when the number of operators skilled for all or at least a number of operations is sufficiently low.

2.1 Solution Graph

A feasible schedule may also be viewed as a set of partial orderings for the processing of tasks in their jobs, machines and assisting operators. These orderings can be represented by means of a solution graph where nodes represent tasks and arcs express processing orders. To visualize all orderings, a fictitious task for each operator and two more arcs termed *start* and *end* are included. Every task is connected to the next one in each partial ordering and the corresponding arc labeled with the processing time of the task at the outgoing node. Figure 1 shows a solution graph for a problem instance with 3 jobs, 3 machines and 2 operators. A feasible solution graph has no cycles and the makespan is given by the cost of the largest cost path from node *start* to node *end*.

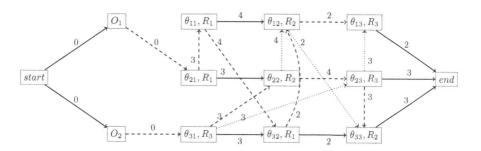

Fig. 1. A solution graph for a problem instance with 3 jobs, 3 machines and 2 operators. Operator 1 is skilled to assist all tasks but θ_{31} and θ_{33}, and the operator can assist all tasks but θ_{32} and θ_{12}. The makespan is 14 and corresponds to the cost of the critical path (*start* O_1 θ_{21} θ_{11} θ_{32} θ_{12} θ_{33} *end*).

3 Genetic Algorithm for the JSPSO

In this section, we review the main components and characteristics of the genetic algorithm used in this work, namely the coding schema, the genetic operators, the decoding algorithm and the general structure of the genetic algorithm.

3.1 Coding Schema

We propose here a coding schema for the JSPSO where a chromosome consists of two permutations of symbols. The first one is a permutation with repetition as it was proposed in [3]. This sequence is a permutation of the set of operations that represents a tentative ordering to schedule them, each one being represented by its job number. For example, the sequence (2, 1, 1, 3, 2, 3, 1, 2, 3) is a feasible sequence for a problem with 3 jobs and 3 machines and represents the task ordering $(\theta_{21}, \theta_{11}, \theta_{12}, \theta_{31}, \theta_{22}, \theta_{32}, \theta_{13}, \theta_{23}, \theta_{33})$. As it was shown in [8], where this encoding was considered for the classic JSP, permutations with repetition have a number of interesting properties; for example, they tend to represent orders of tasks as they appear in good schedules.These properties were also good for other variants of the JSP, like for example the JSO [6]; so, it is expected that they will be good for the JSPSO as well.

The second sequence is the *operator sequence*; it is given by a permutation with repetition of the symbols $1 \ldots p$ of size $n \times m$, without restriction on the number of appearances of each symbol, and represents operator preferences. Notice that some operators could appear several times in the chromosome and also some operator may be absent. For example the following two sequences represent a feasible chromosome for the instance considered in Figure 1.

$$\begin{aligned} (2,1,3,3,2,1,2,1,3) \\ (2,1,1,2,2,1,2,1,1) \end{aligned} \tag{1}$$

This encoding should be understood in such a way that if task u appears before task v in the first permutation, then u should be preferably scheduled before v. At the same time, the second permutation represents priorities for the operators to be allocated to tasks. In order to avoid that all tasks see the same order of operators, the first candidate operator for a task will be that in the same position as the task in the first permutation and then the remaining ones are taken following the chromosome as a circular structure.

Given this interpretation of the encoding, the above chromosome represents the tentative task orderings $(\theta_{21}, \theta_{11}, \theta_{32})$, $(\theta_{12}, \theta_{22}, \theta_{33})$ and $(\theta_{31}, \theta_{23}, \theta_{13})$ on the machines 1, 2 and 3 respectively. Also, it represents the allocation preference of operator 1 to the tasks θ_{11}, θ_{12} and θ_{13}, as the symbol 1 appears in the same position as the task in the second sequence and operator 1 is one of the operators skilled to assist these tasks. Analogously it represents the preference of operator 2 for the tasks θ_{22} and θ_{23}. However, for the remaining tasks the operator in the same position as the task is not skilled to assist the task and so an iteration over the second sequence, starting from the same position as the task, has to be done until an skilled operator is found. So, for example, operator 2 has the largest preference for θ_{32} as this is the third operator in its sequence after two 2 and operator 2 is not skilled to assist θ_{32}, and θ_{33} show preference by operator 2 as this is the second, after one 1, following the circular structure of the permutation of operators. As mentioned, it may happens that for a given task, no skilled operator is contained in the operator sequence. In this case, the chosen operator will be the one that allows the task to be allocated the lowest starting time, in accordance with the strategy used by the decoding algorithm; in case of tie, the operators are taken in order to make the decoding algorithm deterministic. Here, it is important to remark that the chromosome shows preferences by the actual processing orders and operator assignments will depend on the decoding algorithm.

3.2   Decoding Algorithm

Considering the above encoding, different decoding strategies may be used. In scheduling, a decoding strategies are usually designed by means of schedule builders or schedule generation schemes. There are two main categories of schedule builders: serial and parallel. Serial schedule builders iterates over a list of tasks and each one is scheduled "as soon as possible", while parallel schedule builders select the available tasks in successive time periods and schedule some of these tasks on the resources available. Serial schedule builders are less computationally expensive than parallel ones, but at the same time they often build schedules of lower quality.

We opted to use a single serial schedule generation scheme with appending. So, the decoding algorithm iterates over the tasks from left to right in the chromosome sequence. Each task will be assisted by the first operator skilled in the task operator sequence and the task scheduled after all previously scheduled tasks at the earliest possible time.

If this decoding algorithm is applied to the chromosome in eq. 1 it builds the schedule of Figure 1. This is clear as this simple schedule builder translates the tentative orderings of the tasks and their operator preferences to the schedule. This direct translation can be done thanks to the structure of the chromosome.

After building a schedule, we consider the option to code back the schedule into the chromosome. This means that every gene in the chromosome is assigned the assisting operator as the first candidate operator. In this way, it is expected that more characteristics of the parents may be translated to the offsprings.

3.3   Genetic Operators

For chromosome mating, we propose to extend the Job-based Order Crossover (JOX) described in [3] for permutations with repetition to deal with the two sequences of the chromosome. Given two permutations with repetition, JOX selects a random subset of jobs and copies their genes to the offspring in the same positions as they are in the first permutation, then the remaining genes are taken from the second permutation so as they maintain their relative ordering. We clarify how JOX works in the next example. Let us consider the following two permutations with repetition

Permutation1 (**2** 1 1 3 **2** 3 1 **2** 3)   Permutation2 (3 3 1 2 1 3 2 2 1)

If the selected subset of jobs from the first permutation just includes the job 2, the generated permutation is

Offspring (**2** 3 3 1 **2** 1 3 **2** 1).

Hence, operator JOX maintains for each machine a subsequence of operations in the same order as they are in Permutation1 and the remaining in the same order as they are in Permutation2. This fact, together with the strategy of the decoding algorithm, makes it clear that an offspring inherits the partial orders among the tasks of the jobs taken from each permutation, while some of the partial orders among tasks coming from different permutations may be different. Also, at difference of other crossovers, each gene in the offspring represents the same task as it represents in the permutation it comes from.

Extending JOX to dealing with the operator sequences is simple: in the offspring, each gene maintains the first candidate operator it has in the chromosome it comes from. In this way, if this operator is skilled to assist the task, then the assisting operator of this task will be the same in the schedule build from both the offspring and the parent the gene comes from. So if for each task the first candidate operator is the same in both parents, and all of these operators are skilled for their corresponding tasks, in the schedule build from the offspring the operator allocated to each task will be the same as in both parents. For the sake of diversity, this fact must be taken into account when devising mutation operators.

We consider a single mutation (SM) which swaps two consecutive positions at random, the swapping including the task gene and the operator gene. This seems to be appropriate as it may produce small changes. Moreover, to get a variety of distributions of operators in the operator sequences, we will also use an operator mutation (OM) which will change the value in a location of the operator sequence at random. When a chromosome is mutated, one of SM or OM is chosen with probability 0.5.

3.4   Genetic Algorithm Structure

We use here a rather conventional genetic algorithm with generational replacement. In order to avoid premature convergence, we opted not to use the classic roulette wheel selection combined with unconditional replacement. Instead, in the selection phase all chromosomes are organized into pairs at random, then each pair undergoes crossover and mutation. After this, the offsprings are

---

Algorithm 1. Genetic Algorithm.

Data: A JSPSO problem instance $\mathcal{P}$ and a set of parameters
$(P_c, P_m, \#gen, \#popsize, CodingBack)$
Result: A feasible schedule for $\mathcal{P}$
Generate and evaluate the initial population $P(0)$;
for $t=1$ to $\#gen-1$ **do**
  Selection: organize the chromosomes in $P(t-1)$ into pairs at random ;
  Recombination: mate each pair of chromosomes and mutate the two offsprings in accordance with P_c and P_m;
  Evaluation: evaluate the resulting chromosomes and code back the schedules into the chromosomes if the option $CodingBack$ is chosen;
  Replacement: make a tournament selection among every two parents and their offsprings to generate $P(t)$;
end
return *the best schedule built so far*;

---

evaluated and the schedules coding back into the chromosomes, if this option is chosen. Finally, the new population is obtained by means of tournament selection keeping the best two individuals among every two parents and their two offsprings. The algorithm requires 5 parameters: crossover and mutation probabilities (P_c and P_m), number of generations (#gen), the population size (#popsize) and the CodingBack option. Algorithm 1 shows the main steps of the genetic algorithm.

4 Experimental Study

The purpose of the experimental study is to assess the performance of the proposed GA and to compare the coding back (C) and non coding back (NC) options. We considered a set of instances derived from the well-known FT10 instance for the classic JSP with $n = 100$ tasks and $q = 10$ machines. These instances were generated considering different values for p (5, 7 and 9 operators) and different probabilities, Pr, that an operator can assist one task (0.2 and 0.6). Five instances were generated from each pair (Pr, p) and five more from each p taking Pr as 0.2 or 0.6 at random for each task. So we have 45 instances in all. These instances are denoted 2_5_1, 2_5_2, ..., 2/6_5_1, etc.

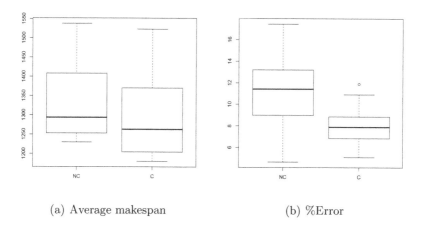

(a) Average makespan   (b) %Error

Fig. 2. Average makespan and errors in percentage of the solutions reached by the GA with the C and NC options

We have chosen a rather conventional parameter setting: $P_c = 1$[1], $P_m = 0.1$, #popsize = 100 and #gen = 1000. The GA was run 30 times for each instance

---

[1] Taking $P_c < 1$ makes that some pairs of chromosomes are not mated and in this case the offsprings are the same as their parents if they are not mutated, so the best parent is chosen twice in the replacement phase, what may contribute to premature convergence.

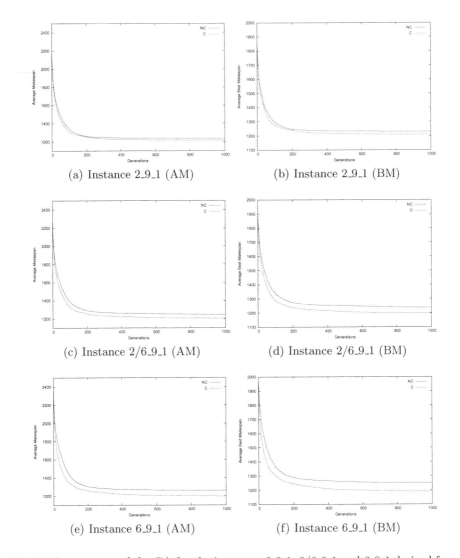

(a) Instance 2_9_1 (AM) (b) Instance 2_9_1 (BM)

(c) Instance 2/6_9_1 (AM) (d) Instance 2/6_9_1 (BM)

(e) Instance 6_9_1 (AM) (f) Instance 6_9_1 (BM)

Fig. 3. Convergence of the GA for the instances 2_9_1, 2/6_9_1 and 6_9_1 derived from the FT10, coding (C) and non-coding (NC) back options. Each plot represents the evolution of the mean makespan and best solution of the population averaged for 30 runs.

and in each run the evolution of the best and average makespan was recorded. The target machine was Intel Core i7-3770K 3.50 GHz. 12 GB RAM and the algorithm was coded in C++.

Firstly, we evaluated the average quality of the schedules. To this end, we consider the average errors in percentage terms (%Error) of the makespan w.r.t. the best makespan obtained for each instance. Figure 2 shows the distribution

Table 1. Summary of results from GA on the 45 instances averaged for each subset of 5 instances. Times are given in seconds.

Sets		Best	NC				C			
Pr	p	Known	Best	Avg.	Time	Err.	Best	Avg.	Time	Err.
2	5	1378,00	1391,00	1495,12	8,70	8,52	1394,80	1494,38	9,33	8,47
2	7	1225,20	1244,00	1321,31	8,82	7,85	1230,00	1308,78	9,56	6,83
2	9	1108,80	1132,80	1238,12	9,06	11,69	1121,00	1206,73	9,78	8,85
6	5	1279,60	1314,00	1409,94	17,41	10,20	1284,00	1371,44	18,10	7,19
6	7	1136,60	1197,20	1300,21	19,34	14,41	1140,40	1235,17	20,17	8,68
6	9	1103,60	1140,00	1253,09	21,38	13,55	1103,80	1184,40	22,16	7,32
2/6	5	1327,40	1342,40	1439,28	12,99	8,44	1329,20	1420,65	13,59	7,03
2/6	7	1160,20	1196,00	1290,93	14,43	11,28	1164,60	1261,41	15,46	8,73
2/6	9	1094,20	1150,20	1242,15	15,05	13,53	1094,20	1196,08	15,81	9,32
Average		1201,51	1234,18	1332,24	14,13	11,05	1206,89	1297,67	14,89	8,05

of errors in percent and mean makespan averaged for all 45 instances. As we can observe, schedules obtained with the C option are clearly better than those obtained with NC.

Then, we considered the evolution of the GA. Figure 3 shows the convergence patterns of the average makespan (AM) and the best makespan (BM) for three instances. As we can observe, in all cases the C option produces a better convergence pattern than the C option.

To further clarify the differences we show in Table 1 more detailed results on teh 45 instances averaged for each group of 5 instances. These results confirm that the C option is the best choice as the makespan with C is about 2.7% better than it is with the NC option, and also the percentage error with C is 3% less than it is with NC. As we can observe, in all cases the values from C are better than those from NC, being the time taken quite similar.

In order to enhance the conclusions we have conducted some statistical analysis following [5]. Particularly, we have used paired Wilcoxon tests samples. We have used three different alternative hypothesis: that the errors, makespan, and best solution in average are smaller from C than they are from NC. The results of these tests indicated that the null hypothesis is rejected in all cases at a high level of significance with p_{value} of $3,58E-08$, $5,78E-08$ and $1,12E-03$ respectively. So, these analysis confirm that the coding back option is the best choice.

5   Conclusions

In this paper we have proposed a genetic algorithm (GA) to solve the job shop scheduling problem with skilled operators (JSPSO). We have demonstrated that the well-known permutation with repetition coding schema, which represents natural orderings for tasks, can be naturally adapted to this problem. To do this, we have combined a permutation with repetition with a permutation of operators to represent operator preferences. We have used a single serial schedule builder to evaluate chromosomes which translate the chromosome task ordering into the schedule and assigns each task the first skilled operator in its preference list. We have seen that this encoding/decoding combination allows for a proper convergence of the GA. Moreover, we have considered coding the characteristics of the schedules back into the chromosomes. This means that the assisting operator of each task is moved to the first position in its preference list. The experimental study showed that the GA convergence clearly improves with the coding back option: not only it is quicker at the beginning, but also it takes longer time in getting stabilized, and so the GA reaches eventually much better schedules.

As future work, we will consider more sophisticated decoding algorithms such as for example parallel schedule builders and we will combine the GA with local search, as done for example in [6] for the JSO. This will require devising some neighborhood structures. To this end, we propose to extend some classic structures based on changing the processing ordering of tasks in the critical path [4], [7] and also we will consider changing the assisting operator of some critical task to devise a new class of neighborhoods structures.

Acknowledgements. This research has been supported by the Spanish Government under research project TIN2013-46511-C2-2-P.

References

1. Agnetis, A., Flamini, M., Nicosia, G., Pacifici, A.: A job-shop problem with one additional resource type. J. Scheduling 14(3), 225–237 (2011)
2. Agnetis, A., Murgia, G., Sbrilli, S.: A job shop scheduling problem with human operators in handicraft production. International Journal of Production Research 52(13), 3820–3831 (2014)
3. Bierwirth, C.: A generalized permutation approach to job shop scheduling with genetic algorithms. OR Spectrum 17, 87–92 (1995)
4. Dell' Amico, M., Trubian, M.: Applying tabu search to the job-shop scheduling problem. Annals of Operational Research 41, 231–252 (1993)
5. García, S., Fernández, A., Luengo, J., Herrera, F.: Advanced nonparametric tests for multiple comparisons in the design of experiments in computational intelligence and data mining: Experimental analysis of power. Information Sciences 180, 2044–2064 (2010)
6. Mencía, R., Sierra, M.R., Mencía, C., Varela, R.: A genetic algorithm for job-shop scheduling with operators enhanced by weak lamarckian evolution and search space narrowing. Natural Computing 13(2), 179–192 (2014)
7. Van Laarhoven, P., Aarts, E., Lenstra, K.: Job shop scheduling by simulated annealing. Operations Research 40, 113–125 (1992)
8. Varela, R., Serrano, D., Sierra, M.: New codification schemas for scheduling with genetic algorithms. In: Mira, J., Álvarez, J.R. (eds.) IWINAC 2005. LNCS, vol. 3562, pp. 11–20. Springer, Heidelberg (2005)

Preliminary Design of the Real-Time Control Software for the Adaptive Optics of AOLI

Carlos Colodro-Conde[3]([✉]), Luis F. Rodríguez-Ramos[2], Isidro Villó[3],
Craig Mackay[1], Rafael Rebolo[2,5], Jonathan Crass[1],
Juan J. Fernández-Valdivia[6], David L. King[1], Lucas Labadie[4],
Roberto López[2], Alejandro Oscoz[2], Antonio Pérez-Garrido[3], Marta Puga[2],
José M. Rodríguez-Ramos[6], and Sergio Velasco[2]

[1] Institute of Astronomy, University of Cambridge, Cambridge, UK
[2] Instituto de Astrofísica de Canarias, La Laguna, Spain
and Departamento de Astrofísica, Universidad de La Laguna, La Laguna, Spain
[3] Universidad Politécnica de Cartagena, Cartagena, Spain
carlos.colodro@upct.es
[4] I. Physikalsiches Institut, Universität zu Köln, Köln, Germany
[5] Consejo Superior de Investigaciones Científicas, Seville, Spain
[6] Departamento de Física Fundamental y Experimental, Electrónica y Sistemas,
Universidad de La Laguna, La Laguna, Spain

Abstract. This paper describes the preliminary design of the software that will perform real-time control of the adaptive optics section of a new astronomical instrument that will combine the benefits of low order adaptive optics with lucky imaging. Such combination will allow astronomers to obtain near diffraction limited images in the visible spectrum on large ground-based telescopes using faint natural guide stars.

Keywords: Adaptive optics · Real-time control · Astronomical instrumentation

1 Introduction

The Adaptive Optics Lucky Imager (AOLI) [1,2,3] is a new instrument under development to demonstrate near diffraction limited imaging in the visible on large ground-based telescopes. This instrument is designed to combine the techniques of AO and Lucky Imaging into a single instrument. The AO (Adaptative Optics) component of the system employs a geometric or a non-linear curvature WFS (wavefront sensor), two Electron Multiplying CCDs (EMCCDs) to maximise its sensitivity to natural guide stars and a deformable mirror (DM) manufactured by ALPAO [4] to apply wavefront corrections in real-time. It is initially designed specifically for use on the William Herschel Telescope (WHT), and it can be adapted to work with the 10.4m Gran Telescopio Canarias (GTC) after some modifications. It is estimated that 17.5-18th magnitude natural guide stars will be usable as reference objects on the WHT, with the limiting magnitude for the GTC being around 1 magnitude fainter.

© Springer International Publishing Switzerland 2015
J.M. Ferrández Vicente et al. (Eds.): IWINAC 2015, Part II, LNCS 9108, pp. 51–60, 2015.
DOI: 10.1007/978-3-319-18833-1_6

The present work focuses on describing the preliminary design of the real-time control (RTC) software, which runs in the AO (Adaptive Optics) computer. Its purpose is to read the output of the wavefront fitting software (which will run in the same computer) and close the control loop by actuating over the DM, in such a way that the wavefront aberrations due to a turbulent atmosphere are minimized in real-time. A diagram of the AO computer processing pipeline is shown in Figure 1.

This paper is structured as follows. Section 2 describes the architecture of the RTC software. Section 3 presents the output of a simulated control run. Finally, Section 4 draws the main conclusions.

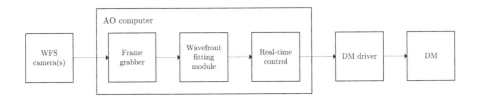

Fig. 1. AO computer

2   Software Architecture

The RTC software must be designed with performance and flexibility in mind. It is required that the software can be adapted to work with DMs with an arbitrary number of actuators. Additionally, it is required that the software accepts two different output formats coming from the WFS: a 2D surface representing the measured wavefront in the spatial domain or a set of Zernike coefficients. These coefficients can also be expanded to 2D surfaces by applying the equations that define the Zernike polynomials [5].

For software developments with a certain degree of complexity, it is almost mandatory to have a clear, modular design. Of course, AOLI RTC software will follow this principle. Modularity is good not only because it eases development, validation and maintenance, but also because it will help to accelerate parts of the code depending of their specific performance requirements.

The RTC subsystem is in fact composed of three software packages: RTC-SW (real-time control software), RTC-SUP (real-time control support) and RTC-SIM (real-time control simulators). The RTC-SW package is the real-time control itself, that is, the one that will try to minimize wavefront aberrations during science observations. The aim of the RTC-SUP is to calibrate, characterize and configure some parts of the system before starting science observations. Finally, RTC-SIM is intended to be used only during the development stage, so as to help to validate the rest of the software from an algorithmic perspective, without having to access the real hardware.

Each of the packages listed above will be described in more detail throughout the following subsections.

2.1   RTC-SW

The RTC-SW is the part of the RTC that will run during science observations, trying to correct the aberrations produced by atmospheric turbulences in real time. The overall architecture of the RTC-SW, depicted in Figure 2, consists on a configurable processing pipeline.

The highest constraint for the RTC-SW is that it must be executed as fast as possible, in order to perform a control that can correct the atmospheric turbulence with a small error over time. It is required that the RTC-SW responds in less than 500 microseconds, that is, below 5% of the sampling time of the WFS (100 Hz). The most timing-critical blocks can be accelerated with the help of one or more GPUs (by means of OpenCL [6] or CUDA [7]). Another option is to use math libraries that try to get the most of a specific processor model (for example, MKL [8] for Intel CPUs), but these type of approaches are typically slower. In any case, most of the calculations will be done in their matrix form so as to facilitate the implementation with any of the mentioned accelerating procedures. The target programming language for the RTC-SW will be C/C++, although prototypes will be programmed in MATLAB in order to facilitate the validation of the algorithms.

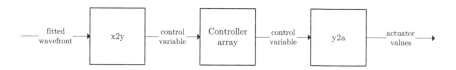

Fig. 2. RTC-SW overall architecture

The function of each block is described below.

x2y. This block converts the input wavefront to the actual variable that will be controlled. The name $x2y$ is because the input (x) can be either surface samples (s) or Zernike modes (z), and the output (y) can be surface samples (s), Zernike modes (z) or actuator values (a). This block will be implemented by multiplying the input wavefront (represented as a vector) with a conversion matrix that must have been pre-calculated by means of the RTC-SUP. If no conversion is needed ($x = y$), this block does nothing.

Controller Array. The actual control function is implemented here. There will be one PID (proportional-integral-derivative) loop for each control variable. For example, if the format of the control variable is surface samples and there are 10000 controllable surface samples, there will be 10000 control loops implemented in this block, each one with its own K_p, K_i and K_d parameters. All of these parameters are optional, so one can implement a PI (proportional-integral) control by disabling the K_d vector. Operations in this block will be vector-based.

y2a. This last component is analogue to the $x2y$ block described before. It converts the control variable (y) to actuator values (a), which will be passed directly to the DM software driver. If the control variables are actuator values ($y = a$), this block does nothing.

2.2 RTC-SUP

The RTC-SUP will implement the characterization and configuration functions that have to be performed before running the RTC-SW.

As opposed to the RTC-SW, the RTC-SUP is not required to have high processing speeds. The focus should be put more on providing a friendly graphical user interface that performs all the necessary off-line functions. These functions will be programmed in the MATLAB language, which will ease the RTC-SUP development of the RTC-SUP due to the fact that MATLAB is a high-level programming language with lots of built-in functions and GUI components.

The main functions of the RTC-SUP are explained below.

Static Characterization. This procedure consists on stimulating each one of the actuators of the DM and read the response measured by the WFS, whose output format can be either surface samples or Zernike modes depending on the nature of the WFS fitting process. The RTC-SUP will automatically select each actuator, inject a given current to it, and record the values outputted by the WFS. The result of the static characterization is either an $a2s$ matrix (for a surface WFS) or an $a2z$ matrix (for a Zernike modes WFS). Any of these two matrices is enough to fully characterize the system from a static point of view. The rest of the matrices that could be needed to configure the control loop can be calculated mathematically, with no interaction with the physical system.

Dynamic Characterization. The aim of this function is to characterize the AO system dynamically so that a proper PID control can be designed by setting the values of the K_p, K_i and K_d parameters. The dynamic characterization consist on introducing a sequence of actuations to the DM, read the resulting WFS measurements and finally obtain the impulse response of the system, which fully characterizes it from a dynamic point of view. In fact, there will be one measured impulse response for each variable under control (actuators, surface samples or Zernike modes).

2.3 RTC-SIM

The aim of the RTC-SIM package is to reduce the risks of not having access to the real hardware from the first stages of development. It will consist on a set of simulators that can be easily integrated within the RTC-SW and the RTC-SUP, using a software interface very similar to the one provided by the real elements. Unlike the RTC-SW and the RTC-SUP, the RTC-SIM package cannot be run as a standalone program.

The programming language will be MATLAB. This means that the RTC-SIM cannot be integrated in the final version of the RTC-SW, which will be programmed in C/C++. Nonetheless, it will be possible to do so with the RTC-SW prototypes, which are to be written in MATLAB.

The following subsections describe the characteristics of each simulator available in RTC-SIM.

Incident Wavefront Simulator. This software simulator will emulate an incident wavefront with atmospheric turbulence that varies over time. One or more layers of the atmosphere will be simulated following a time-evolving Kolmogorov model [10,11], with configurable parameters. Figure 3 shows an example run of the wavefront simulator, with parameters $R_0 = 0.3$, $L_0 = 50$, $\alpha = 0.02$ and 3 simulated turbulence layers. The parameters R_0 and L_0 describe the spatial distribution of the turbulence, while α determines its variation speed.

The existence of a wavefront simulator allows testing the response of the system under the presence of astronomical seeing without the need of having AOLI integrated with the telescope. This software simulator should be replaced in a future stage with its hardware counterpart, that is, the AOLI calibration system.

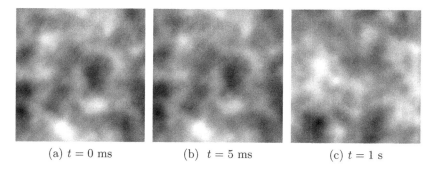

(a) $t = 0$ ms   (b) $t = 5$ ms   (c) $t = 1$ s

Fig. 3. Example output of the incident wavefront simulator

DM Simulator. The DM simulator will provide a software interface similar to that expected from the actual DM drivers, which basically consist of one or more initialization functions and an actuation function (for sending displacements to the DM). The DM simulator will take the output of the wavefront simulator and calculate the shape of the reflected wavefront, as a function of the currents injected to the actuators.

The fact of not having acess to the real DM in the first stages of development makes it necessary to generate an artificial influence function and frequency response. These two elements will define the behaviour of the DM simulator fully, both statically and dynamically.

The synthetic influence will be obtained by solving the Poisson equation over the mirror membrane, after applying the simplifications explained in [12]. Simplifications are acceptable because the aim is just to obtain a representative simulator of a generic DM. Figure 4 shows some of the calculated influence functions for a simulated DM with 227 actuators.

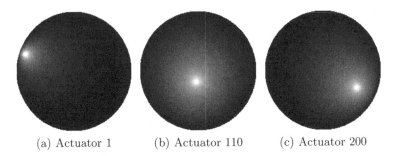

(a) Actuator 1   (b) Actuator 110   (c) Actuator 200

Fig. 4. Example of calculated influence functions

On the other hand, the frequency response can be generated as a Chebyshev first order low-pass filter, with the cut frequency set as specified in the ALPAO datasheets. One of the characteristic of Chebyshev filters is that they have a ripple, so resonance effects will be taken into account this way. Figure 5 plots an example step response generated with the following Chebyshev parameters: gain = 1, order = 1, cutoff frequency = 1.7 KHz and ripple = 2 dB.

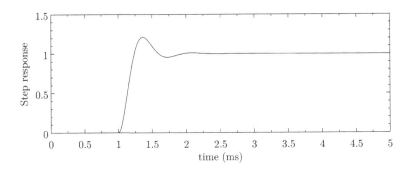

Fig. 5. Generated step response for the DM actuators

After the real DM is characterized, it will be possible to integrate the output of this characterization in the DM simulator, thus improving its similarity with reality.

WFS Simulator. This simulator outputs a fitted wavefront, represented either as surface samples or Zernike modes. For that purpose, it will take the reflected wavefront obtained by the DM simulator and integrate it over the specified period of simulated time, just like a real camera would do.

3   Results

This section presents the results obtained after programming and integrating all the pieces of software described in Section 2. In this first battery of tests, the RTC-SW was implemented in MATLAB in combination with the RTC-SIM package, so as to be able to validate the control algorithm. The RTC-SUP and RTC-SIM packages were written completely in MATLAB as well.

The exposure time of the simulated WFS was established to 10 ms. The latency for the wavefront fitting and control processes were set to 8 ms and 6 ms, respectively. It is appropriate to specify a delay for these software processes because they have a considerable computational cost (they require a high number of mathematical operations).

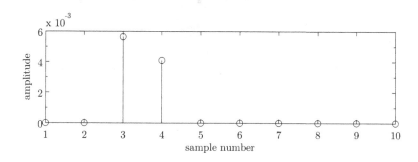

Fig. 6. Example impulse response measured with the RTC-SUP software

Before doing a test run of the RTC-SW, the conversion matrices and PID parameters have to be established first. The influence functions were measured with the static characterization tool of the RTC-SUP, and the results matched with the ones that were synthetically generated for the DM simulator (see Figure 4. The conversion matrices were calculated from the obtained influence functions, setting surface samples as the target control variable. Then, the impulse response was measured using the dynamic characterization tool of the RTC-SUP. Among the different measurement methods that were programmed, the authors chose to use Maximum Length Sequences (MLS) [9]. The result of this measurement for one of the surface samples is shown in Figure 6.

Note that it is not possible to measure the actual dynamic response of the mirror membrane. The reason is that the mirror reacts in a matter of 2 milliseconds (see Figure 5), and the acquisition rate of the WFS (the only available measurement from the AO system) is 10 milliseconds. In spite of not being able to measure the actual response of the DM, dynamic characterization is still of utmost importance, due to the fact that it describes the AO system just as it is seen by the RTC. This is the response that truly matters, because the PID parameters are adjusted according to this information. For this control run, the

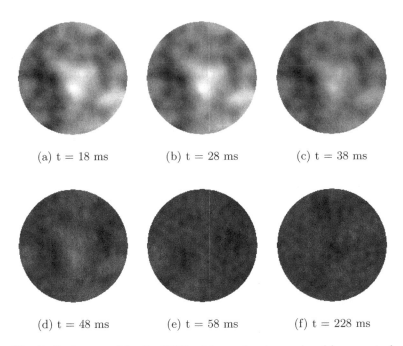

(a) t = 18 ms   (b) t = 28 ms   (c) t = 38 ms

(d) t = 48 ms   (e) t = 58 ms   (f) t = 228 ms

Fig. 7. Surfaces read by the WFS while performing a closed loop control

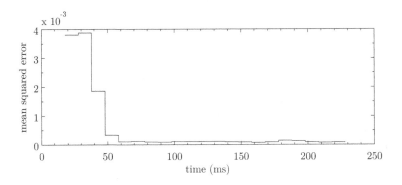

Fig. 8. Measured MSE while performing a closed loop control

PID parameters were designed to control surface samples with a response time of 55 milliseconds and a phase margin of 60°, with little ringing. For the example response shown in Figure 6, the PID parameters would be $K_p = 57.9$, $K_i = 3807.4$ and $K_d = 0$ (in fact, that would make it a PI control).

The recorded results of a test run of the RTC-SW are represented in Figure 7. The wavefronts read by the WFS (which was configured as a surface sensor) confirm that the RTC-SW is effectively removing a big component of the atmospheric turbulence. The overall response time also matches the one with which the individual PID controllers were designed.

Figure 8 further analyzes the obtained results by plotting the measured MSE (mean squared error) over time. It can be seen in this figure that after the loop has reached its steady state, the MSE is very close to zero (an MSE equal to zero would mean a perfect correction). The improvement factor when compared to the initial MSE (i.e., before turning on the control loop, first sample in the graph) is around 43x.

4   Conclusions

The main conclusion of this paper is that the software developed for controlling the adaptive optics of AOLI is valid from an algorithmic point of view. In the future, this software will have to be integrated with the rest of parts of the instrument, and a full validation campaign will have to be performed. The software-only validation that was done for this work is considered useful because it has minimized the risks of an unsuccessful final validation. The additional time that was spent developing the software simulators is justified because a problem detected during the final validation campaign would have a high impact in the project schedule.

It is important to note that performances obtained in Section 3 are not representative of the ones that will be obtained with the real AOLI, because lots of simulation parameters have been established with a certain degree of arbitrariness. In these simulations, better MSEs would have been obtained if the incident wavefront simulator had been configured with a lower variation speed (α), so that the control loop would have found it easier to correct such variations. Another way to improve the results is to lower the simulated latencies, enabling the design of faster (and still stable) control loops. In fact, the values established for the latencies represent a rather poor case.

The next steps in the RTC software development plan involve a pre-integration with the hardware (again, to minimize the risks of an unsuccessful validation), porting the RTC-SW to C/C++ with one of the acceleration methods explained in Section 2.1 and finally developing a GUI for the RTC-SW and RTC-SUP. At that stage, the RTC software will be ready to be validated with the real AOLI.

Acknowledgements. This work was supported by the Spanish Ministry of Educación, Cultura y Deporte under the grant FPU12/05573, and by the Spanish Ministry of Economía project ESP2013-48362-C2-2-P.

References

1. High spatial resolution optical imaging of the quintuple T Tauri system LkHα 262-263 with AOLI, Velasco, S., et al. MNRAS (2015) (in prep.)
2. Mackay, C., et al.: High-resolution imaging in the visible on large ground-based telescopes. SPIE Astronomical Telescopes+ Instrumentation. International Society for Optics and Photonics (2014)

3. Crass, J., et al.: The AOLI low-order non-linear curvature wavefront sensor: a method for high sensitivity wavefront reconstruction. SPIE Astronomical Telescopes+ Instrumentation. International Society for Optics and Photonics (2012)
4. Camet, S., Curis, J.-F., Rooms, F.: Deformable mirror having force actuators and distributed stiffness. U.S. Patent No. 8,794,773 (August 5, 2014)
5. Noll, R.J.: Zernike polynomials and atmospheric turbulence. JOsA 66(3), 207–211 (1976)
6. Khronos OpenCL Working Group. The opencl specification (2008)
7. Nvidia, C.U.D.A.: C programming guide. NVIDIA Corporation, Santa Clara (2014)
8. Intel Corporation. Intel Math Kernel Library (Intel MKL) (retrieved February 11, 2014)
9. Rife, D.D., Vanderkooy, J.: Transfer-function measurement with maximum-length sequences. Journal of the Audio Engineering Society 37(6), 419–444 (1989)
10. Tatarskii, V.I.: Wave propagation in turbulent medium. Wave Propagation in Turbulent Medium, by Valerian Ilich Tatarskii. Translated by RA Silverman, 285 p. Published by McGraw-Hill (January 1961)
11. Glindemann, A., Lane, R.G., Dainty, J.C.: Simulation of time-evolving speckle patterns using Kolmogorov statistics. Journal of Modern Optics 40(12), 2381–2388 (1993)
12. Claflin, E.S., Bareket, N.: Configuring an electrostatic membrane mirror by least-squares fitting with analytically derived influence functions. JOSA A 3(11), 1833–1839 (1986)

Graph-Based Learning on Sparse Data for Recommendation Systems in Social Networks

J. David Nuñez-Gonzalez[1(✉)] and Manuel Graña[1,2]

[1] Computational Intellligence Group
of the University of the Basque Country (UPV/EHU), Vizcaya, Spain
jdnunez001@gmail.com
[2] ENGINE project at the Wroclaw University of Tecnology (WrUT),
Wroclaw, Poland

Abstract. Graph-based supervised learning is a useful tool to model data supported by powerful algebraic techniques. This paper proposes a novel approach using Graph-based supervised learning to handle the problem of building Recommendation Systems in Social Networks. Specifically, the main contribution of this paper is to propose two ways to construct Recomendations Systems based on Colaborative Filtering. The first way consists in building a feature matrix from the Web of Trust of each user. The second way builds a feature matrix based on similiraties among users according to the criterion of the system without taking into consideration the Web of Trust of the target user.

1 Introduction

The structure of a social network is often represented by a graph, where nodes correspond to users and edges to their social relationships. Two users are linked by an edge if they are friends. The possibility of giving ratings about other users or items is a common feature of social networks, which is very useful for recommendation systems.a Recommendation System generates proposals of items that may receive a high rating by a user. Those ratings can include a statement of the trust of one user on another. This specific form of social structure is the Web of Trust (WoT), where the edges between nodes are explicitly constructed, and the graph is directly related to the ratings given by the users. Let us consider the subconscious intelligence embedded in the social system [1]. From this point of view, we are interested on the construction of implicit graphs using available information on the users (attributes of each node) and looking for similarities among user attribute values. After a revision of the State of the Art, the main goal of this paper is to propose two ways to build custom classifiers for each target user in order to make recommendations. Specifically we propose, on the one hand, a Recommendation Systems based on collaborative filtering using the WoT of the target user to build a custom classifier. On the other hand, we propose an implicit Recommendation Systems building a custom regressor based on colaborative filtering defined on similarities among users. One specific difficulty is that the input data is sparse, that is, each input vector contains a large percentage of zero values, therefore both proposals are treated with sparse data tools.

J.M. Ferrández Vicente et al. (Eds.): IWINAC 2015, Part II, LNCS 9108, pp. 61–68, 2015.
DOI: 10.1007/978-3-319-18833-1_7

Contents of the paper The structure of paper is as follows: Section 2 revise other applications of graph-based learning techniques. Section 3 reports our approach to apply graph-based learning in Social Networks. Section 4 gives our conclusions and plans for future work.

2 State of the Art

2.1 Graph Learning Similarity Metrics

Let us recall some definitions. A graph is a collection of n vertices (or nodes) and m edges that is denoted as $G(V, E)$. An adjacency matrix A with $n \times n$ size represents the nodes of the graph and existing links. Thus, an element $a_{ij} \in A$ from the adjacency matrix will be $a_{ij} = 1$ if exists an edge that links both vertices i and j. Sometimes edges have an associated weight, which may be a value or set of values . So, the weight matrix W (with size $n \times n$) represents those weights being each element of the matrix $w_{ij} \in W/w_{ij} \in \mathbb{R}$. The degree of a vertex is the number of edges that links each vertex with others. The degree of matrix is called D which is in $n x n$ size and each element $d_{ij} \in D/d_{ij} \in \mathbb{R}$. Thus, the matrix L that represents Laplacian Graph is defined as $L = D - A$.

Networks also involve different kind of measures. [14] gives some of them as degree centrality, eigenvector centrality, katz centrality, pagerank, closeness centrality and betweenness inter alia. Eigenvector centrality is defined by [6] as $x_i = \kappa_1^{-1} \sum_j A_{ij} x_j$ where κ_i are the eigenvalues of A and κ_1 is the largest of them, A_{ij} is an element of the adjacency matrix and x is the centrality of vertex j. So, the centrality x_i of vertex i is proportional to the sum of the centralities of i's neighbors.

Applications. Graph-based learning is already being used in other contexts. For instance, [9] proposes graph-based semisupervised learning graph classifier based on kernel smoothing. A Sequential Predictions Algorithm (SPA) are used as a graph-based algorithm which propagates labels across vertices. [17]gives a graph-based multiprototype competitive learning and its applications to solve partitioning nonlinearly separable datasets. [18] proposes a clustering-based graph laplacian framework for value function approximation in reinforcement learning by subsampling in Markov decision processes (MDPs) with continuous state spaces. In patter recognition [7] proposes learning graph matching. In graph matching, patterns are modeled as graphs and pattern recognition amounts to finding a correspondence between the nodes of different graphs. The used method for learning graph matching is bistochastic normalization, a state-of-the-art quadratic assignment relaxation algorithm. [11] proposes graph-based semisupervised learning (GSSL) as paradigm for modeling the manifold structures that may exist in massive data sources in highdimensional spaces.

[2,3] proposes some work related with Singular Value Descomposition (SVD) for dimensionality reduction strategy, eigenanatomy in the context of neuroimaging data. Another work involves sparse canonical correlation analysis (SCCAN)

that identifies generalizable, structural MRI-derived cortical networks that relate to five distinct categories of cognition. Those techniques, SVD and sparsification are contextualized in our work.

2.2   Sparse Data and Classification

Sparse matrix is a matrix in which most of the elements are zero. With large matrix traditional methods to store the array in memory of a computer or for solving systems of linear equations need a lot of memory and processing time. They are designed specific algorithms for these purposes when the matrices are dispersed. Many tools for sparse data and classification are given on Web to be used in Matlab. This reference[1] is just an example which would be our tool. Sparse representation (SR) is a principle that a signal can be approximated by a sparse linear combination of dictionary atoms. Is formulated as follows:

$$\mathbf{b} = x_1\mathbf{a}_1 + ... + x_k\mathbf{a}_k + \varepsilon = A\mathbf{x} + \varepsilon$$

where $A = [a_1, ..., a_k]$ is called dictionary, a_i is a dictionary atom, $\mathbf{x}$ is a sparse coefficient vector, and ε is an error term. A, x, k are the model parameters. Sparse representation solves the problem of finding a compact signal in terms of a linear combination of atoms in a dictionary. Each signal can be reconstructed by only one linear combination of atoms. Thus, a dictionary can be learned with the following expression:

$$\min_{\mathbf{x}} \frac{1}{2} \parallel \mathbf{b} - A\mathbf{x} \parallel_2^2 + \lambda^T \parallel\mathbf{x}\parallel_1 \mid \mathbf{x} \geq 0$$

Applications Several works focus their interest in solving sparsification problems. We summarize some works about it. For instance, [13] relates sparse coding and Multilayer Perceptron (MLP) by converting sparse code into convenient vectors for MLP input for classification of any sparse signals.[4] proposes a Bayesian framework 3-D human pose estimation from monocular images to get a posterior distribution for the sparse codes and the dictionaries from labeled training data. [5] gives a method that constructs a series of sparse linear SVMs to generate linear models in order to reduce data dimensionality.[8] gives a survey of algorithms and results to induce grammars from sparse data sets. [10] shows models of biological neurons to create sparse representations with true zeros for naturally sparse data. [12] focuses on the large-scale matrix factorization problem solving it via optimization algorithm, based on stochastic approximations. [15] proposes transfer learning for image classification with sparse prototype representations. [16] combines a general Bayesian framework with Relevance Vector Machines in order to obtain sparse solutions to regression and classication tasks utilising models linear in the parameters.

---

[1] https://sites.google.com/site/sparsereptool/

3   Proposed Methods

As mentioned before, we are interested on building classifiers in two ways. First one take into account the opinion of the target user. Distances among user are calculated to build the features matrix of second way, based on the knowledge of the system. Next Sections show these ideas in more detail.

3.1   Raw Data

Epinions [2] gives two datasets: the trust network dataset and the rating dataset. The first one represents the graph of trust among users, thus, the directed graph $G_u(U, T)$ represents users as nodes and trust links as edges. A link will exist from user u_i to another user u_j if the first user trust the second one. On the dataset, each link among two users is represent in this way: $< u_i, u_j >$. The graph of the rating dataset could be defined as $G_r(U \cup (C \times I), E, R)$ where nodes are users and items (products) and edges are links between users and items. Users and items are linked when a user gives a rating about an item. Edges have a a weight corresponding to the rating. Thus, the information on the dataset is represented in this way: $< (u, (c, i)), r >$ that means that a user u_i gives a rating r to the item i of the category c. There are 27 categories, because of the importance of the domains in Recommendation Systems, we define the set of whole items as $I = \{I_1, ..., I_{27}\}$ according with the number of categories. Having the trust network graph and having the 27 graphs of ratings, we are able to build classifiers. Each I_i is the rating matrix for the items of the category i. Ratings are values in the range $\{1..5\}$. Zero means that no rating is given for an item.

3.2   Feature Matrix Based on Web of Trust

Algorithm 1. Algorithm for extraction of target user Web of Trust
---
For each target user u_i in $G_u(U, T)$
  if $(\exists t_{ij} \in T)$ then
    $[trustedUsers]_{u_i} \leftarrow u_j$
  endif

Focusing on the Web of Trust we can get the explicit graph which represents the trust values from a target user u_i to another set of users on the community $U_j = \{u'_1, ..., u'_n\} | t_{uu'} = 1$. The algorithm shown above ilustrates this idea.

---
[2] http://www.public.asu.edu/~{}jtang20/datasetcode/truststudy.htm

3.3   Feature Matrix Based on Similarities

Algorithm 2. Algorithm for extraction of target user similar users based on ratings

---

For each target user u_i in G_r $(\{\{U \cup I\}, R\})$
  For each I_i matrix rating
    $\psi \Lambda \phi^T = SVD(I_i)$
    $\Phi = \phi * \Lambda$ /*Each row of Φ are eigenvectors from a user i
    for each user   /*Get distances
      $d(u_i, u_j) = \|\Phi_i - \Phi_j\|$
    end
  end
end
Select α most similiar users in D_u
$[SimilarUsers]_{u_i} \leftarrow \alpha$ similar users

This time we can get implicit graph based on the knowledge collected in the system. We will operate the rating matrix using Singular Value Decomposition (SVD) that is a method for identifying and ordering the dimensions along which data points exhibit the most variation. This is an used tool in recommender systems to predict people's item ratings. We know that using SVD, rating matrix I can be states as follows:

$$I = \psi \lambda \phi^T$$

where ψ is the matrix of eigenvectors of $I(I)^T$, then ϕ is the matrix of eigenvectors of $(I)^T I$ and we just transpose it to get ϕ^T, and finally λ is the matrix of square roots of non-zero eigenvalues ordered in decreasing order. Having an attribute matrix of nxp we would add as many zero-columns as necessary in λ to keep the proper dimensions to allow multiplication of $\psi \lambda \phi^T$. Thus, computing eigenvectors $\Phi = \phi \lambda$ we can easily get distances matrix D_u of nodes (users) calculating the distance between vectors. By this way, we define distance between users as follows:

$$d(u_i, u_j) = norm\|\Phi_i - \Phi_j\| = \sqrt{\sum (\Phi_i - \Phi_j)^2}$$

Once we have distance matrix, we can choose a set of users U_d that are the closest users to the target user u_i. Those users will conform the most similar users and they are be used to perform the features matrix as well as we will described in the next Section, but instead of having the array of trusted users we have the array of similar users.

Previous shown algorithm is used to get the array of similar users. As said before, we use SVD to get eigenvectors from users to suddenly use norm function in Matlab which give the Euclidean distance between two vectors. For a user u_i the α similar users will be choosen to perform the feature matrix.

3.4   Proposed Experiment

Once we have the users that we are interested in, we build the feature matrix being rows items and columns users ratings. We mantain rating values of users that we are interested in. Others, we turn to value 0. Thus, we have to predict rating values from a user target. We use Matlab to get the array of users that we are interested in. Because of the high dimension of matrix, we select 5000 users and 5000 items for experimentation. Items with no rating and users that have not rated any item are removed from the feature matrix.

A Success predicting the ratings of the i-th target u_i user happens when the predicted rating is $r'_u = r_u$, otherwise it is a Failure. The accuracy of the prediction for the u_i is computed as the ratio $\frac{success(u_i)}{success(u_i)+failure(u_i)}$. In general, the accuracy for the whole social network will be computed as as

$$A = \frac{\sum_{i=1}^{n} success(u_i)}{\sum_{i=1}^{n} success(u_i) + \sum_{i=1}^{n} failure(u_i)}$$

Because of the range of the ratings we can not assume that all failures are the same, in other words, if we have to predict a rating of '2' marks, making a prediction of '3' marks is an smaller error that making a prediction of '5' marks. Therefore, we have a regression problem instead of a classification problem. The performance measures taken into account accordingly are: Mean Absolute Error (MAE), Root Mean Squared Error (RMSE), Relative absolute Error (RAE) and Root Relative Squared Error (RRSE).

Tables 1 and 2 show the results of built regressors based on Web of Trust users information, and on distances in hte user attribute space, respectively. We have tested six different regressors provided in the open source and free Weka software[3]:

- Linear Regression which uses the Akaike criterion for model selection, and is able to deal with weighted instances. Best results with this classifier are obtained without attribute selection.
- Multilayer Perceptron, which is the most famous artificial neural network that uses backpropagation to classify instances.
- SMOreg classifier which is an implementation of Support Vector Regression.
- K-nearest neighbours classifier with K=1 gives the best results in this data.
- Random Tree uses constructs a tree that considers K randomly chosen attributes at each node.
- Finally, Additive Regression which is a meta classifier that enhances the performance of a regression base classifier. Each iteration fits a model to the residuals left by the classifier on the previous iteration. Prediction is accomplished by adding the predictions of each classifier. The Additive Regression version in this paper is Support Vector Regression with a Polynomial Kernel, and Support Vector Machines as optimization function.

All experiments are tested with fold cross validation option. Features extrated from Web of Trust give better results than features based on users distances. In the first case, in general, the Mean Absolute Error is over 0.30-0.36 marks. In the second one, in general, the Mean Absolute Error is over 0.56-0.74 marks. The best classifiers in both experiments is Additive Regression.

---

[3] http://www.cs.waikato.ac.nz/ml/weka

Table 1. Results of features extracted from Web of Trust

	MAE	RMSE	RAE	RRSE
Linear Regression	0.37	0.79	32.13%	56.87%
Multilayer Perceptron	0.59	0.94	51.67%	67.42%
Support Vector Regression	0.81	0.34	30.09%	57.96%
KNN	0.36	0.79	31.60%	56.84%
Additive Regression	*0.30*	*0.96*	*25.91%*	*68.69%*
Random Tree	0.36	0.79	31.60%	56.84%

Table 2. Results of features extracted from user distances

	MAE	RMSE	RAE	RRSE
Linear Regression	0.74	1.49	44.47%	79.11%
Multilayer Perceptron	0.97	1.64	58.15%	87.23%
Support Vector Regression	*0.56*	*1.14*	*33.43%*	*60.41%*
KNN	1.25	1.39	85.10%	84.32%
Additive Regression	*0.56*	*1.14*	*33.43%*	*60.41%*
Random Tree	0.74	1.49	44.47%	79.11%

4 Conclusion and Future Work

We propose methods for Recommendation Systems in Social Networks based on
Colaborative Filtering. Those methods follow the way of the Web of Trust built
from users trust values and the way of the intelligent layer of the system. To
represent this idea we propose a feature extraction based on similarities among
users. Better results are obtained with the Web of Trust that is provided by
users. As future research lines, we want to work on the sparse methods in order
to propose improvements to be applicated on Recommendation Systems.

Acknowledgements. This research has been partially funded by EU through
SandS project, grant agreement no 317947. The GIC has been supported by
grant IT874-13 as university research group category A. The ENGINE project
is granted by the European Commision by grant 316097.

References

1. Graña, M., Apolloni, B., Fiasché, M., Galliani, G., Zizzo, C., Caridakis, G.,
   Siolas, G., Kollias, S., Barrientos, F., San Jose, S.: Social and smart: Towards an instance of subconscious social intelligence. In: Iliadis, L., Papadopoulos, H., Jayne, C.
   (eds.) EANN 2013, Part II. CCIS, vol. 384, pp. 302–311. Springer, Heidelberg (2013)
2. Avants, B., Dhillon, P., Kandel, B.M., Cook, P.A., McMillan, C.T., Grossman, M.,
   Gee, J.C.: Eigenanatomy improves detection power for longitudinal cortical change
   (cited By 3). In: Ayache, N., Delingette, H., Golland, P., Mori, K. (eds.) MICCAI
   2012, Part III. LNCS, vol. 7512, pp. 206–213. Springer, Heidelberg (2012)

3. Avants, B.B., Libon, D.J., Rascovsky, K., Boller, A., McMillan, C.T., Massimo, L., Coslett, H.B., Chatterjee, A., Gross, R.G., Grossman, M.: Sparse canonical correlation analysis relates network-level atrophy to multivariate cognitive measures in a neurodegenerative population. NeuroImage 84, 698–711 (2014)
4. Babagholami-Mohamadabadi, B., Jourabloo, A., Zarghami, A., Kasaei, S.: A bayesian framework for sparse representation-based 3-d human pose estimation. IEEE Signal Processing Letters 21(3), 297–300 (2014)
5. Bi, J., Bennett, K.P., Embrechts, M., Breneman, C.M., Song, M., Guyon, I., Elisseeff, A.: Dimensionality reduction via sparse support vector machines. Journal of Machine Learning Research 3, 2003 (2003)
6. Bonacich, P.: Power and centrality; A family of measures. American Sociological Review, 52 (1987)
7. Caetano, T.S., McAuley, J.J.: Li Cheng, Quoc V. Le, and A.J. Smola. Learning graph matching. IEEE Transactions on Pattern Analysis and Machine Intelligence 31(6), 1048–1058 (2009)
8. Cicchello, O., Kremer, S.C.: Inducing grammars from sparse data sets: A survey of algorithms and results. J. Mach. Learn. Res. 4, 603–632 (2003)
9. Culp, M., Michailidis, G.: Graph-based semisupervised learning. IEEE Transactions on Pattern Analysis and Machine Intelligence 30(1), 174–179 (2008)
10. Glorot, X., Bordes, A., Bengio, Y.: Deep sparse rectifier neural networks. In: Gordon, G.J., Dunson, D.B. (eds.) Proceedings of the Fourteenth International Conference on Artificial Intelligence and Statistics (AISTATS 2011), vol. 15, pp. 315–323. Journal of Machine Learning Research - Workshop and Conference Proceedings (2011)
11. Liu, W., Wang, J., Chang, S.-F.: Robust and scalable graph-based semisupervised learning. Proceedings of the IEEE 100(9), 2624–2638 (2012)
12. Mairal, J., Bach, F., Ponce, J., Sapiro, G.: Online learning for matrix factorization and sparse coding. J. Mach. Learn. Res. 11, 19–60 (2010)
13. Mayoue, A., Barthelemy, Q., Onis, S., Larue, A.: Preprocessing for classification of sparse data: Application to trajectory recognition. In: 2012 IEEE Statistical Signal Processing Workshop (SSP), pp. 37–40 (August 2012)
14. Newman, M.: Networks: An Introduction. Oxford University Press, Inc., New York (2010)
15. Quattoni, A., Collins, M., Darrell, T.: Transfer learning for image classification with sparse prototype representations. In: IEEE Conference on Computer Vision and Pattern Recognition, CVPR 2008, pp. 1–8 (June 2008)
16. Tipping, M.E.: Sparse bayesian learning and the relevance vector machine. J. Mach. Learn. Res. 1, 211–244 (2001)
17. Wang, C.-D., Lai, J.-H., Zhu, J.-Y.: Graph-based multiprototype competitive learning and its applications. IEEE Transactions on Systems, Man, and Cybernetics, Part C: Applications and Reviews 42(6), 934–946 (2012)
18. Xu, X., Huang, Z., Graves, D., Pedrycz, W.: A clustering-based graph laplacian framework for value function approximation in reinforcement learning. IEEE Transactions on Cybernetics 44(12), 2613–2625 (2014)

Self-sampling Strategies for Multimemetic Algorithms in Unstable Computational Environments

Rafael Nogueras and Carlos Cotta[(⊠)]

Dept. Lenguajes y Ciencias de la Computación, Universidad de Málaga,
ETSI Informática, Campus de Teatinos, 29071 Málaga, Spain
ccottap@lcc.uma.es

Abstract. Optimization algorithms deployed on unstable computational environments must be resilient to the volatility of computing nodes. Different fault-tolerance mechanisms have been proposed for this purpose. We focus on the use of dynamic population sizes in the context of island-based multimemetic algorithms, namely memetic algorithms which explicitly represent and evolve memes alongside solutions. These strategies require the eventual creation of new solutions in order to enlarge island populations, aiming to compensate the loss of information taking place when neighboring computing nodes go down. We study the influence that the mechanism used to create these new individuals has on the performance of the algorithm. To be precise, we consider the use of probabilistic models of the current population which are subsequently sampled in order to produce diverse solutions without distorting the convergence of the population and the progress of the search. We perform an extensive empirical assessment of those strategies on three different problems, considering a simulated computational environment featuring diverse degrees of instability. It is shown that these self-sampling strategies provide a performance improvement with respect to random reinitialization, and that a model using bivariate probabilistic dependencies is more effective in scenarios with large volatility.

1 Introduction

One of the greatest advantages of metaheuristics, and in particular of population-based variants thereof, is their amenability for deployment on parallel and distributed environments [2]. Consider for example the so-called island model of evolutionary algorithms [24], whereby multiple populations evolve in parallel and occasionally exchange information. Distributing these islands among different computing nodes can lead to notably improved solutions and remarkable reductions in the computational time required to reach them [1]. Although this strategy has been successfully exploited since the late 1980s, emerging computational environments such as peer-to-peer (P2P) networks [13] and volunteer computing networks [22] are currently bringing both new opportunities and new challenges. Regarding the latter, it must be noted that the nature of these

© Springer International Publishing Switzerland 2015
J.M. Ferrández Vicente et al. (Eds.): IWINAC 2015, Part II, LNCS 9108, pp. 69–78, 2015.
DOI: 10.1007/978-3-319-18833-1_8

computational platforms is inherently dynamic due to the volatility of computational resources attached to them (i.e., computing nodes can enter and abandon the system in a dynamic way – the term *churn* has been coined to denote this phenomenon). For this reason, it is essential that algorithms running on this kind of environments are adequately suited to work under these conditions. For example, it has been shown that the instability of the computational environment can lead to the loss of the current incumbent solution in island-based evolutionary algorithms (EAs) [8]; it will also pose additional difficulties to the progress of the search due to the loss of good solutions, genetic diversity, etc. Although EAs are intrinsically robust at a fine-grain scale [10], algorithms must nevertheless be fault-aware in some sense in order to cope with highly volatile systems. In this line, different fault-management strategies have been proposed – see [16]. In essence, these strategies can be seen as corrective, exhorting their effect when a computing node re-enters the system after having left it previously. As an alternative to these strategies, we can think of reactive strategies whereby the algorithm self-adapts to the changing environment on the fly.

An example of such strategies can be found in the model for dynamic resizing of populations defined by Nogueras and Cotta – see Sect. 2. Unlike other fault-management strategies, this scheme is intrinsically decentralized and emergent, suiting computational scenarios without global control, and does not require external persistent storage. By using this model, a rather constant population size is kept on a global scale, having the size of individual populations fluctuate to compensate for node activations/deactivations. Notice in this sees that while populations can be easily reduced by truncation and/or distribution of individuals, enlarging them requires the creation of new solutions. In this work we focus on this aspect and study different probabilistic modeling strategies for this purpose. We apply these in the context of island-based model of multimemetic algorithms (MMAs) [11], an extension of memetic algorithms [15] in which computational representations of problem solving strategies (neighborhood definitions for a local search operator in this case) are explicitly stored and evolved as a part of solutions, much in the line of the concept of memetic computing [20]. We consider the deployment of these techniques on a simulated computational environment that allows experimenting with different churn rates. We report the results of a broad empirical assessment of the strategies considered in Sect. 4. We close the paper with conclusions and an outline of future work in Sect. 5.

2   Algorithmic Model

We have n_ι panmictic (i.e., unstructured) islands, each of them running a MMA in which memes are attached to individuals and evolve alongside them. These memes are represented as pattern-based rewriting rules following the model by Smith [23]. Memes are subject to mutation and are transferred from parent to offspring via local selection (offspring inherit the meme of the best parent). We refer to [18] for details. Besides the use of memes, the MMA run on each island can be otherwise regarded as a steady-state evolutionary algorithm using

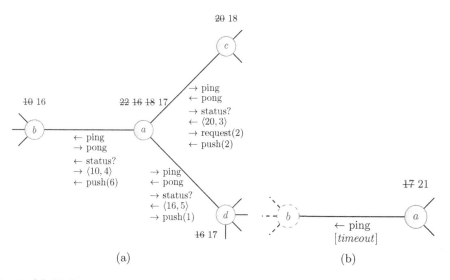

Fig. 1. (a) Node a attempts to balance with nodes b, c and d in that order. The numbers next to each node indicate the population sizes (crossed-out values correspond to previous sizes). (b) Node a attempts to balance with node b which is now inactive. The information from the previous balancing attempt, namely that node b had size 16 and 4 active neighbors, is recalled in order to enlarge the population of node a by 16/4 individuals.

tournament selection, one-point crossover, bit-flip mutation, and replacement of the worst parent. From a global point of view, the islands work in parallel and are interconnected as a scale-free network, a pattern of connectivity observed in many real-world systems, and in particular in peer-to-peer systems. We use Barabási-Albert model [3] in order to generate these networks. This network topology is used for the purposes of information exchange via migration. This operation is asynchronously performed with a certain probability at the end of each cycle. When performed, we use random selection of migrants and deterministic replacement of the worst individuals in the receiving island [17].

To model the instability of the computational environment we assume each island of the MMA runs on one out of n_ι nodes. These nodes are volatile and become inactive or active on a time-dependent basis (much like P2P networks or volunteer computing platforms). To be precise we use a Weibull distribution to model the dynamics of the system – see [12, 16]. In response to such variability of computational resources, the MMA tries to dynamically resize the islands so as to counteract the loss of information taking place when a computing node goes down. To this end, each island performs a handshake with its neighbors at the beginning of each evolutionary cycle and tries to balance the size of their population by exchanging individuals (see Fig. 1a). In the event of a new computing node going up, this same process is used for it to absorb a fraction of its neighbors' population (using a default population size C_1 if no neighbor can donate solutions). If during this handshaking process a node detects that a neighbor that was active has just become inactive, it enlarges its own population to compensate the loss of that island (see Fig. 1b).

For this purpose new individuals have to be created. Next section studies how to approach this task.

3   Self-sampling Strategies

Dynamic population resizing strategies such as those described before only capture the quantitative aspect of the process when it comes to enlarge a population without having a donor to provide solutions. The simplest approach to solve the qualitative question of how to produce those required individuals is to resort to blind search: new solutions, as many as needed, are generated from scratch, essentially using the same mechanism by which the initial population was created. While this can be regarded as an appropriate method to boost diversity, introducing fresh information in the population, it certainly has the drawback of not keeping up with the momentum of the search. In fact, the new random individuals are to a certain extent dragging backwards the population in terms of global convergence to promising regions of the search space. Needless to say, this effect will be more or less strong depending upon the frequency at which the algorithm has to resort to this mechanism, which will be directly linked to the severity of churn.

As an alternative to such a random reinitialization from scratch, some smarter strategies based on probabilistic modeling could be used. The underlying idea would be to estimate the joint probability distribution $p(\boldsymbol{x})$ representing the population of the island to be enlarged; subsequently, that distribution would be sampled as many times as necessary in order to produce the new individuals. By doing so, two objectives are pursued: (1) the momentum of the search is kept since the new individuals would be representative of the current state of the population (as indicated by the probabilistic model) and (2) diversity would be still introduced since the new individuals can be different to individuals already in the population.

To approach the probabilistic modeling we assume the population is a matrix $[pop_{ij}]$ where each row $pop_{i\cdot}$ represents an individual and each column $pop_{\cdot j}$ represent a variable (a genetic/memetic symbol). We consider two alternatives to model this population: univariate and bivariate models. In univariate models, variables are assumed to be independent and therefore the joint probability distribution $p(\boldsymbol{x})$ is factorized as

$$p(\boldsymbol{x}) = \prod_{j=1}^{n} p(x_j).$$

This is the model used by simple estimation of distribution algorithms (EDAs) such as UMDA [14], in which $p(x_j)$ is estimated as

$$p(x_j = v) = \frac{1}{\mu} \sum_{i=1}^{\mu} \delta(pop_{ij}, v),$$

where $\mu = |pop|$ is the size of the population, pop_{ij} is the value of the j-th variable of the i-th individual in pop, and $\delta(\cdot,\cdot)$ is Kronecker delta ($\delta(a,b) = 1$ if $a = b$ and $\delta(a,b) = 0$ otherwise).

On the other hand, bivariate models assume relations between pairs of variables. Thus, $p(\boldsymbol{x})$ is factorized as

$$p(\boldsymbol{x}) = p(x_{j_1}) \prod_{i=2}^{n} p(x_{j_i}|x_{j_{a(i)}}),$$

where $j_1 \cdots j_n$ is a permutation of the indices $1 \cdots n$, and $a(i) < i$ is the permutation index of the variable which x_{j_i} depends on. This model is used by EDAs such as MIMIC [5] and COMIT [4]. We focus on a mechanism analogous to the latter EDA and hence we pick j_1 as the variable with the lowest entropy $H(X_k)$ in the population, and then we pick j_i ($i > 1$) as the variable (among those not yet selected) that minimizes $H(X_k|X_{j_s}, s < i)$. Each variable will therefore depend on a variable previously selected hence leading to a tree-like dependence structure. As a final detail, it must be noted that we compute separate models for both genotypes and memes.

4   Experimental Analysis

The experiments have been done using an MMA with $n_\iota = 32$ islands whose initial size is $\mu = 16$ individuals, interconnected utilizing a scale-free topology generated with Barabási-Albert model using parameter $m = 2$. We consider crossover probability $p_X = 1.0$, mutation probability $p_M = 1/\ell$ –where ℓ is the genotype length– and migration probability $p_{\mathrm{mig}} = 1/80$. As indicated in Sect. 2, memes are rewriting rules. Following [18], their length varies from $l_{\min} = 3$ up to $l_{\max} = 9$ and is mutated with probability $p_r = 1/l_{\max}$. To control churn, we consider that the Weibull distribution describing node availability is defined by the shape parameter $\eta = 1.5$ (implying an increasing hazard rate since it is larger than 1) and by six scale parameters $\beta = -1/\log(p)$ for $p = 1 - (Kn_\iota)^{-1}$, $K \in \{1,2,5,10,20,\infty\}$ describing scenarios of different volatility: from no churn ($K = \infty$) to very high churn ($K = 1$). As an approximate interpretation, notice that under an hypothetically constant hazard rate these scale parameters would correspond to an average of one island going down/up every K cycles. Parameter C_1 used during eventual island reinitialization from scratch is set to $2\mu = 32$. We perform 25 simulations for each algorithm and churn scenario, each of them comprising $maxevals = 50000$ evaluations. We denote by $\mathrm{LBQ_{rand}}$, $\mathrm{LBQ_{umda}}$ and $\mathrm{LBQ_{comit}}$ the MMAs using random reinitialization of individuals, or probabilistic modeling with univariate (UMDA-like) or bivariate (COMIT-like) models respectively. As a reference we also consider an MMA without any balancing at all which we denote as noB. We have considered three test functions, namely Deb's trap (TRAP) function [6] (concatenating 32 four-bit traps), Watson et al.'s Hierarchical-if-and-only-if (HIFF) function [25] (using 128 bits) and Goldberg et al.'s Massively Multimodal Deceptive Problem (MMDP) [7] (using 24 six-bit blocks).

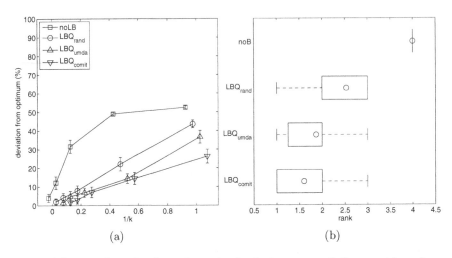

(a)   (b)

Fig. 2. (a) Mean deviation from the optimal solution across all three problems for each algorithm as a function of the churn rate. (b) Distribution of ranks for each algorithm. As usual the boxes comprise the 2nd and 3rd quartiles of the distribution, the whiskers extend to the most extreme data points not considered outliers and the circle marks the mean rank.

Table 1. Results of Holm Test ($\alpha = 0.05$) using LBQ_{comit} as control algorithm

i	algorithm	z-statistic	p-value	α/i
1	LBQ_{umda}	5.657e–01	2.858e–01	5.000e–02
2	LBQ_{rand}	1.980e+00	2.386e–02	2.500e–02
3	noB	5.091e+00	1.779e–07	1.667e–02

A global view of the results is provided by Fig. 2a. This figure shows the relationship between the mean distance to the optimum attained by each algorithm –averaged for the three problems– as a function of the churn rate (complete numerical data is provided in Table 2). Not surprisingly the distance to the optimum increases for increasing churn for all algorithms. It is however interesting to note how the different algorithms do this at a different rate. In the case of noB there is an abrupt performance degradation as we move to the right of the X axis (increasing churn) but this degradation is much more gently (rather linear actually) for variants with balancing. Among these, variants with probabilistic modeling perform distinctly better than LBQ_{rand}, exhibiting a lower slope, that is, a smaller degradation for increasing churn. This is more clearly seen in Fig. 2b in which a box plot of the distribution of relative ranks (1 for the best, 4 the worst) of each algorithm on each problem and churn scenario is shown. The advantage of LBQ variants over noB as well as the advantage of $LBQ_{[umda|comit]}$ over LBQ_{rand} is clearly depicted. To ascertain the significance of these differences, we have conducted the Quade test [21], obtaining that at least one algorithm is significantly different to the rest (p-value ≈ 0). Subsequently we have conduct a post-hoc test, namely Holm test [9] (see Table 1) using LBQ_{comit} as control algorithm. It is shown to be significantly different to

Table 2. Results (25 runs) in terms of deviation from the optimal solution of the different MMAs on the three problems considered. The median $(\tilde{x})$, mean $(\bar{x})$ and standard error of the mean $(\sigma_{\bar{x}})$ are indicated. The three symbols next to each entry indicate whether differences are statistically significant ($\bullet$) or not ($\circ$). The first symbol correspond to a comparison between the corresponding algorithm and the fault-less ($k = \infty$) version; the second one reflects a comparison with $\mathrm{LBQ_{rand}}$ and the third one is a comparison with the algorithm that provides the best results for the corresponding problem and value of k (marked with $\star$).

strategy	K	TRAP $\tilde{x}$	TRAP $\bar{x} \pm \sigma_{\bar{x}}$	HIFF $\tilde{x}$	HIFF $\bar{x} \pm \sigma_{\bar{x}}$	MMDP $\tilde{x}$	MMDP $\bar{x} \pm \sigma_{\bar{x}}$
–	∞	0.00	0.55 ± 0.18	0.00	1.00 ± 1.00	1.50	2.08 ± 0.33
noB	20	1.25	1.65 ± 0.39 $\bullet\bullet\bullet$	0.00	4.88 ± 2.05 $\circ\circ\circ$	5.99	5.51 ± 0.77 $\bullet\bullet\bullet$
	10	8.75	8.72 ± 1.09 $\bullet\bullet\bullet$	0.00	12.30 ± 3.11 $\bullet\circ\circ$	13.48	15.25 ± 1.03 $\bullet\bullet\bullet$
	5	27.50	28.59 ± 1.49 $\bullet\bullet\bullet$	44.44	39.61 ± 3.28 $\bullet\bullet\bullet$	25.13	26.55 ± 0.69 $\bullet\bullet\bullet$
	2	48.12	47.49 ± 0.71 $\bullet\bullet\bullet$	61.98	61.51 ± 0.43 $\bullet\bullet\bullet$	38.45	38.02 ± 0.51 $\bullet\bullet\bullet$
	1	51.88	52.35 ± 0.57 $\bullet\bullet\bullet$	64.76	64.17 ± 0.27 $\bullet\bullet\bullet$	41.12	40.93 ± 0.54 $\bullet\bullet\bullet$
$\mathrm{LBQ_{rand}}$	20	0.00	0.50 ± 0.26 $\circ\circ\bullet$	0.00	2.83 ± 1.64 $\circ\circ\circ$	3.00	2.56 ± 0.38 $\circ\circ\circ$
	10	0.00	0.45 ± 0.19 $\circ\circ\star$	0.00	7.28 ± 2.32 $\bullet\circ\circ$	4.49	4.71 ± 0.55 $\bullet\circ\bullet$
	5	5.00	5.22 ± 0.75 $\bullet\circ\bullet$	0.00	9.67 ± 2.55 $\bullet\circ\star$	8.99	8.72 ± 0.66 $\bullet\circ\bullet$
	2	21.88	22.83 ± 1.25 $\bullet\circ\bullet$	21.88	21.71 ± 3.39 $\bullet\circ\bullet$	21.80	21.48 ± 0.87 $\bullet\circ\bullet$
	1	44.38	44.15 ± 1.04 $\bullet\circ\bullet$	51.39	50.17 ± 1.79 $\bullet\circ\bullet$	36.78	36.38 ± 0.57 $\bullet\circ\bullet$
$\mathrm{LBQ_{umda}}$	20	0.00	0.00 ± 0.00 $\bullet\bullet\star$	0.00	2.44 ± 1.35 $\circ\circ\star$	1.50	1.89 ± 0.45 $\circ\circ\circ$
	10	0.00	0.60 ± 0.24 $\circ\circ\circ$	0.00	9.94 ± 2.81 $\bullet\circ\circ$	4.49	3.93 ± 0.45 $\bullet\circ\bullet$
	5	1.88	2.82 ± 0.57 $\bullet\bullet\star$	0.00	10.44 ± 2.36 $\bullet\circ\circ$	7.49	7.25 ± 0.57 $\bullet\circ\bullet$
	2	17.50	15.55 ± 1.02 $\bullet\bullet\bullet$	0.00	6.77 ± 2.27 $\bullet\bullet\star$	21.14	20.56 ± 0.83 $\bullet\circ\bullet$
	1	38.75	38.76 ± 1.22 $\bullet\bullet\bullet$	35.07	34.01 ± 3.20 $\bullet\bullet\bullet$	37.80	37.28 ± 0.88 $\bullet\circ\bullet$
$\mathrm{LBQ_{comit}}$	20	0.00	0.10 ± 0.07 $\bullet\circ\circ$	0.00	3.28 ± 1.54 $\circ\circ\circ$	1.50	1.66 ± 0.37 $\circ\circ\star$
	10	0.00	0.70 ± 0.28 $\circ\circ\circ$	0.00	5.83 ± 2.17 $\bullet\circ\star$	3.00	2.47 ± 0.40 $\circ\bullet\star$
	5	3.75	3.60 ± 0.62 $\bullet\circ\circ$	0.00	12.28 ± 2.80 $\bullet\circ\circ$	5.99	5.20 ± 0.52 $\bullet\bullet\star$
	2	10.62	10.88 ± 0.82 $\bullet\bullet\star$	19.44	17.89 ± 3.08 $\bullet\circ\bullet$	14.98	14.23 ± 0.83 $\bullet\bullet\star$
	1	27.50	26.92 ± 1.42 $\bullet\bullet\star$	21.53	18.95 ± 3.54 $\bullet\bullet\star$	32.14	32.88 ± 0.75 $\bullet\bullet\star$

both noB and $\mathrm{LBQ_{rand}}$ at $\alpha = 0.05$ level. The difference between $\mathrm{LBQ_{umda}}$ and $\mathrm{LBQ_{comit}}$ is not statistically significant at this level on a global scale. A more fine-grained perspective is shown in Table 2. In most cases $\mathrm{LBQ_{comit}}$ provides the best results and it is better (with statistical significance) than $\mathrm{LBQ_{umda}}$ for the most extreme churn scenarios. This is further illustrated in Fig. 3 in which the evolution of best fitness (for the TRAP function) is shown for variants using probabilistic modeling. The performance is statistically indistinguishable for low and moderate churn (up to $K = 5$) but there is a clear superiority (statistically significant) of $\mathrm{LBQ_{comit}}$ in the two most severe scenarios. This can be explained by the fact that (1) population enlarging is less frequently demanded in scenarios with low churn, and hence the particular strategy chosen bears a smaller effect on performance and (2) the better accuracy of the bivariate model for modeling the population seems to be decisive to keep the progress of the search in scenarios with severe churn.

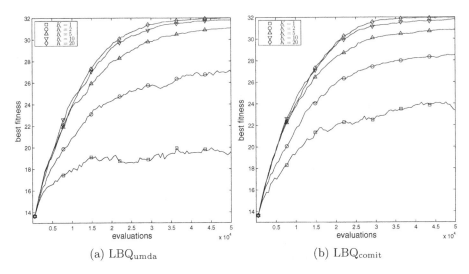

(a) LBQ$_{umda}$ (b) LBQ$_{comit}$

Fig. 3. Evolution of best fitness in TRAP for LBQ$_{umda}$ and LBQ$_{commit}$ depending on the churn rate

5   Conclusions

Resilience is a key feature optimization techniques must exhibit in order to be successfully deployed on unstable computational environments. To this end, we have studied in this work the effect that the use of self-sampling strategies have on the performance of island-based multimemetic algorithms endowed with dynamic population balancing mechanisms. These self-sampling procedures are aimed to produce new solutions (used to enlarge the population of an island when needed) similar but not identical to other solutions in a certain population, so that the distortion on the search caused by churn is kept at a minimum. This is done by building a probabilistic model of the current population and subsequently sampling it. We have shown that the use of these strategies can effectively improve the performance of the base algorithm, maintaining better the momentum of the search particularly in situations of severe churn. We have considered two different approaches for building the probabilistic model, based on a univariate and bivariate dependencies respectively. While both approaches provide globally comparable results, the latter seems again superior when churn is high, a fact which is interpreted in light of the need of more accurate population models in such highly unstable scenarios. There are many directions for future work. In the short term we plan to extend the collection of environmental scenarios considered by including other network parameters as well as dynamically-rewired network topologies [19]. Considering more complex probabilistic models (i.e., multivariate models) is another objective which we plan to approach in the mid-term.

Acknowledgements. This work is partially supported by MICINN project ANY-SELF (TIN2011-28627-C04-01) and EphemeCH (TIN2014-56494-C4-1-P), by Junta de Andalucía project DNEMESIS (P10-TIC-6083) and by Universidad de Málaga, Campus de Excelencia Internacional Andalucía Tech.

References

1. Alba, E.: Parallel evolutionary algorithms can achieve super-linear performance. Information Processing Letters 82, 7–13 (2002)
2. Alba, E.: Parallel Metaheuristics: A New Class of Algorithms. Wiley-Interscience (2005)
3. Albert, R., Barabási, A.L.: Statistical mechanics of complex networks. Review of Modern Physics 74(1), 47–97 (2002)
4. Baluja, S., Davies, S.: Using optimal dependency-trees for combinatorial optimization: Learning the structure of the search space. In: 14th International Conference on Machine Learning, pp. 30–38. Morgan Kaufmann (1997)
5. Bonet, J.S.D., Isbell Jr., C.L., Viola, P.: MIMIC: Finding optima by estimating probability densities. In: Mozer, M., Jordan, M., Petsche, T. (eds.) Advances in Neural Information Processing Systems, vol. 9, pp. 424–430. The MIT Press (1996)
6. Deb, K., Goldberg, D.E.: Analyzing deception in trap functions. In: Whitley, L.D. (ed.) Second Workshop on Foundations of Genetic Algorithms, pp. 93–108. Morgan Kaufmann, Vail (1993)
7. Goldberg, D.E., Deb, K., Horn, J.: Massive multimodality, deception, and genetic algorithms. In: Parallel Problem Solving from Nature – PPSN II, pp. 37–48. Elsevier, Brussels (1992)
8. Hidalgo, J.I., Lanchares, J., Fernández de Vega, F., Lombraña, D.: Is the island model fault tolerant? In: Proceedings of the 9th Annual Conference Companion on Genetic and Evolutionary Computation, GECCO 2007, pp. 2737–2744. ACM, New York (2007)
9. Holm, S.: A simple sequentially rejective multiple test procedure. Scandinavian Journal of Statistics 6, 65–70 (1979)
10. Jiménez Laredo, J.L., Bouvry, P., Lombraña González, D., Fernández de Vega, F., García Arenas, M., Merelo Guervós, J.J., Fernandes, C.M.: Designing robust volunteer-based evolutionary algorithms. Genetic Programming and Evolvable Machines 15(3), 221–244 (2014)
11. Krasnogor, N., Blackburne, B.P., Burke, E.K., Hirst, J.D.: Multimeme algorithms for protein structure prediction. In: Guervós, J.J.M., Adamidis, P.A., Beyer, H.-G., Fernández-Villacañas, J.-L., Schwefel, H.-P. (eds.) PPSN 2002. LNCS, vol. 2439, pp. 769–778. Springer, Heidelberg (2002)
12. Liu, C., White, R.W., Dumais, S.: Understanding web browsing behaviors through weibull analysis of dwell time. In: Proceedings of the 33rd International ACM SIGIR Conference on Research and Development in Information Retrieval, SIGIR 2010, pp. 379–386. ACM, New York (2010)
13. Milojičić, D.S., Kalogeraki, V., Lukose, R., Nagaraja, K., Pruyne, J., Richard, B., Rollins, S., Xu, Z.: Peer-to-peer computing. Tech. Rep. HPL-2002-57, Hewlett-Packard Labs (2002)
14. Mühlenbein, H., Paaß, G.: From recombination of genes to the estimation of distributions I. Binary parameters. In: Ebeling, W., Rechenberg, I., Voigt, H.-M., Schwefel, H.-P. (eds.) PPSN 1996. LNCS, vol. 1141, pp. 178–187. Springer, Heidelberg (1996)

15. Neri, F., Cotta, C., Moscato, P.: Handbook of Memetic Algorithms. SCI, vol. 379. Springer, Heidelberg (2013)
16. Nogueras, R., Cotta, C.: Studying fault-tolerance in island-based evolutionary and multimemetic algorithms. Journal of Grid Computing (2015), doi:10.1007/s10723-014-9315-6
17. Nogueras, R., Cotta, C.: An analysis of migration strategies in island-based multimemetic algorithms. In: Bartz-Beielstein, T., Branke, J., Filipič, B., Smith, J. (eds.) PPSN 2014. LNCS, vol. 8672, pp. 731–740. Springer, Heidelberg (2014)
18. Nogueras, R., Cotta, C.: On meme self-adaptation in spatially-structured multimemetic algorithms. In: Dimov, I., Fidanova, S., Lirkov, I. (eds.) NMA 2014. LNCS, vol. 8962, pp. 70–77. Springer, Heidelberg (2015)
19. Nogueras, R., Cotta, C.: Self-balancing multimemetic algorithms in dynamic scale-free networks. In: Mora, A.M., Squillero, G. (eds.) EvoApplications 2015. LNCS, vol. 9028, pp. 177–188. Springer, Heidelberg (2015)
20. Ong, Y.S., Lim, M.H., Chen, X.: Memetic computation-past, present and future. IEEE Computational Intelligence Magazine 5(2), 24–31 (2010)
21. Quade, D.: Using weighted rankings in the analysis of complete blocks with additive block effects. Journal of the American Statistical Association 74, 680–683 (1979)
22. Sarmenta, L.F.: Bayanihan: Web-based volunteer computing using java. In: Masunaga, Y., Tsukamoto, M. (eds.) WWCA 1998. LNCS, vol. 1368, pp. 444–461. Springer, Heidelberg (1998)
23. Smith, J.E.: Self-adaptation in evolutionary algorithms for combinatorial optimisation. In: Cotta, C., Sevaux, M., Sörensen, K. (eds.) Adaptive and Multilevel Metaheuristics. SCI, vol. 136, pp. 31–57. Springer, Heidelberg (2008)
24. Tanese, R.: Distributed genetic algorithms. In: 3rd International Conference on Genetic Algorithms, pp. 434–439. Morgan Kaufmann Publishers Inc., San Francisco (1989)
25. Watson, R.A., Hornby, G.S., Pollack, J.B.: Modeling building-block interdependency. In: Eiben, A.E., Bäck, T., Schoenauer, M., Schwefel, H.-P. (eds.) PPSN 1998. LNCS, vol. 1498, pp. 97–106. Springer, Heidelberg (1998)

On the Influence of Illumination Quality in 2D Facial Recognition

Ángel Sánchez[✉], José F. Vélez, and A. Belén Moreno

Departamento de Informática y Estadística, Universidad Rey Juan Carlos,
c/Tulipán s/n, 28933 Móstoles (Madrid), Spain
angel.sanchez@urjc.es

Abstract. Detecting automatically whether a facial image is greatly affected or not by the illumination conditions, allows us in some cases to discard those deteriorated images for further recognition tasks or, in other cases, to apply a preprocessing method only to the images that really need it. With this aim, our paper presents a study on the isolated influence of illumination quality of 2D images in facial recognition. First, a fuzzy inference system is designed to be as an objective and automatic method to evaluate the illumination quality of facial patterns. Then, we estimate the best recognition result for the same images using different image classification methods. By combining both the illumination quality with the corresponding recognition results for same face images, and computing the regression line of this set of patterns, we detect a nearly-linear regression trend between illumination quality and recognition rate for the images tested. This result can then be used as a quality measure of patterns in 2D facial recognition, and also for deciding whether it is worth or not using these facial images in recognition tasks.

Keywords: Illumination quality · Facial variations · 2D facial recognition · Anthropometric feature points · Fuzzy inference system

1 Introduction

Degradation of visual quality in digital images can be caused by a large variety of distortions during adquisition, processing, compression, storage or transmission processes [5]. In the applications where images are viewed by humans, the method for quantifying visual image quality (IQ) is the subjective evaluation. However, this kind of evaluation can be time-consuming and expensive. In consequence, it would be more practical to develop quantitative and objective measures for automatically determining IQ assessment in agreement with human quality judgments [14]. Different IQ analysis algorithms and measures have been proposed and compared in recent years [12][14]. According to Zou et al. [16], the methods addressing the illumination variation problem can be classified into two groups: "passive" and "active" methods, respectively. The "passive" approaches attempt to overcome the problem by considering the visible spectrum images where face appearance has been changed by the illumination variations. This group includes methods applying illumination variation models in linear/non-linear subspaces, those using illumination invariant features (i.e. which are obtained with Gabor-like filters), algorithms that introduce photometric normalization and methods based on 3D

© Springer International Publishing Switzerland 2015
J.M. Ferrández Vicente et al. (Eds.): IWINAC 2015, Part II, LNCS 9108, pp. 79–87, 2015.
DOI: 10.1007/978-3-319-18833-1_9

morphable models. In the "active" approaches the ilumination variation problem is overcomed by using active image techniques to capture facial patterns in consistent illumination conditions or using images with invariant illumination. This group of "active" methods includes also the use of 3D face information, and/or additional face images in other non-visible spectra (i.e. thermal infrared ones). Moreover, many works apply a "two-step" framework [15] addressing the illumination variation problem, where first an appropriate illumination preprocessing is performed and then a facial feature extraction is applied for capturing the effect2D face representation in the recognition task under varying illumination.

Illumination changes in a scene produce many complex effects on the image of the objects [3]. Different methods and measures to determine how faces were affected by the illumination conditions have also been proposed [11] [1]. Such variations in the appearance of faces could be even much larger than those caused by personal identity [16]. Adini, Moses and Ullman [3] studied these important image variations due to changes in illumination direction. Many 2D and 3D standard face datasets consider the different types of facial variabilities in order to evaluate the merits of the analyzed face recognition algorithms [2] [10]. However, and up to our knowledge, there are no studies that quantify how the isolated illumination degree of patterns affects to the 2D recognition of these faces. In this paper, we evaluate the illumination effect without considering other possible intrinsic conditions (such as pose, expressions or partial occlusions of the faces) or extrinsic ones (such as indoor/outdoor scenarios, different camera orientations or varying scales) in the facial recognition results. For our purpose, it is important the selection of an appropriate database for the tests where the illumination conditions in patterns could be isolated from other types of facial variations. Such is the case of the Yale Face Database B [4] which also presents a large variability of illumination conditions in the images of subjects contained in it.

In this work, we used only a simple facial pattern representation: a vector formed by the ordered concatenation of local window regions around a set of facial feature points. Different classification techniques like K-NN, SVM and two types of MLP neural networks were tested. As the question addressed in this study is: "How does it affect the global face illumination alone to the recognition of people?", we searched an objective IQ assessment method that would avoid the further processing of images presenting low quality for facial recognition. Different from other works, we neither compensate nor correct the illumination effect by making some assumptions about the light source [4] nor carry out any illumination normalization stage [6] like in the retinex-based methods [9].

2   Proposed Approach for Studyding the Isolated Influence of Illumination in Facial Recognition

This section describes the architecture of our approach to analyze the isolated influence of illumination in facial recognition. Next, we present the two independent subsystems to separately determine the recognition and illumination quality estimations for each test image. Finally, both subsystems are combined to produce a collection of two-dimensional facial test patterns. Statistical regression analysis is used to estimate the relationship between illumination quality and recognition result variables.

2.1   System Architecture

Fig. 1 shows the complete system designed to automatically determine how illumination quality and facial recognition are related for the 2D test images. Two subsystems, which work separately with the facial images, are involved: the recognition estimation and the illumination estimation subsystems, respectively. After that, both components are combined in order to determine a trend relation between illumination and recognition using a regression analysis approach.

In the 2D facial model considered, a collection of twenty relevant anthropometric facial feature points (see Fig. 2 with the acronyms of selected points), corresponding to a subset of those ones defined by the standard MPEG-4 [7], were automatically extracted from each facial image. For each of these points, a texture squared-region 21×21 size centered on the respective feature points was selected to create the input pattern vectors for each 2D face image. The appropriate size of the regions was determined through experimentation.

2.2   Recognition Estimation Subsystem

The input feature vectors describing each face were built by concatenating, in a fixed order, the texture region pixels for the facial points in Fig. 2. We performed the experiments with three different types of classification methods: K-Nearest Neighbors (K-NN), Support Vector Machines (SVM) and MultiLayer Perceptron Neural Networks (MLP), respectively. Moreover, we have tested two configurations for a two-hidden-layer MLP networks: the fully connected MLP (FC-MLP), where all neurons of the input layer are connected to all neurons of the first hidden layer, and the non-fully connected MLP one (NFC-MLP), where the input neurons are grouped into blocks (i.e. each block correspond to a facial feature region) and these are separately connected to different and disjoint sets of neurons in the first hidden layer. We also performed the k-fold cross-validation method for all tested classifiers. Some details of best parameter configurations for each algorithm are presented in the Experiments section.

2.3   Illumination Estimation Subsystem

To automatically determine the illumination quality of a given 2D facial image I, we computed the corresponding normalized 8-bin histogram for each feature point window, as shown in Fig. 3.

After that, we calculate the average illumination I_{av} and contrast C for each window as given by the following equations.

$$I_{av} = \frac{1}{8} \sum_{1}^{8} I_{hist} \qquad C = \frac{I_{max} - I_{min}}{I_{max} + I_{min}} \qquad (1)$$

where I_{hist} is the frequency value of each histogram bin, and I_{max} and I_{min} are the respective maximum and the minimum values of the 8-bin histogram used. To determine the illumination quality of each feature window, we first fuzzify the values of the two respective input variables I_{av} and C. Triangular-shaped fuzzy sets for the input

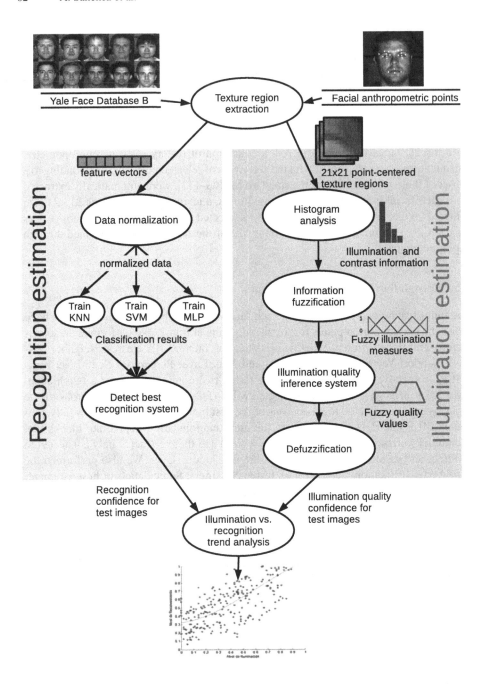

Fig. 1. Proposed methodology to analyze the influence of illumination in 2D facial recognition

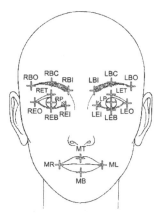

Fig. 2. Facial model with the anthropometric points considered

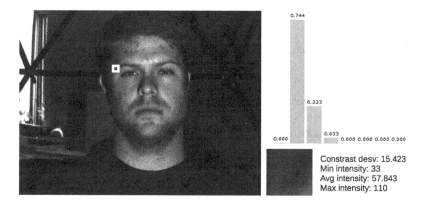

Constrast desv: 15.423
Min intensity: 33
Avg intensity: 57.843
Max intensity: 110

Fig. 3. Corresponding histogram and feature values extracted from the outer-corner right-eyebrow region (around the RBO point of Fig. 2) in a sample 2D face image

variables were used. Then, a zero-order Takagi-Sugeno (TS-0) fuzzy inference system [13] was created by considering the fuzzy rules presented in Table 1.

The output variable in this table represents the illumination quality and it classifies each feature window into six possible cathegories which are labelled from 'A' to 'F', respectively. Class 'A' corresponds to the "best" type of illumination in a feature window, while class 'F' is the "worst" one. The defuzzification process is carried out using the weighted average method [8]. Each feature window is then scored to a value which is the highest for class 'A' and the lowest for class 'F'. In particular, class 'A' was assigned a score of 5, class 'B' was given score of 4, and so on, and finally class 'F' was scored with 0. In this way, the objective quality of each given 2D facial image, represented as a sequence of 20 feature-centered windows, is a normalized integer value between 0 and 100. This "percentage" value is computed as the sum of the scores of all considered facial windows in the model presented by Fig. 2).

Table 1. Fuzzy inference rules for determining the image quality from illumination and contrast input variables

Illumination\Contrast	Very Low	Low	Medium	High	Very High
Very Bad	F	F	F	F	F
Bad	F	F	E	D	D
Fair	E	D	C	C	B
Good	D	C	B	B	A
Very Good	D	C	B	A	A

3   Experiments

This section summarizes: (a) the main features of the database used in the experiments, (b) the recognition results achieved with the different classification methods tested, and finally (c) how is the correlation between both illumination quality with facial recognition variables.

3.1   Yale Face Database B

The Yale face database B [4] is well suited for illumination-varying 2D facial recognition experiments. It contains 5,760 images taken from 10 subjects under 576 viewing conditions (that is, 9 poses and 64 illumination conditions per subject's pose). Images in the database are divided into 5 subsets based on the angle of the light source directions. This set of images was captured using a purpose-built illumination rig which is fitted with 64 computer controlled strobes. The 64 images of each subject in a particular pose were almost-simultaneously acquired by the system, so there could appear only a very small change in head pose for these 64 images. Among all the images, there is one with ambient illumination and it was captured without a strobe going off. Since we were only concerned in this work with the illumination of the facial images (and not with the pose changes), only the 640 frontal-pose images of the 10 subjects were used in our experiments (since these images represent the 64 possible illumination conditions under study).

3.2   Facial Recognition Results

Experiments were repeated for the four different types of classifiers considered: K-NN, SVM, FC-MLP and NFC-MLP, respectively. The best facial recognition results for each configuration of parameters of each pattern classification method are shown in Table 2. This Table displays the parameter values and time consuming (using an Intel I7 processor, at 2.2GHz and 4 GB of RAM) for the classification techniques evaluated using Yale Face Database B. Classifiers were tested both using single-validation and cross-validation approaches. The best value for cross-validation is shown for each specific classifier in the corresponding position of this Table. It can be noticed that best classification results for the test patterns in this database were achieved using a fully-connected multilayer perceptron (FC-MLP) classifier.

Table 2. Comparative table shown best results for each classifier tested

Classifiers		Parameters and time consuming	Simple validation results (%)	Parameters and time consuming	Cross validation results (%)
K-NN		Train 70.76 % Test 29.24 % k = 4 dim: 8820 16''	86.31	10-fold k=4 1' 20''	81.53
SVM		Train 70.76 % Test 29.24 % Kernel RBF (default) 2' 37''	79.47	10-fold Kernel RBF (default) 30' 45''	75.22
MLP	FC-MLP	Train 70.76 % Test 29.24 % Layers: [400 30 10] lr =0.001 goal = 0.01 tansig 14' 22'' (740 iterations)	100.00	5-fold Layers: [400 30 10] lr = 0.001 goal = 0.01 logsig 18' 52'' (163 mean iterations)	97.86
	NFC-MLP	Train 60 % Test 40% Layers: [400 30 10] lr =0.01 goal = 0.001 logsig 6' 03 (291 iterations)	93.23	5-fold Layers: [400 30 10] lr = 0.001 goal = 0.01 logsig (161 mean iterations) 20' 8''	97.38

3.3   Illumination Quality Versus Facial Recognition Results

The results produced by the FC-MLP (i.e. the best classifier tested) for every test pattern were combined with their corresponding estimated illumination qualities (which were calculated as explained in the previous Section). Fig. 4 shows this combination of recognition and illumination estimations for all of the 260 test patterns. One can see that there is a linear relationship between both variables. By computing the regression line for all the points in this chart (red line in Fig. 4), we obtained $y = 0.764x + 0.272$ and the angle α that forms this line with the horizontal axis is $37.3°$ showing a near-linear tendence, and that "when the illumination quality of 2D facial images increases, the corresponding recognition result achieved by the FC-MLP neural classifiers also increases linearly".

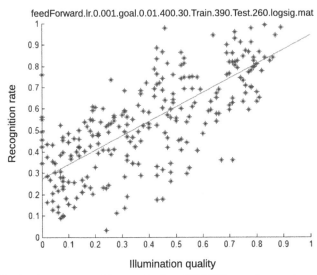

feedForward.lr.0.001.goal.0.01.400.30.Train.390.Test.260.logsig.mat

Fig. 4. Illumination vs. recognition results for a FC-MLP neural network configuration

4 Conclusion

This paper presented a study on the isolated influence of illumination quality in 2D facial recognition. First, a TS-0 fuzzy inference system to objective and automatically evaluate the illumination quality of each facial pattern is presented. Then, we estimate the facial recognition results for the best of several classification methods tested (in our case, the FC-MLP algorithm). By combining both the illumination quality estimations with the recognition results achieved for the same test patterns, and then calculating the regression line using these data, we detected a nearly-linear relation between both illumination and recognition variables. This result can be then used as an illumination quality measure of facial images and if it would be worth to use them for further recognition tasks. As future work, the following aspects need to be investigated: (a) to study the influence of the feature window size in the facial recognition results and (b) to analyze the respective discriminative capacity in the recognition task of each feature window contained in the proposed facial model.

Acknowledgements. This work has been funded by the Spanish MICINN project TIN2011-29827-C02-01.

References

1. Abaza, A., Harrison, M.A., Bourlai, T., Ross, A.: Design and evaluation of photometric image quality measures for effective face recognition. IET Biometrics 3, 314–324 (2014)
2. Abate, A.F., Nappi, M., Riccio, D., Sabatino, G.: 2D and 3D face recognition: A survey. Pattern Recognition Letters 28, 1885–1906 (2007)

3. Adini, Y., Moses, Y., Ullman, S.: Face recognition: the problem of compensating for changes in illumination direction. IEEE Trans. on Pattern Analysis and Machine Intelligence 19, 721–732 (1997)

4. Georghiades, A., Belhumeur, P.N., Kriegman, D.: From Few to Many: Illumination Cone Models for Face Recognition under Variable lighting and Pose. IEEE Trans. on Pattern Analysis and Machine Intelligence 23, 643–660 (2001)

5. Gonzalez, R.C., Woods, R.E.: Digital Image Processing, 3rd edn. Prentice Hall (2008)

6. Gross, R., Baker, S., Matthews, I., Kanade, T.: Face Recognition across Pose and Illumination. In: Li, S.Z., Jain, A.K. (eds.) Handbook of Face Recognition, pp. 193–216. Springer (2004)

7. Koenen, R.: MPEG-4 Overview, International Organisation for Standarisation, ISO/IEC JTC1/SC29/WG11 No. 4668 (2002), http://www.oipf.tv/docs/mpegif/overview.pdf

8. Leekwijck, W.V., Kerre, E.E.: Defuzzification: criteria and classification. Fuzzy Sets and Systems 108, 159–178 (1999)

9. Park, Y.K., Min, B.C., Kim, J.K.: A New Method of Illumination Normalization for Robust Face Recognition. In: Martínez-Trinidad, J.F., Carrasco Ochoa, J.A., Kittler, J. (eds.) CIARP 2006. LNCS, vol. 4225, pp. 38–47. Springer, Heidelberg (2006)

10. Phillips, P.J., et al.: Overview of the face recognition grand challenge. In: Hebert, M., Kriegman, D. (eds.) Proc. IEEE Computer Society Conference on Computer Vision and Pattern Recognition (CVPR 2005), pp. 947–954 (2005)

11. Rizo-Rodrıguez, D., Mendez-Vazquez, H., Garcıa-Reyes, E.: An Illumination Quality Measure for Face Recognition. In: Ercil, A. (ed.) Proc. 20th International Conference on Pattern Recognition (ICPR 2010), pp. 1477–1480. IEEE Press (2010)

12. Sheikh, H.R., Sabir, M.F., Bovik, A.C.: An evaluation of recent full reference image quality assessment algorithms. IEEE Trans. on Image Processing 15, 3440–3451 (2006)

13. Sugeno, M.: Industrial Applications of Fuzzy Control. Elsevier Publishing Company (1985)

14. Wang, Z., Bovik, A.C., Sheikh, H.R., Simoncelli, E.P.: Image Quality Assessment: From Error Visibility to Structural Similarity. IEEE Trans. Image Processing 13, 600–612 (2004)

15. Zhang, J., Xie, X.: A study on the effective approach to illumination-invariant face recognition based on a single image. In: Zheng, W.-S., Sun, Z., Wang, Y., Chen, X., Yuen, P.C., Lai, J. (eds.) CCBR 2012. LNCS, vol. 7701, pp. 33–41. Springer, Heidelberg (2012)

16. Zou, X., Kittler, J., Messer, K.: Illumination Invariant Face Recognition: A Survey. In: Bowyer, K.W. (ed.) Proc. First IEEE Intl. Conf. on Biometrics: Theory, Applications, and Systems (BTAS 2007), pp. 1–8 (2007)

A Binary Cuckoo Search Algorithm for Solving the Set Covering Problem

Ricardo Soto[1,2,3], Broderick Crawford[1,4,5], Rodrigo Olivares[1(✉)],
Jorge Barraza[1], Franklin Johnson[6], and Fernando Paredes[7]

[1] Pontificia Universidad Católica de Valparaíso, Chile
{ricardo.soto,broderick.crawford}@ucv.cl, jorge.barraza.c@mail.pucv.cl
[2] Universidad Autónoma de Chile, Chile
[3] Universidad Central de Chile, Chile
[4] Universidad Finis Terrae, Chile
[5] Universidad San Sebastián, Chile
[6] Universidad de Playa Ancha, Chile
franklin.johnson@upla.cl
[7] Escuela de Ingeniería Industrial, Universidad Diego Portales, Chile
fernando.paredes@udp.cl

Abstract. The non-unicost set covering problem is a classical optimization benchmark that belongs to the Karp's 21 NP-complete problems. In this paper, we present a new approach based on cuckoo search for solving such problem. Cuckoo search is a modern nature-inspired metaheuristic that has attracted much attention due to its rapid convergence and easy implementation. We illustrate interesting experimental results where the proposed cuckoo search algorithm reaches several global optimums for the non-unicost instances from the OR-Library.

Keywords: Set Covering Problem · Metaheuristics · Cuckoo Search

1 Introduction

The Set Covering Problem (SCP) is a classic problem which belongs to the class of NP-complete problems where the input is given for multiple items or set of data items that have something in common [11]. There are many applications for this type of problems such as facility location, selection files in a database, simplification of Boolean expressions, and line balancing production, among others [13]. Various algorithms have been proposed to solve it, which have been extensively reported in the literature. Exact algorithms are mostly based on branch-and-bound and branch-and-cut algorithms [1]. The problem is that these algorithms can only solve instances with limited size. To tackle this concern, much research in recent years has been focused on the development of new heuristics able to find optimal solutions in a reasonable amount of time. Classical greedy algorithms are quite fast and easy to implement, but are unable to generate high quality solutions due to their deterministic nature [8]. An improved greedy algorithm incorporating memory and random components got promising

© Springer International Publishing Switzerland 2015
J.M. Ferrández Vicente et al. (Eds.): IWINAC 2015, Part II, LNCS 9108, pp. 88–97, 2015.
DOI: 10.1007/978-3-319-18833-1_10

results [12]. Compared with classical greedy algorithms, Lagrangian relaxation-based heuristics are much more effective such as the ones proposed in [7,4]. As top-level general search strategies, metaheuristics have also been applied to solve SCPs. Some examples in this context include genetic algorithms [2], simulated annealing [3], tabu search [6], and so on. A larger description of efficient algorithms devoted to SCPs can be found in [5].

In this paper, a new approach based on binary cuckoo search involving pre-processing phases for the SCP is presented. Cuckoo Search (CS) [14,15] is a recently developed, population-based metaheuristic based on the principle of the brood parasitism of some cuckoo species. So far, it has been shown that the CS algorithm is very efficient in dealing with multimodal, global optimization problems; being also easy to tune and implement. As far as we know, CS has not been applied yet to solve SCPs. We illustrate experimental results on a set of OR-Library instances where CS appears as a good candidate for tackling this problem.

This paper is organized as follows: In Sect. 2, we formally present the SCP and describe the pre-processing method employed. In Sect. 3, we provide a overview of CS followed by the proposed approach. Finally, we present the experimental results, conclusions, and future work.

2 Problem Description

The Set Covering Problem (SCP) can be formally defined as follows:
Let $A = (a_{ij})$ be an n-row, m-column, zero-one matrix. We say that a column j covers a row i if $a_{ij} = 1$. Each column j is associated with a nonnegative real cost c_j. Let $I = \{1, ..., n\}$ and $J = \{1, ..., m\}$ be the row set and column set, respectively. The SCP calls for a minimum cost subset $S \subseteq J$, such that each row $i \in I$ is covered by at least one column $j \in S$. A mathematical model for the SCP is

$$Minimize\ f(x) = \sum_{j=1}^{m} c_j x_j \tag{1}$$

subject to

$$\sum_{j=1}^{m} a_{ij} x_j \geq 1, \quad \forall i \in I \tag{2}$$

$$x_j \in \{0, 1\}, \quad \forall j \in J \tag{3}$$

The purpose is to minimize the sum of the costs of the selected columns, where $x_j = 1$ if the column j is in the solution and $x_j = 0$ otherwise. Each row i is covered at least by one column j. This is guaranteed by SCP constraints.

2.1   Pre-processing

Preprocessing is a popular method to accelerate the solving process by reducing the instance sizes. A number of preprocessing methods for the SCP have been proposed in the literature [10]. In this study, two of those found to be most effective are used.

Column Domination: If a column j whose rows I_j can be covered by other columns with a total cost lower than c_j, we say that column j is dominated and can be removed. As expected, this reduction is ineffective for unicost problems.

Column Inclusion: If a row is covered by only one column after the above domination, this column is definitely included in an optimal solution.

3   Overview of Cuckoo Search Algorithm

In the last years, nature-inspired metaheuristics have been placed among the best algorithms for solving optimization problems. In this work, we solve the SCP by using a relatively new metaheuristic called Cuckoo search (CS). CS is inspired by the obligate brood parasitism of some cuckoo species whose main characteristic is to let their eggs in nests from other bird species. CS was developed using three fundamental rules, which are described in the following.

1. A cuckoo egg represents a solution to the problem and it is left in a randomly selected nest, a cuckoo only can left one egg at a time.
2. Nests holding the higher quality eggs will pass to the next generations.
3. The nest owner can discover a cuckoo egg with a probability $p_a \in [0, 1]$. If this occurs, the nest owner can left his nest and build other nest in other location. The number of total nests is a fixed value.

The generation of a new solution is performed by using Lévy flight as follows.

$$sol_i^{t+1} = sol_i^t + \alpha \oplus \text{Levy}(\beta) \tag{4}$$

where sol_i^{t+1} is the solution in iteration $t + 1$, and $\alpha \geq 0$ is the step size, which is associated to the range of values that the problem needs (scale value), being determined by upper and lower bounds [15].

$$\alpha = 0.01(U_b - L_b) \tag{5}$$

The step length in the generation of a new solution is given by the Lévy flight distribution.

$$\text{Lévy} \sim u = t^{-\beta}, \quad (1 < \beta < 3) \tag{6}$$

The Lévy flight distribution has an infinite variance with an infinite mean. Here, the steps essentially form a random walk process with a power-law step-length distribution with a heavy tail. A few solutions generated by Lévy walk will be neighborhoods of the best solution achieved. This will increase velocity of the local search. However, a considerable number of produced solutions by CS should be sufficiently distant from the best solution achieved. This fraction of p_a nests are generated using a random walk step. This avoid the CS to be trapped in a local optimum.

4   Description of the Proposed Approach

In this section, we describe the CS algorithm proposed to solve the SCP. We employ a binary representation for the SCP solution as depicted in Figure 1, where $x_j = 1$ if the column j belongs to the solution and $x_j = 0$ otherwise. We employ the simple CS form where each nest holds only one egg.

Fig. 1. Binary solution representation

Algorithm 1. Cuckoo Search via Lévy Flights

Initialise CS parameters;
Produce the first generation of n nests;
while $t < MaxGeneration$ **do**
    Obtain a random cuckoo/generate a solution by Lévy Flight distribution;
    Quantify its fitness F_j and then compare with old Fitness F_i;
    if $F_j > F_i$ **then**
        Substitute j as the new best solution;
    end if
    A fraction of worse nest $p(a)$ will be abandoned and new nests will be made;
    Maintain the better solutions (or nests with high quality);
    Rank the better solutions and find the best one;
end while
Postprocess results and visualization;

Algorithm 1 depicts the proposed procedure, at the beginning the CS parameters are initialized (α, β, p_a, size for the initial population and maximum number of generation) and an initial population of n nests is generated. Then, a while loop manages the CS actions which are self-explanatory. The objective function of the SCP is employed to compute the fitness of each solution. Solutions are produced by using the Lévy flight distribution according to Eq. 4. From this process, a real number between 0 and 1 is generated, which needs to be discretized. To this end, we proceed as follows:

$$x_j = \begin{cases} 1 & \text{if } r < x'_j \quad \textbf{or} \quad x'_j > U_b \\ 0 & \text{if } r \geq x'_j \quad \textbf{or} \quad x'_j < L_b \end{cases}$$

where x'_j holds the value to be discretized for variable x_j of the SCP solution, and r is a normal distributed random value. Then, the new solution produced is subjected to evaluation. If it is not feasible, it is repaired using the heuristic feasibility operator described in Sect. 4.1. Finally, the best solutions are memorized and the generation count is incremented or the process is stopped if the termination criteria has been met.

4.1 Heuristic Feasibility Operator

CS, as various metaheuristics, may provide solutions that violate the constraints of the problem. For instance, in the SCP, a new solution that has not all his rows covered, clearly violates a subset of constraints. In order to provide feasible solutions the algorithm needs additional operators. To this end, we employ a heuristic operator that achieves the generation of feasible solutions, and additionally eliminates column redundancy. This additional step allows the CS to be more efficient [2].

To make all feasible solutions we will calculate a ratio based on the sum of all the constraint matrix rows covered by a column j.

$$\frac{Cost\ of\ a\ column\ j}{number\ of\ covered\ rows\ by\ column\ j} \tag{7}$$

The unfeasible solution are repaired by covering the columns of the solution that had the lower ratio. After this, a local optimization step is applied, where column redundancy is eliminated. A column is redundant when it can be deleted and the feasibility of the solution is not affected.

Let:

> I = The set of all rows,
>
> J = The set of all columns,
>
> α_i = The set of columns that cover row i, $i \in I$,
>
> β_j = The set of rows covered by column j, $j \in J$,
>
> S = The set of columns in a solution,
>
> U = The set of uncovered rows,
>
> w_i = The number of columns that cover row i, $i \in I$.

(i) Initialize $w_i := |S \cap \alpha_i|$, $\forall i \in I$

(ii) Initialize $U := \{i | w_i = 0, \forall i \in I\}$

(iii) For each row i in U (in increasing order of i):

   (a) Find the first column j in increasing order of j in α_i that minimizes

   $c_j / |U \cap b_j|$

   (b) Add j to S and set $w_i := w_i + 1$, $\forall i \in b_j$. Set $U := U - b_j$.

(iv) For each column j in S (in decreasing order of j), if $w_i \geq 2$, $\forall i \in \beta_j$,
    set $S := S - j$ and set $w_i := w_i - 1$, $\forall i \in \beta_j$.

(v) S is now a feasible solution for the SCP that contains no redundant
    columns.

The recognition of the rows that are not covered are in Steps (i) and (ii).
Steps (iii) and (iv) are "greedy" heuristics. In step (iii) the columns with lower
ratios are added to the solution. Finally, in the step (iv) the redundant columns
with higher costs are deleted while the solution is feasible.

5   Experiments and Results

The CS Algorithm for solving SCP was implemented in Java and launched in a
3.0 GHz quad core with 8 GB RAM machine running MS Windows 7 Ultimate.
The configuration of the CS is as follows: $t = 5000$, $n = 15$, $\beta = \frac{3}{2}$, $\alpha = 0.01$, $p_a =$
0.25, $U_b = 1$, and $L_b = 0$, for instance set : 4, 5, 6, A, B, C and D. For instances
set: NRE, NRF, NRG and NRH, we increment population to $n = 50$, since the
higher hardness of such instances. We test 65 SCP non-unicost instances from
OR-Library (available at http://www.brunel.ac.uk/~mastjjb/jeb/info.html), all
of them were executed 30 times. Each instance was preprocessed by using rules
defined in Sect. 2. Table 1 shows detailed information about tested instances,
where "Density" is the percentage of non-zero entries in the SCP matrix.

Table 1. Test instances

Instance set	No. of instances	n	m	Cost Range	Density(%)	Optimal solution
4	10	200	1000	[1,100]	2	Known
5	10	200	2000	[1,100]	2	Known
6	5	200	1000	[1,100]	5	Known
A	5	300	3000	[1,100]	2	Known
B	5	300	3000	[1,100]	5	Known
C	5	400	4000	[1,100]	2	Known
D	5	400	4000	[1,100]	5	Known
NRE	5	500	5000	[1,100]	10	Unknown
NRF	5	500	5000	[1,100]	20	Unknown
NRG	5	1000	10000	[1,100]	2	Unknown
NRH	5	1000	10000	[1,100]	5	Unknown

Table 2, 3 and 4 show the results obtained using the proposed CS algorithm.
Column 1 depicts the instance number, m' represents the number of constraint
matrix columns of the preprocessed instance. Column 3 depicts the best known
optimum value for the instance. Columns 4 and 5 provide the best and worst optimum value reached from all executions. The RPD(%) represents the difference
between the best known optimum value and the best optimum value reached by
CS in terms of percentage, which is computed as follows:

$$RDP = \frac{(Z - Z_{opt})}{Z_{opt}} \times 100$$

where Z_{opt} is the best known optimum value and Z is the best optimum value reached by CS. Finally, the average value of 30 executions is depicted.

Table 2. Computational results for instances 4, 5 and 6

Instance	m'	Optimum	Min.	Max.	RPD(%)	Average
4.1	169	429	430	439	0.233	432.53
4.2	212	512	512	531	0	517.83
4.3	225	516	517	536	0.193	523.5
4.4	200	494	494	507	0	499.7
4.5	215	512	512	536	0	519.37
4.6	229	560	560	567	0	562.97
4.7	194	430	430	442	0	434.47
4.8	215	492	492	503	0	497.5
4.9	243	643	643	670	0	655.14
4.10	200	514	514	535	0	522.4
5.1	220	253	253	262	0	256.06
5.2	262	302	304	318	0.657	310.77
5.3	215	226	226	238	0	230.73
5.4	225	242	242	247	0	244,03
5.5	185	211	212	222	0.471	215,3
5.6	211	213	213	219	0	215,97
5.7	220	293	293	302	0	295,9
5.8	245	288	288	299	0	291
5.9	230	279	279	284	0	280.43
5.10	229	265	265	273	0	268.67
6.1	212	138	140	149	1.428	141.85
6.2	243	146	146	151	0	148.6
6.3	237	145	145	153	0	149.53
6.4	200	131	131	135	0	133.1
6.5	249	161	161	169	0	166.17

Table 3. Computational results for instances A, B, C and D

Instance	m'	Optimum	Min.	Max.	RPD(%)	Average
A.1	383	253	254	264	0.393	257.8
A.2	387	252	256	261	1.562	258.57
A.3	391	232	233	239	0.429	235.6
A.4	378	234	237	246	1.265	240.03
A.5	387	236	236	238	0	237.31
B.1	453	69	69	74	0	71.83
B.2	459	76	76	83	0	78.77
B.3	498	80	80	82	0	80.83
B.4	488	79	79	84	0	81.17
B.5	460	72	72	73	0	72.4

Table 3. *(Continued)*

Instance	m'	Optimum	Min.	Max.	RPD(%)	Average
C.1	519	227	228	238	0.438	231.8
C.2	561	219	221	225	0.904	223.037
C.3	588	243	247	254	1.619	250.07
C.4	550	219	221	231	0.904	225.05
C.5	541	215	216	224	0.462	220
D.1	639	60	60	65	0	61.27
D.2	679	66	66	70	0	68.17
D.3	693	72	73	79	1.369	75.03
D.4	652	62	62	65	0	63.57
D.5	627	61	61	65	0	63.1

Table 4. Computational results for instances NRE, NRF, NRG, and NRH

Instance	m'	Optimum	Min.	Max.	RPD(%)	Average
NRE.1	831	29	29	29	0	29
NRE.2	953	30	31	31	3.225	31
NRE.3	878	27	28	28	3.571	28
NRE.4	927	28	30	30	6.666	30
NRE.5	970	28	28	28	0	28
NRF.1	726	14	14	14	0	14
NRF.2	672	15	15	15	0	15
NRF.3	752	14	15	15	6.666	15
NRF.4	698	14	15	15	6.666	15
NRF.5	659	13	14	14	7.142	14
NRG.1	2076	176	176	176	0	176
NRG.2	1942	154	156	156	1.282	156
NRG.3	2003	156	169	169	7.692	169
NRG.4	1974	168	170	170	1.176	170
NRG.5	2022	168	170	170	1.176	170
NRH.1	2796	63	64	64	1.562	64
NRH.2	2745	63	64	64	1.562	64
NRH.3	2765	59	62	62	4.838	62
NRH.4	2779	58	59	59	1.694	59
NRH.5	2703	55	56	56	1.785	56

In Fig. 2, we show the evolution of the instances with a faster convergence of cost minimization. The rapid convergence of these instances demonstrates that CS can be competitive with others metaheuristics to solve the SCP. In instance SCPB.3 the convergence is achieved between iterations 50 and 100, while for instances SCPB.4 and SCPB.5 convergence is achieved before. We have also observed that the optimization process applied after the repair function greatly helps the CS convergence, being as well an important step to improve the cost value of each instance. The experimental results also exhibit the robustness of the approach, which is able to reach reasonable good results by only increasing population size when the instance becomes harder.

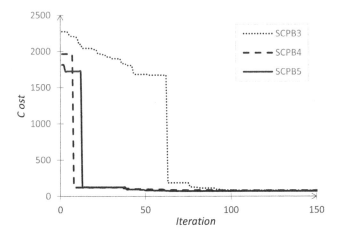

Fig. 2. Evolution of mean best values for SCPB.3, SCPB.4 and SCPB.5

6 Conclusions

In this paper we have presented a new CS algorithm for solving SCPs. The metaheuristic is quite simple to implement and can be adapted to binary domains by only using lower and upper bounds. We have tested 65 non-unicost instances from OR-Library where several global optimum values where reached (RPD=0). The results have also exhibited the rapid convergence and robustness of the proposed algorithm which is capable to obtain reasonable good results by only varying population size when the instance becomes harder. As future work, we plan to experiment with additional modern metaheuristic and to provide a larger comparison of modern techniques to solve SCPs. The integration of autonomous search [9] to the presented approach would be another direction of research to follow as well.

References

1. Balas, E., Carrera, M.C.: A dynamic subgradient-based branch-and-bound proce- dure for set covering. Oper. Res. 44(6), 875–890 (1996)
2. Beasley, J.E., Chu, P.C.: A genetic algorithm for the set covering problem. Eur. J. Oper. Res. 94(2), 392–404 (1996)
3. Brusco, M.J., Jacobs, L.W., Thompson, G.M.: A morphing procedure to supple- ment a simulated annealing heuristic for cost- and coverage-correlated set-covering problems. Ann. Oper. Res. 86, 611–627 (1999)
4. Caprara, A., Fischetti, M., Toth, P.: A heuristic method for the set covering prob- lem. Oper. Res. 47(5), 730–743 (1999)
5. Caprara, A., Toth, P., Fischetti, M.: Algorithms for the set covering problem. Ann. Oper. Res. 98, 353–371 (2000)

6. Caserta, M.: Tabu search-based metaheuristic algorithm for large-scale set covering problems. In: Doerner, K., Gendreau, M., Greistorfer, P., Gutjahr, W., Hartl, R., Reimann, M. (eds.) Metaheuristics. Operations Research/Computer Science Interfaces Series, vol. 39, pp. 43–63. Springer, US (2007)

7. Ceria, S., Nobili, P., Sassano, A.: A Lagrangian-based heuristic for large-scale set covering problems. Math. Program. 81(2), 215–228 (1998)

8. Chvatal, V.: A greedy heuristic for the set-covering problem. Math. Oper. Res. 4(3), 233–235 (1979)

9. Crawford, B., Castro, C., Monfroy, E., Soto, R., Palma, W., Paredes, F.: Dynamic Selection of Enumeration Strategies for Solving Constraint Satisfaction Problems. Romanian Journal of Information Science and Technology 15(2), 106–128 (2013)

10. Fisher, M.L., Kedia, P.: Optimal solution of set covering/partitioning problems using dual heuristics. Management Science 36(6), 674–688 (1990)

11. Garey, M.R., Johnson, D.S.: Computers and Intractability: A Guide to the Theory of NP-Completeness. W. H. Freeman & Co., New York (1990)

12. Lan, G., DePuy, G.W.: On the effectiveness of incorporating randomness and memory into a multi-start metaheuristic with application to the set covering problem. Comput. Ind. Eng. 51(3), 362–374 (2006)

13. Vasko, F.J., Wolf, F.E.: Optimal selection of ingot sizes via set covering. Oper. Res. 35(3), 346–353 (1987)

14. Yang, X.-S., Deb, S.: Cuckoo search via Lévy Flights. In: Proc. of World Congress on Nature & Biologically Inspired Computing (NaBIC 2009), pp. 210–214. IEEE Publications, USA (December 2009)

15. Yang, X.S.: Bat algorithm and cuckoo search: a tutorial. In: Yang, X.S. (ed.) Artificial Intelligence, Evolutionary Computing and Metaheuristics. SCI, vol. 427, pp. 421–434. Springer, Heidelberg (2013)

State Machine Based Architecture to Increase Flexibility of Dual-Arm Robot Programming

Héctor Herrero[1]([✉]), Jose Luis Outón[1], Urko Esnaola[1], Damien Sallé[1],
and Karmele López de Ipiña[2]

[1] Tecnalia Research and Innovation, Industry and Transport Division,
San Sebastián, Spain
{hector.herrero,joseluis.outon,urko.esnaola,damien.salle}@tecnalia.com
[2] Department of Systems Engineering and Automation,
UPV/EHU (Basque Country University), San Sebastián, Spain
karmele.ipina@ehu.es

Abstract. This paper introduces a state machine based architecture with the aim of increasing flexibility of dual-arm robot programming. The proposed architecture allows absolute control of the execution, easing coordination of the arms if necessary. This work attempts to deal with dual-arm robotic programming challenges, providing a robust and reliable core which is able to interconnect different software modules where each one provides different capabilities. A pilot station is under development at Airbus Operations plant in Puerto Real, Spain.

1 Introduction

An analysis of the current situation in manufacturing plants allows to highlight 3 major trends:

- An ever increasing customization of products and short lifecycle. Which requires an increase in the Flexibility of production means (1 unique system must handle all the product diversity and operations)
- A strong variation in production volumes. Which requires an increase in the Reconfigurability of production (1 system for one product/task within recombinable production lines)
- Limited access to skilled operators due to ageing workforce, changes in education and an ever faster technology development. Which requires new solutions to Assist operators and provide collaborative work environments.

The research addressed in this paper focuses on the first trend: the need for highly flexible robotic systems. Despite of a large effort in the research community, large companies as well as SMEs still don't have appropriate software tools and solutions to react rapidly and at costs compatible with an interesting return of investment for the automation of their processes. The direct consequence is

© Springer International Publishing Switzerland 2015
J.M. Ferrández Vicente et al. (Eds.): IWINAC 2015, Part II, LNCS 9108, pp. 98–106, 2015.
DOI: 10.1007/978-3-319-18833-1_11

that mainly production operations are performed manually, with high operation costs that endanger those companies with respect to lower-wages countries. This research is thus oriented at developing and providing a software ecosystem that allows for a rapid and efficient programming of production processes, providing the required flexibility. Even if this approach is generic and applicable to industrial manipulators, this paper will be focused on dual-arm robotic operations.

The dual-arm robots provide more dexterity, in addition to the advantage that they can be used in the existing workstations. Due to these arguments the dual-arm robot implantation is growing year by year, not only in large multinationals, but also in small and medium enterprises (SMEs). Theoretically, sector experts say investments for robot implantation are amortized in 1-2 years, but this affirmation can not be extrapolated to applications with dual-arm robots and specially to short production series or many changes prone environment, neither to industrial processes which need human-robot collaboration or special environment supervision, and to many other cases in which specific solutions are needed.

The growing of dual-arm systems [1] is resulting in a lot of efforts made by robotic researchers to manage them. Programming, coordinating and supervising bi-manual robots is a need that is increasingly being demanded by the community.

In this paper we present the challenges that can be identified for dual-arm robotic programming (Section 2). To ease deployment of this kind of applications we propose an architecture that allows controlling execution very simply and that considerably facilitates programming through skill based organization of robot primitives (Section 3). To understand advantages of the proposed architecture, a riveting use case is presented (Section 4). Finally, we present conclusions and future work (Section 5).

2   Dual-Arm Robotic Programming Challenges

One of the points that differs the dual-arm robots from traditional robots is that two different scenarios can be presented [1]:

1. Non-coordinated manipulation.
2. Coordinated manipulation.

In the case of non-coordinated manipulation each arm works in a different process, i.e., one arm does not have to worry about the status of the other arm, except for self collision detection. For this setting it is necessary an architecture which does not obstruct the execution with unjustified synchronization or waits.

In the second scenario, both arms perform different parts of the same task. This setup can be divided into goal-coordinated and bi-manual coordination. In the case of goal-coordinated manipulation the arms are not interacting with each other but, they have some elements in common, e.g. two arms palletizing items in the same box. For the bi-manual manipulation, it is necessary to control and coordinate synchronously the whole robot, e.g., two arms manipulation processes or processes in which one arm is holding something while the other one is doing assembly operations. For these configurations it is necessary the communication or coordination between arms in order to assure the correct behavior of the whole system.

The traditional robot programming is still not very flexible, thus the dual-arm programming suffers the same problems. In the industry smaller and smaller series are ordered, and as a consequence costs of reprogramming the robots grow. Even though there are usually different parts, the process is very similar, i.e., assembling parts with different types of screw. In this case the assembly operation is the same, only the screw size, type or position is changing. Grouping the robot basic movements (primitives) according to tasks or skills is an alternative that many authors have followed [2][3][4].

One of the most relevant issues in dual-arm robotic programming, especially for industrial applications, is the lack of both powerful and easy to use graphical user interfaces [5]. An easy to configure GUI, which allows the previously mentioned skill based programming, will enable operators to program and maintain the industrial processes. This, in addition to the workers feel themselves part of the automation process, will also contribute to reduce the costs of the robotic systems deployment.

Another important topic concerning dual-arm robotics that will not be dealt here is related with the scenario mentioned above. In the case of processes with independent task for each arm, the effect of the inertia generated by the other arm should be considered. This is a common problem in dual-arm robots in which the arms are connected to a central torso. This phenomenon may produce a loss of accuracy in some instants.

3   State Machine Based Execution Coordination

State machines for execution control can face dual-arm challenges. These tools are commonly used for general-purpose processes and, in particular, they have been extensively adopted by the robotic community. State-machines are an easy way for describing behaviors and for modeling how components react to different external and internal stimuli [6]. In this area there are different implementation alternatives, e.g., there are many projects using Orocos rFSM. rFSM is a small and powerful state-chart implementation designed for coordinating complex systems such as robots [7]. SMACH [8] is another implementation of state machines. It can be defined as task-level architecture for rapidly creating complex robot behavior.

The proposed architecture in this paper has focused on both alleviating problems related with coordinated and non-coordinated manipulation tasks and preparing the way for the ongoing development of skills based programming to ease the use of dual-arm robots.

3.1   Technology

Different open source software was chosen to implement this architecture. Considering the large and active community behind it, ROS is used like middleware [9].

ROS provides the necessary ecosystem to manage complex applications involving trajectory planning with collision detection, pick and place, perception, simulation and much more. In this paper SMACH is used to state machine implementation. SMACH allows not only controlling execution, but also designing complex hierarchical state machines; this feature is especially helpful in order to develop further phases of the project. The proposed architecture is designed to continue developing works related with formalizing skills and automatic code generation, following the same proposal as the BRICS project [10], which is part of EU FP7.

Regarding the hardware, TECNALIA owns a Kawada Nextage Open robot [11]. It has two arms attached to a rotatory torso with 6 degrees of freedom per arm and a stereo vision equipped head with two degrees of freedom, 15 DOF altogether managed by a single controller. This robot is connected to ROS through a bridge developed through collaboration between Tokyo University's JSK Laboratory [12], TORK [13] and TECNALIA. OpenRTM middleware developed by AIST [14] also is used to interconnect the robot with ROS. This combination of components allows using all ROS capabilities. At this point, it should be emphasized that thanks to the use of ROS, the presented architecture is hardware vendor independent: different robotic hardware will be used just utilizing appropriate interface between ROS and the robot controller.

3.2 Core

One of the first requirements that was identified was introspection, which is a tool able to provide current execution state continuously, allowing us to manage possible errors and improving the recovery of them. In Fig. 1 the proposed architecture is outlined. The proposed architecture consists of two state machines, one per arm, with some common states. These common states are used for coordinated manipulation. The system starts from a *Ready* state and keeps changing to different states that can be seen as available abilities or capacities of the robot. Note that some states have not been included in order to simplify the diagram. These states are *Pause/Stop, Error handling* and *Finish*.

The proposed work in this paper allows supervising the environment and permits cancelling or adapting plans according to sensor values and perception system information.

3.3 States

Each state has been implemented as a module that generally is independent from the core. Only a few modules have been defined as basic and required. These special modules are *Articular/Cartesian movements, Full body coordinated motion* and *Trajectory execution*. All available modules for this version are shown in Fig. 1. It should be emphasized that according to the requirements of the different applications, the available states can be updated by incorporating new capabilities or removing others which will not be used.

In order to understand the proposed architecture, Table 1 summarizes the different states and their utility. Each state may contain a more or less complicated

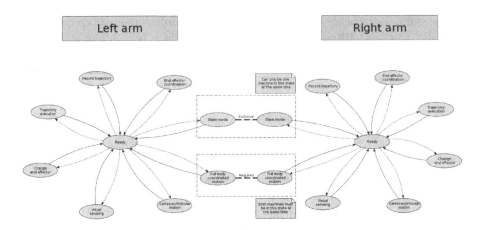

Fig. 1. Proposed state machine based architecture to control dual-arm robots

Table 1. Summary of states

State	Description
Ready	Robot is ready for receive new instruction
Articular/Cartesian motion	Manage robot movements both in cartesian space and articular space
Full body coordinated motion	Allows to control both arms coordinately. Two arms must be in this state to start coordinated motion
Trajectory execution	Executes trajectories, provided by a trajectory planner or previously stored in a database
Change end-effector	Include necessary operations for changing end-effectors
End-effector coordination	Manage end-effector operations, and coordinate movements if a combined operation is required
Record trajectory	Allows to record trajectories with a trajectory planner or by teach by demonstration
Visual servoing	Coordinates robot motions with perception provided information
Slave mode	Puts robot in bi-manual coordinated manipulation mode, one arm actuates as master and the other one as slave.

structure according to its purpose. On the one hand, for example, the *Change-end effector* state only contains a few arm movements and some simple pneumatic operations for the end-effector exchange. On the other hand, the *Articular/Cartesian motion* state is highly general, i.e., this state contains all the required code to manage motions both in cartesian and articular spaces.

3.4   Overall Architecture

As illustrated in Fig. 2, the proposed state machine interconnects the previously commented skill programming (Section 3) and the robotic lower level control system.

As it is detailed in the next section, the skills are composed by primitives which are translated to states. On the one hand, the skill execution engine triggers state changes at the low level. On the other hand, in the case of Kawada Nextage Open the states are connected to the robotic system through an OpenRTM bridge. But it should not be forgotten that ROS allows hardware independence, and changing the bridge properly another robotic system can be used (for example, Orocos or Fast Research Interface [15] to interface a Kuka LWR with the proposed architecture).

This combination of a skill application framework and a low level state machine allows us to considerably improve the flexibility, hardiness, easier programming, hardware independence and environment control, resulting in a more industry oriented solution.

In order to illustrate these abstract concepts, an example of a manipulation task is detailed in Section 5.

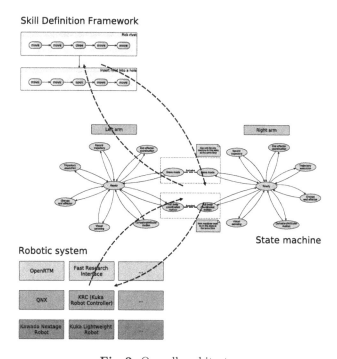

Fig. 2. Overall architecture

4   Validation On Riveting Use Case

This development is performed in collaboration with Airbus Operations (Puerto Real plant in Spain). One of the most relevant tasks in the aerostructure assembly is the installation of rivets. So picking a rivet and introducing it into a drilling has been selected as a manipulation skill example. This skill is composed of the following steps:

1. Pick and extract a rivet from a tray
2. Insert a rivet into a drilling

We can decompose this skill into robot primitives. In this case, the required primitives are *move*, *close* and *open*. The meaning of these primitives can be intuited easily: moving the robot to the provided position, and opening and closing the gripper, respectively.

Once the skills are decomposed, the resulting primitives are the ones that are executed by the proposed state machine architecture. Each state is processing the primitive callbacks, and handling errors if they take place. Thus, the error handling is simpler and managed specifically in each state or module. With the example that is being analysed, the sequence of the machine state is shown in Fig. 3.

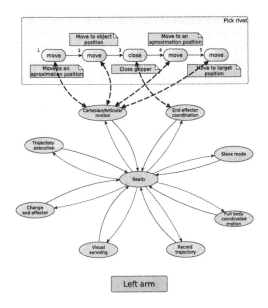

Fig. 3. State machine sequence in a pick rivet skill

5   Conclusions and Future Work

In this paper, a revision of the most common robotic architectures has been done. The trend of using distributed systems versus monolithic systems has been highlighted. At application level, deliberative and reactive architectures have been mentioned. At a higher level of abstraction, agent and state machine based architectures have been reviewed.

Considering the architectures found in the bibliography and the challenges that are emerging from the dual-arm robot programming, the need of a novel architecture is increasingly evident in order to increase the flexibility of dual-arm robots. Thus, this would allow an easier and faster deployment of industrial dual-arm robotics cells.

Fig. 4. Pilot station at Airbus facilities using proposed architecture

To improve the control and coordination of anthropomorphic robots, state machine based architectures have been introduced. This approach allows us to increase the robustness and reliability of the whole system. The proposed architecture is designed to act as a basis for easier programming methodologies. At present, a pilot station is under deployment at the Airbus Operations plant in Puerto Real, Spain (Fig. 4).

In future work, following the efforts made by other authors [16][17], we will further investigate how to define and implement new skills easily and quickly, and especially their integration into an intuitive and graphical interface.

Regarding the state machine based architecture, if the proposed approach is used, the non-coordinated manipulation scenario is more controlled and allows an easier and faster deployment of new applications. In the future, the focus will be set on the coordinated manipulation with the intention of easing this kind of tasks.

Acknowledgment. The authors would like to thank Juan Francisco García Amado for managing the pilot cell at the Airbus facilities. We greatly appreciate the support and supervision of Karmele Lopéz de Ipiña, Professor in the department of System Engineering and Automation at the University of the Basque Country. We also would like to thank Kei Okada from JSK Laboratory for his support with Nextage robot.

References

1. Smith, C., Karayiannidis, Y., Nalpantidis, L., Gratal, X., Qi, P., Dimarogonas, D.V., Kragic, D.: Dual arm manipulation a survey. Robotics and Autonomous systems 60(10), 1340–1353 (2012)
2. Thomas, U., Hirzinger, G., Rumpe, B., Schulze, C., Wortmann, A.: A new skill based robot programming language using uml/p statecharts. In: 2013 IEEE International Conference on Robotics and Automation (ICRA), pp. 461–466. IEEE (2013)
3. Sen, S., Sherrick, G., Ruiken, D., Grupen, R.A.: Hierarchical skills and skill-based representation. In: Lifelong Learning (2011)
4. Zhou, J., Ding, X., Qing, Y.Y.: Automatic planning and coordinated control for redundant dual-arm space robot system. Industrial Robot: An International Journal 38(1), 27–37 (2011)
5. Poppa, F., Zimmer, U.: Robotui-a software architecture for modular robotics user interface frameworks. In: 2012 IEEE/RSJ International Conference on Intelligent Robots and Systems (IROS), pp. 2571–2576. IEEE (2012)
6. Alonso, D., Vicente-Chicote, C., Pastor, J.A., Álvarez, B.: Stateml: From graphical state machine models to thread-safe ada code. In: Kordon, F., Vardanega, T. (eds.) Ada-Europe 2008. LNCS, vol. 5026, pp. 158–170. Springer, Heidelberg (2008)
7. Orocos rfsm, http://people.mech.kuleuven.be/mklotzbucher/rfsm/readme.html
8. Smach, http://wiki.ros.org/smach
9. Ros, http://www.ros.org/
10. Bischoff, R., Guhl, T., Prassler, E., Nowak, W., Kraetzschmar, G., Bruyninckx, H., Soetens, P., Haegele, M., Pott, A., Breedveld, P., et al.: Brics-best practice in robotics. In: 2010 41st International Symposium on Robotics (ISR) and 2010 6th German Conference on Robotics (ROBOTIK), pp. 1–8. VDE (2010)
11. Kawada nextage, http://nextage.kawada.jp/en/
12. Jouhou system kougaku laboratory, Tokyo university, http://www.jsk.t.u-tokyo.ac.jp/
13. Tokyo opensource robotics kyokai association, http://opensource-robotics.tokyo.jp/?lang=en
14. Openrtm-aist middleware, http://openrtm.org/
15. Fast research interface library, http://cs.stanford.edu/people/tkr/fri/html/
16. Andersen, R.H., Solund, T., Hallam, J.: Definition and initial case-based evaluation of hardware-independent robot skills for industrial robotic co-workers. In: ISR/Robotik 2014; Proceedings of the 41st International Symposium on Robotics, pp. 1–7 (2014)
17. Vanthienen, D., De Laet, T., Decré, W., Smits, R., Klotzbücher, M., Buys, K., Bellens, S., Gherardi, L., Bruyninckx, H., De Schutter, J.: Itasc as a unified framework for task specification, control, and coordination, demonstrated on the pr2. In: IEEE International Conference on Mechatronics and Robotics (2011)

A Cellular Automata Model for Mobile Worm Propagation

A. Martín del Rey$^{(\boxtimes)}$, A. Hernández Encinas, J. Martín Vaquero,
A. Queiruga Dios, and G. Rodríguez Sánchez

Department of Applied Mathematics,
University of Salamanca, Salamanca, Spain
{delrey,ascen,jesmarva,queirugadios,gerardo}@usal.es

Abstract. The mobile devices, and especially the smartphones, are exposed to the malicious effects of malware. In this sense the study, simulation and control of epidemic processes due to malware is an important issue. The main goal of this work is to introduce a new mathematical model to study the spread of a mobile computer worm. Its dynamic is governed by means of two cellular automata based on logic transition functions. Some computer simulations are shown and analyzed in order to determine how a mobile worm might spread under different conditions.

Keywords: Mobile malware · Computer worms · Cellular automata · Mathematical modeling

1 Introduction

In recent years, there has been a growing uptake in smartphones and other mobile devices which has been due, undoubtedly, to their increasing connectivity and extensive capabilities. Nowadays, people start to use smartphones for sensitive transactions such as online shopping and banking, and consequently there is an unstoppable rise of the number of immediate threats to mobile devices.

As is well-known, (mobile) malware is one of the most important threats targeting smartphones. There are several types of mobile malware: trojans, worms, computer viruses, adware, spyware, etc. and the effects of such specimens range from brief annoyance to device crashes and identity left. Moreover, mobile malware can be a serious threat to business as well: an attacker can use an infected smartphone as a gateway or proxy to obtain access into a restricted business network.

Malware commonly targets Google's Android operating system. This fact is mainly due to the following: (1) Its leader position: Android holds 80% of the total market share in tablet and smartphone devices; (2) Its open-source nature; (3) The loose security policies for apps which make possible the distribution of malware through both the Android Play Store and the unofficial stores. In this sense, the 97% of malware threats target the Android operating system.

Taking into account this scenario, it is very important to design methodologies to study and simulate epidemic processes caused by mobile malware especimens. To achieve this goal mathematical modeling plays an important role.

© Springer International Publishing Switzerland 2015
J.M. Ferrández Vicente et al. (Eds.): IWINAC 2015, Part II, LNCS 9108, pp. 107–116, 2015.
DOI: 10.1007/978-3-319-18833-1_12

Unfortunately not many mathematical models dealing with this issue have been appeared. The great majority of these works are based on continuous mathematical tools such as systems of ordinary differential equations (see [1,2,6]) or recurrence relations ([4]). These are well-founded and coherent models from the mathematical point of view, offering a detailed study of the main characteristics of their dynamic: stability, equilibrium, etc. Nevertheless they exhibit some important drawbacks: They do not take into account the local interactions between the smartphones and they are unable to simulate the individual evolution of each mobile device. These could be rectified simply if we use discrete individual-based models as cellular automata. As far we know, there are not many discrete mathematical models for mobile malware based on cellular automata: the most important study the mobile worm spread and are due to Peng *et al.* ([5]) and to Martín *et al.* ([3]), and they consider the bluetooth connections as the transmission vector.

The model by Peng *et al.* is a compartmental model where the population is divided into susceptible, exposed, infectious, diagnosed and recovered smartphones. Every cell stands for a regular geographical area which is occupied by at most one smartphone, and these smartphones are fixed in the corresponding cell (that is, movements between cells are not permitted). Moreover, the local transition function depends on two parameters: the infected factor and the resisted factor and any other individual characteristics are not taken into account.

On the other hand, the model by Martín *et al.* is also a compartmental model (considering susceptibles, carriers, exposed and infected devices) based on the use of two cellular automata: one of them rules de global dynamic of the system (that is, it computes de total number of the devices belonging to each compartment at every step of time), and the other one governs the local dynamic (that is, the evolution of each smartphone). In this model the geographical area where the smartphones are placed is tessellated into several square tiles that stand for the cells of the global CA. In addition, the smarphones placed in a global cell stand for the cells of the local CA. Moreover, the smartphones can move freely from one (global) cell to another, and there could be more than one smartphone placed in a (global) cell.

The main goal of this work is to improve the model introduced in [3]. Specifically, the following improvements are done: (1) Not only the bluetooth connections are considered as transmission vector but also the instant messaging apps; in this sense the local CA is suitably modified in order to consider both infection paths. (2) A new compartment is considered: the recovered smartphones (those infected devices where the worm has been successfully detected and removed and are endowed with an immunity period). (3) Modified boolean functions ruling the transition from susceptible to exposed or carrier are presented: they include the connectivity factor of neighbor cells and also modified parameters in order to obtain more realistic simulations. Specifically, in this novel model the infection probabilities due to Bluetooth connections and instant messaging apps depends on the transmission rate, the resistant coefficient and the number of the connected smartphones. (4) Moreover, some simulations have been performed with a modified version of the local transition function including a memory coefficient. This parameter considers the number of reinfections of each smartphone to train it to decrease the probability of a new infection.

The rest of the paper is organized as follows: the mathematical background on cellular automata is introduced in section 2; in section 3 the proposed model is presented, and some illustrative simulations are shown in section 4. Finally, the conclusions and further work are introduced in section 5.

2 Cellular Automata

A cellular automaton (CA for short) is a particular type of agent-based model formed by a finite number of memory units called cells which are uniformly arranged in a two-dimensional space. Each cell is endowed with a particular state that changes synchronously at every step of time according to a local transition rule ([7]). In a more precise way, a CA can be defined as the 4-uplet $\mathcal{A} = (\mathcal{C}, \mathcal{S}, V, f)$, where $\mathcal{C} = \{(i,j), 1 \leq i \leq r, 1 \leq j \leq c\}$ is the cellular space formed by a two-dimensional array of $r \times c$ cells.

At every step of time, every cell $(i,j) \in \mathcal{S}$ is endowed with a state $x_{ij}[t] \in \mathcal{S}$. The neighborhood of each cell is defined by means of an ordered and finite set of indices $V \subset \mathbb{Z} \times \mathbb{Z}$, $|V| = m$ such that for every cell (i,j) its neighborhood $V_{(i,j)}$ is the following set of m cells:

$$V_{(i,j)} = \{(i+\alpha_1, j+\beta_1), \ldots, (i+\alpha_m, j+\beta_m) : (\alpha_k, \beta_k) \in V\}. \qquad (1)$$

Usually Moore neighborhoods are considered ([7]):

$$V_{(i,j)} = \{(i-1, j-1), (i-1, j), (i-1, j+1)(i, j-1), (i, j), \qquad (2)$$
$$(i, j+1), (i+1, j-1), (i+1, j), (i+1, j+1)\}.$$

The states of all cells changes at every discrete step of time according to a local transition function $f \colon \mathcal{S}^m \to \mathcal{S}$, whose variables are the states of the neighbor cells at the previous step of time; then:

$$x_{ij}[t+1] = f(x_{i+\alpha_1, j+\beta_1}[t], \ldots, x_{i+\alpha_m, j+\beta_m}[t]) \in \mathcal{S}. \qquad (3)$$

As the number of cells of the CA is finite, boundary conditions must be considered in order to assure the well-defined dynamics of the CA. One can state several boundary conditions depending on the nature of the phenomenon to be simulated, although in this work we will consider null boundary conditions, that is: if $(i,j) \notin C$ then $x_{ij}[t] = 0$.

As is introduced above, the standard paradigm for cellular automata states that the cellular space is rectangular and the local interactions between the cells are limited to the spatially nearest neighbor cells. Nevertheless one can consider a different topology for the cellular space depending on the phenomenon to be simulated. Consequently, the notion of cellular automaton on graph G emerges: it is a CA whose cellular space is defined by means of a graph G such that each cell stands for a node of G and there is an edge between the nodes u, v, if the cell associated to the node u belongs to the neighborhood of the cell associated to node v (that is, the connections between the nodes define the neighborhoods).

3   Description of the Model

3.1   General Assumptions

In this work we propose a model to simulate mobile worm spreading which is based on the use of two CA: one of them rules de global dynamic of the system and follows the standard paradigm (the topology is defined by an homogeneous rectangular lattice), whereas the other one governs the local dynamic and its topology depends on a certain graph G. The following general assumptions are made in the model:

(1) The population of the smartphones is placed in a finite geographical area where they can move freely.

(2) The smartphones are divided into five classes or compartments: susceptibles S (those smartphones that are susceptible to the infection), carriers C (the smartphones that carry the computer worm but it can not be activated), exposed E (the mobile devices that have been successfully infected but the malware remains latent), infectious I (those smartphones where the computer worm is activated and ready to propagate), and finally, recovered R (those infected smartphones where the worm has been successfully removed).

As a consequence, the model is compartmental: susceptible smartphones becomes carriers or exposed when the computer worm reaches them (note that the carrier status is acquired when the malicious code reaches a smartphone based on a different operative system than the worm's target; on the contrary, exposed devices are those which have been reached by a computer worm which could be activated due to the coincidence of the operative systems); exposed smartphones get the infectious status when the worm is activated and can perform its payload; infectious smartphones progress to recovered when the computer worm is detected and eliminated; and finally, when the immune period after the recovery from the malware is finished, the recovered smartphones becomes susceptible again.

(3) Bluetooth connections and instant messaging apps are the malware transmission vectors.

3.2   The Global Cellular Automaton

As is previously mentioned, the global dynamic of the system is governed by means of a CA defined by an homogenous topology, and denoted by $\mathcal{A}_G$. This global CA gives at every step of time the number of smartphones which are susceptible, carrier, exposed, infectious and recovered in a certain area. The geographical region is tessellated into $r \times c$ constant-size square smallholdings, each one of them stands for a cell of the cellular space of $\mathcal{A}_G$. Let $\mathcal{C}_G = \{(i,j), 1 \leq i \leq r, 1 \leq j \leq c\}$ be the cellular space and set

$$X_{ij}[t] = (S_{ij}[t], C_{ij}[t], E_{ij}[t], I_{ij}[t], R_{ij}[t]), \qquad (4)$$

the state of the cell (i,j) at time t, where $S_{ij}[t]$ is the number of susceptible smartphones, $C_{ij}[t]$ is the number of carrier smartphones, $E_{ij}[t]$ is the number of exposed smartphones, $I_{ij}[t]$ is the number of infectious smartphones, and $R_{ij}[t]$ is the number of recovered devices. The local transition functions are as follows:

$$S_{ij}[t+1] = S_{ij}[t] - S_{ij}^{\text{out}}[t] + S_{ij}^{\text{in}}[t], \quad C_{ij}[t+1] = C_{ij}[t] - C_{ij}^{\text{out}}[t] + C_{ij}^{\text{in}}[t],$$
$$E_{ij}[t+1] = E_{ij}[t] - E_{ij}^{\text{out}}[t] + E_{ij}^{\text{in}}[t], \quad I_{ij}[t+1] = I_{ij}[t] - I_{ij}^{\text{out}}[t] + I_{ij}^{\text{in}}[t],$$
$$R_{ij}[t+1] = R_{ij}[t] - R_{ij}^{\text{out}}[t] + R_{ij}^{\text{in}}[t], \tag{5}$$

where $S_{ij}^{\text{out}}[t]$ stands for the number susceptible smartphones that move from (i,j) to a neighbor cell at time t, and $S_{ij}^{\text{in}}[t]$ represents the number of susceptible smartphones that arrive at the cell (i,j) at time t coming from a neighbor cell.

3.3   The Local Cellular Automaton

The local dynamic of the system (that is, the evolution of the states of the smartphones) is defined by the local cellular automaton $\mathcal{A}_L$. Specifically, it is a CA on a graph $G = G_1 \cup G_2$ which defines the topology of the cellular space $\mathcal{C}_L$ and the neighborhoods. As there are two transmission vectors: bluetooth connections and instant messaging, the subgraph G_1 defines the connections using bluetooth whereas the subgraph G_2 represents the contact network of instant messaging. Then, the neighborhood of each smartphone is $N_1 \cup N_2$, where N_1 (resp. N_2) is associated to G_1 (resp. G_2). The state of the smartphone v at time t is denoted by $x_v[t] \in \mathcal{S}_L = (S, C, E, I, R)$ where S stands for susceptible, C for carrier, E for exposed, I for infectious and R for recovered. The local transition rules are as follows:

(1) *Transition from susceptible to exposed and carrier*: A susceptible smartphone v becomes exposed or carrier when the mobile worm reaches it. This is ruled by two boolean functions: F that models the dynamic considering the bluetooth connections, and G that takes into account the instant messages. Their explicit expressions are as follows:

$$F(v) = B(v) \wedge B(u) \wedge \bigvee_{\substack{u \in N_1(v) \\ x_u[t-1]=I}} A_{uv}, \quad G(v) = \bigvee_{\substack{u \in N_2(v) \\ x_u[t-1]=I}} \tilde{A}_{uv}, \tag{6}$$

where

$$B(w) = \begin{cases} 1, & \text{with probability } b_w(t) \\ 0, & \text{with probability } 1 - b_w(t) \end{cases} \tag{7}$$

and $A_{uv}, \tilde{A}_{uv}$ are the following boolean variables:

$$A_{uv} = \begin{cases} 1, & \text{with probability } a_{uv} \\ 0, & \text{with probability } 1 - a_{uv} \end{cases} \qquad \tilde{A}_{uv} = \begin{cases} 1, & \text{with probability } \tilde{a}_{uv} \\ 0, & \text{with probability } 1 - \tilde{a}_{uv} \end{cases} \tag{8}$$

where $b_v(t)$ is the probability that the smartphone v (resp. u) has the bluetooth enabled at time t, and a_{uv} (resp. $\tilde{a}_{uv}$) is the probability to receive malicious data using the bluetooth connection from the smartphone u (*resp.* the probability to receive a malicious instant message from u). Note that:

$$a_{uv} = \frac{n_{ij}(t)}{n} \cdot d \cdot \alpha_v, \quad \tilde{a}_{uv} = \frac{|N_2(v)|}{n} \cdot d \cdot \alpha_v, \tag{9}$$

where $n_{ij}(t)$ is the number of smartphones placed in the cell (i,j) at time t, d is the transmission rate, and α_v is the resistant coefficient.

Consequently:

$$x_v[t+1] = \begin{cases} E, \text{ if } (x_v[t] = S) \wedge (F(v) = 1 \vee G(v) = 1) \wedge \mathrm{mal}(v) = 1 \\ C, \text{ if } (x_v[t] = S) \wedge (F(v) = 1 \vee G(v) = 1) \wedge \mathrm{mal}(v) = 0 \\ S, \text{ if } (x_v[t] = S) \wedge (F(v) = 0 \wedge G(v) = 0) \end{cases} \tag{10}$$

where

$$\mathrm{mal}(v) = \begin{cases} 1, \text{ if the OS of } v \text{ is the same than the worm's target} \\ 0, \text{ in other case} \end{cases} \tag{11}$$

(2) *Transition from exposed to infectious*: When the virus reaches the host it becomes infectious after a period of time (the latent period t_v^L). Then:

$$x_v[t+1] = \begin{cases} I, \text{ if } (x_v[t] = E) \wedge (\tilde{t}_v > t_v^L) \\ E, \text{ if } (x_v[t] = E) \wedge (\tilde{t}_v \le t_v^L) \end{cases} \tag{12}$$

where $\tilde{t}_v$ stands for the discrete steps of time passed from the acquisition of the mobile worm.

(3) *Transition from infectious to recovered*: An infectious smartphone becomes recovered when the computer worm is detected and removed. It occurs with probability h_v. Then:

$$x_v[t+1] = \begin{cases} I, \text{ if } (x_v[t] = I) \wedge H(v) = 0 \\ R, \text{ if } (x_v[t] = I) \wedge H(v) = 1 \end{cases} \tag{13}$$

where

$$H(v) = \begin{cases} 1, \text{ with probability } h_v \\ 0, \text{ with probability } 1 - h_v \end{cases} \tag{14}$$

(5) *Transition from recovered to susceptible*: When the immunity period t_{IM} is finished, the recovered smartphone will be susceptible again. Then:

$$x_v[t+1] = \begin{cases} S, \text{ if } (x_v[t] = R) \wedge (t_{IM} \le \hat{t}_v) \\ R, \text{ if } (x_v[t] = R) \wedge (t_{IM} > \hat{t}_v) \end{cases} \tag{15}$$

where $\hat{t}_v$ is the discrete steps of time passed from the recovery.

4   Simulations and Discussion

The main characteristic of the simulations done are the following: (a) A bidimensional lattice of 10×10 cells is considered where $n = 100$ smartphones are initially placed at random. (b) The step of time, t, is measured in hours. (c) At every step of time the mobile devices can move from the main cell to a neighbor cell with a certain probability $m_v \in [0,1]$ for each smartphone v. (d) They are considered 5 infected smartphones at time $t = 0$ (the rest of smartphones are initally considered susceptibles). (e) It is emphasized that the values of these parameters are merely illustrative (they are not computed using real data). Several simulations have been computed in the laboratory with different initial parameters but, for the sake of brevity, only few of them are shown in this work (since the simulations performed using similar conditions exhibit similar trends). The parameters used in the simulations are the following (some of them are different for each smartphone varying in a fixed range): $b_v(t) \in [0.5,1], d = 0.5, m_v \in [0,1], \alpha_v \in [0,1], t_v^L \in [1,12], h_v \in [0,1], t_{IM} \in [5,120]$. Note that $t_v^L, t_{IM} \in \mathbb{N}$.

In Figure 1-(a) the global evolution of the different compartments (susceptible -green-, carrier -black-, exposed -orange-, infectious -red-, and recovered -blue-) is shown when it is supposed that the distribution of the operative systems is as follows (taking into account the latest reports of IT companies): 79% of smartphones are based on Android, 16% is based on iOS, the 3% of mobile devices are based on Microsoft Mobile, and the 2% are based on Blackberry OS. In addition, we state that the five infected smartphones at $t = 0$ are Android-based. In Figure 1-(b) the global evolution of the system is presented when all smartphones are based on Android. As is shown, both scenarios exhibit similar patterns: the system evolves to a quasi-endemic equilibrium with periodic outbreaks (where the number of recovered devices decreases, and the number of susceptible, exposed and infected increases).

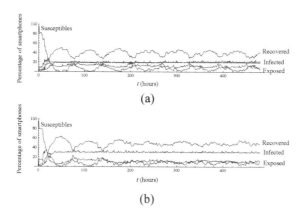

Fig. 1. (a) Heterogeneous population: Android (79%), iOS (16%), Microsoft (3%), and BlackBerry (2%). (b) Homogeneous population (100% Android).

In Figure 2 the evolution of the different compartments is shown when different probabilities of movement, m_v, are considered: in Figure 2-(a) $m_v = 0$ is assumed for all v (that is, the smartphones do not move), $m_v \in [0.4, 0.6]$ for all v is taken in Figure 2-(b), and finally in Figure 2-(c) $m_v = 1$ is considered (all smartphones change the cell at every step of time). Note that as the probability of movement m_v increases, the number of infected, exposed (and recovered) smartphones also increases and the number of susceptible devices decreases.

In Figure 3, some simulations are presented when the immunity period, t_{IM}, varies. Specifically, in Figure 3-(a) the evolution of the different compartments is shown when $t_{IM} = 48$, the dynamic of the system is given in Figure 3-(b) when $t_{IM} = 72$, and finally, in Figure 3-(c) the simulation represents the evolution when $t_{IM} = 120$. In all simulations, the evolution of each compartment can be divided into three phases which are periodically repeated: An initial phase where the number of susceptible smartphones decreases whereas the number of exposed, infected and recovered devices increases; a stable phase where the number of smartphones of each compartment remains constant (or, at least, they change with small variations); and a final phase where the number of recovered smartphones decreases and, in contrast, the number of susceptible, exposed and infected devices increases. Note that the carrier compartment grows in the first phase and after that, it remains constant. The stable phase ends when the immunity period also ends, then as t_{IM} increases the duration of this phase also increases.

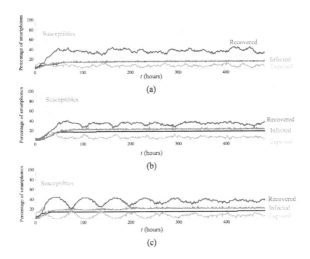

Fig. 2. (a) Evolution without smartphone's movement ($m_v = 0, \forall v$) (b) Evolution with a moderate movement of smartphones ($m_v \in (0.4, 0.6)$) (c) Evolution with a constant movement of smartphones ($m_v = 1, \forall v$)

In the next simulations, the local function which rules the transition from susceptible to exposed or carrier is modified including a new parameter: the

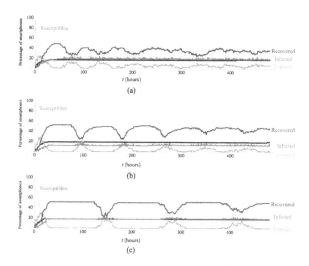

Fig. 3. (a) Evolution with $t_{IM} = 48$ (b) Evolution with $t_{IM} = 72$ (c) Evolution with $t_{IM} = 120$

memory coefficient β_v which stands for the number of reinfections of the smartphone v, that is, $\beta_v = k + 1$ if the mobile device v has been reinfected k times. Consequently, we define:

$$A_{uv} = \begin{cases} 1, \text{with probability } \frac{a_{uv}}{\beta_v} \\ 0, \text{with probability } 1 - \frac{a_{uv}}{\beta_v} \end{cases} \qquad \tilde{A}_{uv} = \begin{cases} 1, \text{with probability } \frac{\tilde{a}_{uv}}{\beta_v} \\ 0, \text{with probability } 1 - \frac{\tilde{a}_{uv}}{\beta_v} \end{cases}$$

$$(16)$$

With this new approach we train the cells of the local CA (that is, the smartphones) in order to obtain experience: the recovery from an infection strengths the resistance of the smartphone against a new infection. The results introduced in Figure 4-(a) show that the use of the new coefficient yields a substantial reduction of the number of infected smartphones (although the trends are similar). Finally in the last simulations we consider G_2 as a complete graph, that is, it is suppose that the contacts (used by instant messaging apps) of every smartphone are the remaining $n - 1$ smartphones. The results presented in Figure 4-(b) show that the number of infected devices increases as the connectivity rate of G_2 increases.

5   Conclusions

In this work a novel individual-based model to simulate the spreading of mobile computer worm is introduced. It improves in different ways the work due to Peng *et al.* ([5]) and Martín *et al.* ([3]). In the model proposed in this work two cellular automata are used to rule the dynamic at macroscopic and microscopic level respectively. The geographical area where smartphones are placed is

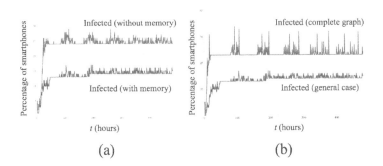

Fig. 4. (a) Evolution of the number of infected smartphones with and without memory. (b) Evolution of the number of infected smartphones when G_2 is a complete graph or a general graph.

tessellated into several square lattice giving the cellular space of the global cellular automata. There can be more than one smartphone in each cell and these smartphones can move from one global cell to another at every step of time. The local dynamic is simulated by means of the local cellular automata whose cells are the individual smarphones. The model facilitates real-time predictions of the evolution of the malware spreading, making it possible to modify the different parameters and control measures.

Acknowledgments. This work has been supported by Ministry of de Economy and Competitiveness (Spain) and by Junta de Castilla y León (Spain).

References

1. Cheng, S.M., Ao, W.C., Chen, P.Y., Chen, K.C.: On Modeling Malware Propagation in Generalized Social Networks. IEEE Commun. Lett. 15(1), 25–27 (2011)
2. Jackson, J.T., Creese, S.: Virus Propagation in Heterogeneous Bluetooth Networks with Human Behaviors. IEEE T. Depend. Secure 9(6), 930–943 (2012)
3. del Rey, A.M., Sánchez, G.R.: A CA model for mobile malware spreading based on bluetooth connections. In: Herrero, A., et al. (eds.) International Joint Conference SOCO 2013-CISIS 2013-ICEUTE 2013. AISC, vol. 239, pp. 619–629. Springer, Heidelberg (2014)
4. Merler, S., Jurman, G. A combinatorial model of malware diffusion via Bluetooth connections. PLos ONE 8, art. no. e59468 (2013)
5. Peng, S., Wang, G., Yu, S.: Modeling the dynamics of worm propagation using two-dimensional cellular automata in smartphones. J. Comput. System Sci. 79(5), 586–595 (2013)
6. Rhodes, C.J., Nekovee, M.: The opportunistic transmission of wireless worms between mobile devices. Physica A 387, 6837–6844 (2008)
7. Wolfram, S.: A New Kind of Science. Wolfram Media Inc., Champaign (2002)

Evolving Cellular Automata to Segment Hyperspectral Images Using Low Dimensional Images for Training

B. Priego, Francisco Bellas, and Richard J. Duro[✉]

Integrated Group for Engineering Research
Universidade da Coruña, A Coruña, Spain
{blanca.priego,fran,richard}@udc.es
http://gii.udc.es

Abstract. This paper describes a hyperspectral image segmentation approach that has been developed to address the issues of lack of adequately labeled images, the computational load induced when using hyperspectral images in training and, especially, the adaptation of the level of segmentation to the desires of the users. The algorithm used is based on evolving cellular automata where the fitness is established based on the use of synthetic RGB images that are constructed on-line according to a set of parameters that define the type of segmentation the user wants. A series of segmentation experiments over real hyperspectral images are presented to show this adaptability and how the performance of the algorithm improves over other state of the art approaches found in the literature on the subject.

Keywords: Hyperspectral image segmentation · Cellular automata · Evolution

1 Introduction

The hyperspectral image segmentation problem has been addressed in the literature through approaches that are mostly extensions of more traditional image processing techniques, such as region-merging [1] or hierarchical segmentation [2]. Some of these approaches, such as morphological levels [3] or Markov random fields [4] [5] make use of fixed window based neighborhoods. Others resort to mathematical morphology based techniques [6] [7] or variations of watershed algorithms [8] [9]. They provide different ways of extending them to multidimensional images, but they always face the ordering problem in high dimensional spaces, leading to difficult issues and very ad hoc solutions when they modify the operators involved. These are mostly fixed algorithms, and, even though through some parameter tuning, especially in the case of watershed based solutions, different types of segmentations may be achieved, it is very hard, if not impossible to adapt to concrete segmentation requirements the users may have.

Some authors, instead of providing a complete algorithm, propose different trainable segmentation and classification algorithms for trainable structures such

© Springer International Publishing Switzerland 2015
J.M. Ferrández Vicente et al. (Eds.): IWINAC 2015, Part II, LNCS 9108, pp. 117–126, 2015.
DOI: 10.1007/978-3-319-18833-1_13

as Artificial Neural Networks. This way the system can learn to perform the segmentation, presumably in an appropriate way, from sets of sample labeled images. Examples of this approach are those presented in [10] [11] or [12]. However, they usually require complex algorithms that lead to computationally intensive processing stages that are very hard to implement in limited computing resources, which are usually the case in many applications where hyperspectrometers are used in the field. Additionally, there is the perennial problem of multidimensional imaging: there are very small numbers of labeled images available and they are not always too reliable in terms of their labels. Thus, it is quite hard to construct a training set that really reflects what the user wants.

Consequently, we have detected a need to provide methodologies that allow this processing to be achieved without having to resort to large training or labeled sets of real images. Thus, Instead of resorting to complex labeled real hyperspectral images, in a previous paper [13], we have presented a method by which the segmentation function has been generalized so that the algorithms can be trained using synthetic RGB images and then directly applied to hyperspectral images. In addition, and in order to preserve the representational power of these high dimensional images the approach presented works with the complete spectral wealth of the images and does not project them onto lower dimensionality spaces. Finally, as the methodology is based on Cellular automata, which is a very distributed processing scheme, the algorithm is very easy to adapt to highly parallel processing architectures, such as GPUs. This algorithm has been called ECAS (Evolutionary Cellular Automata based Segmentation) and its current implementation is ECAS-II. The main objective of this paper is to present the evolution of ECAS and show its capabilities when segmenting hyperspectral images.

The rest of the paper is structured as follows. The next section provides a brief overview of ECAS- II and the philosophy behind its operation. Section 3 is devoted to the presentation of some example segmentations using ECAS-II and a comparison of some of its results to other techniques. Finally, Section 4 provides a series of conclusions on the work carried out.

2   ECAS and ECAS-II

ECAS, in its two versions, is a cellular automata (CA) based algorithm, where a CA is evolved using as set of labeled target images during the evaluation of the candidate CAs in the population. What is original here is that in order to make the process computationally more efficient, we have chosen a distance measure to compare spectra (or states) that is independent of dimensionality. This allows using lower dimensionality training samples, which are much easier to generate and manipulate, in order to obtain CAs that can then be used over higher dimensional multi or hyperspectral images just as long as their general features are similar. The dimensionality independent distance measure we have considered here is the well known *spectral angle*. Thus, the normalized spectral angle,α_{ij}, for a cell with respect to its neighboring cell j is defined as:

$$\alpha_{ij} = \frac{2}{\pi} \cos^{-1} \left(\frac{\sum s_{ij} s_i}{\sqrt{\sum s_{ij}^2} \sqrt{\sum s_i^2}} \right) \quad \alpha_{ij} \in [0, 1] \tag{1}$$

In terms of how the CA operates, first, it is necessary to say that a cell of the automaton is placed over each pixel of the hyperspectral image. The state of the cell (S_i) is given by an N-band spectrum, initially that of its corresponding pixel. Thus, ECAS works over the complete spectra of the image without performing any projections onto lower dimensionalities. Obviously, the state space is continuous and corresponds to the positive vector space $\mathbb{R}^N$. During execution, the CA is iteratively applied. Every iteration it adapts the state (S_i) of each one of the cells on the image through the modification of the N-band vector that corresponds to the spectrum in that location. This modification depends on the set of rules that control the automaton, on the state of the cell over which the automaton is applied (S_i) and, in the original ECAS, on the information of the spectra (state) of the 8 closest neighboring cells.

It is in this aspect that ECAS-II differs from ECAS. In ECAS-II the way the cells are modified is a little more complex as the new state of a cell is calculated taking into account the pixels contained in three different two dimensional windows of size $N_s \times N_s$, where $N_s = \{3, 5, 7\}$ through the determination of three spatial gradients from the spectral angles of the cell and its neighbors in each of these windows.

Two two-dimensional masks $M_{X_{NS}}$ and $M_{Y_{NS}}$ applied to each cell are used to obtain the gradients. Thus, the horizontal and vertical components of the gradient vectors corresponding to cell i, and denoted as $G_{X_{NS_i}}$ and $G_{Y_{NS_i}}$,are:

$$G_{X_{NS_i}} = \sum_{j=1}^{N_S \cdot N_S} \alpha_{i,j} \cdot M_{X_{NS_j}}, \quad G_{Y_{NS_i}} = \sum_{j=1}^{N_S \cdot N_S} \alpha_{i,j} \cdot M_{Y_{NS_j}} \tag{2}$$

being $M_{X_{NS_j}}$ and $M_{Y_{NS_j}}$ the jth elements of the masks.

These three gradient vectors, $\mathbf{G_{NS_i}}$, can be expressed as a modulus, $|\mathbf{G_{NS_i}}|$, and an angle, ϕ_{NS_i}:

$$|\mathbf{G_{NS_i}}| = \sqrt{G_{X_{NS_i}}^2 + G_{Y_{NS_i}}^2}, \quad \phi_{NS_i} = tan^{-1} \left(\frac{G_{Y_{NS_i}}}{G_{X_{NS_i}}} \right) \tag{3}$$

The modulus, $|\mathbf{G_{NS_i}}|$, is related to the intensity of the spectral change following the direction given by the angle, ϕ_{NS_i} for window size N_S . In this algorithm, we consider the values of the three modules and angles of the gradient vector as all the information the CA needs to operate over a cell of the hyperspectral cube.

As a result, the set of transition rules that control the behavior of the CA is made up of M six parameter rules:

$$CA = \begin{pmatrix} |\mathbf{G_{r3_1}}| & |\mathbf{G_{r5_1}}| & |\mathbf{G_{r7_1}}| & \phi_{r5_1} & \phi_{r7_1} & \theta_{r1} \\ \vdots & \vdots & \vdots & \vdots & \vdots & \vdots \\ |\mathbf{G_{r3_k}}| & |\mathbf{G_{r5_k}}| & |\mathbf{G_{r7_k}}| & \phi_{r5_k} & \phi_{r7_k} & \theta_{rk} \\ \vdots & \vdots & \vdots & \vdots & \vdots & \vdots \\ |\mathbf{G_{r3_M}}| & |\mathbf{G_{r5_M}}| & |\mathbf{G_{r7_M}}| & \phi_{r5_M} & \phi_{r7_M} & \theta_{rM} \end{pmatrix} \tag{4}$$

Each iteration of the CA, only one of these M rules is applied over each cell. To decide which, we compare the neighborhood information of the pixel to the first 5 parameters of each one of the M rules and choose the rule that is closest, using as distance the sum of the L^2 norms of the vector differences once the vectors have been rotated and reflected. Only 5 parameters are used because the representation has been chosen so that the three gradient vectors as a group are independent from rotations and reflections (making $\phi_{r3_k} = 0$).

The last parameter of the selected rule contains the information required to update the state of cell i. The state of the cell will be modified by performing a weighted average of its state or spectrum and that of some of its neighbors. If we follow the direction given by $\phi + \theta_{rs}$ for a distance of 1 pixel, where ϕ is the rotation angle of the selected rule, and θ_{rs} is the last parameter of rule s, we will be at a point of the image, P_i. The neighboring cells used to modify the spectrum of cell i will be located at a maximum distance of 1 pixel from P_i.

Following this procedure, the CA is applied iteratively to the whole image producing a new hyperspectral cube every iteration. With the appropriate rules, the final hyperspectral cube will be a segmented version of the original one, where each region will be represented by a narrow range of spectra.

Now, the key to this whole process is to provide a mechanism that permits obtaining these rule sets automatically seeking a particular type of segmentation. As indicated above, in ECAS we have chosen an automatic optimization process in the form of an Evolutionary Algorithm. In this case, the algorithm is Differential Evolution (DE) [14] using as fitness function to be minimized the maximum value of two error measurements: intra-region (e_{intra}) error and inter-region (e_{inter}) error. That is, every generation, the candidate CAs are run over a set of labeled training images and the results they produce are compared to the target segmentation in terms of the two errors indicated above. Obviously, depending on how these training images are constructed, the type of segmentation the resulting CA will provide will be different.

Finally, to produce a provisional classification image, a SVM based pixel wise classification is applied to the original image following the work of Tarabalka et al. [9]. A multi-class pairwise (one versus one) SVM classification of the hyperspectral image is performed using a Gaussian Radial Basis Function (RBF) kernel.

3 Some Experiments

Taking random images as a training set in order to produce a CA based segmenter using the ECAS-II algorithm leads to a segmenter that is general to any type of image. However, if one wants to specify how the segmentation is desired (coarse, detailed, or other) it is necessary to fine tune the training set so that it allows the optimization of the segmenter to the desired type of segmentation, which is basically given by the ground truths provided for the RGB based training set. Thus, to automatically create this training RGB images so that they reflect what is desired over the hyperspectral images, a set of parameters must be given that indicate the type of segmentation required. It is not always easy to choose these parameters, and for this reason, a new tool has been developed. It takes as inputs presegmented hyperspectral images and produces as outputs the corresponding parameters that describe these images (parameters such as number of regions, inter-region spectral distances, intra-region spectral distances, ruggedness of the borders, etc). A second tool, takes these spectral and spatial parameters and outputs random synthetic RGB images that comply with them. The objective of these tools is to provide the evolutionary process with the appropriate RGB dataset for training. This way, ECAS-II can produce the CA transition rules adapted to the nature of the hyperspectral image type and to the sort of segmentation that is required.

In order to test this approach, in this results section, three real hyperspectral scenes have been considered:

- Indiana scene: image recorded by the AVIRIS sensor over the Indian Pines test site in Northwestern Indiana. The image has spatial dimensions of 145 by 145 pixels, and 200-bands per pixel.
- Salinas scene: image captured by the AVIRIS sensor over Salinas Valley, CA, USA, composed of 224 bands and a spatial size of 512×217.
- Pavia University scene: image recorded by the ROSIS-03 satellite sensor. It has 115 bands of and spatial dimensions of 10×340.

Before going into the application of ECAS-II to these scenes and the comparison of the results obtained to those published by other authors, as an illustration of the need to adapt the segmenting CA to the type of image and segmentation required, we are going to perform a small analysis of the spectral features related to the inter and intra-region distances present in these images. These features are related to the spectral homogeneity between pixels that belong to the same region (intra-region spectral distances) and between pixels located at the boundaries between different regions (inter-region spectral distances). The left graph of Fig. 1 plots the histograms corresponding to the inter-region and intra-region spectral distances from pixels belonging to some selected classes of the ground truth provided by experts for the Indiana, Salinas and Pavia scenes. In this case, this ground truth establishes the segmentation level desired by the user and the images themselves determine the type they belong to in terms of spectral and spatial characteristics. The right graph of Fig. 1 plots the most frequent inter-region spectral distance vs. the most frequent intra-region spectral distance for each class in these hyperspectral scenes. It can be observed in both representations that each scene exhibits different combinations of paired intra-region and inter-region spectral distances. It is our hypothesis that a CA that takes this into account will be better adapted to perform an improved segmentation leading to better classification results when passed on to a subsequent classification stage.

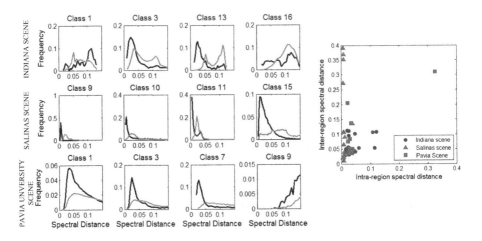

Fig. 1. Left: Intra-region (dark line) and inter-region spectral distances (light line) histograms from pixels belonging to different classes extracted from Indiana, Salinas and Pavia scenes. Right: most frequent inter-region spectral distance vs. most frequent intra-region spectral distance for each class of the considered hyperspectral scenes.

Once the spectral features that represent a given type of hyperspectral image and segmentation process have been extracted, for the operation of ECAS-II, a RGB image dataset is created and used as samples for evaluating the population in the evolutionary algorithm. It is important to emphasize that the dataset used to evolve the CA is very simple, consisting of images made up of three spectral bands and which only share some parameterized spectral features with the real hyperspectral ones.

For the following experiments, three different CAs, denoted as CA#1, CA#2 and CA#3, were evolved using a Differential Evolution (DE) algorithm [14]. CA#1, CA#2 and CA#3 were evolved considering three different RGB image datasets with spectral features similar to those found in the Indiana, Salinas and Pavia scenes respectively. Each CA is made up of 30 rules, and each rule has 6 components, leading to a total of 180 real valued parameters in each chromosome. The Differential Evolution algorithm stops when the maximum number of generations (50) or a lower bound (1e-6) for the error used as fitness value is reached. The population size used was 100 in every case and the crossover and mutation probabilities were set to 0.7 and 0.8 respectively. To compare the performance of ECAS-II to other existing techniques, a classification process was applied over the segmentation results for the University of Pavia image

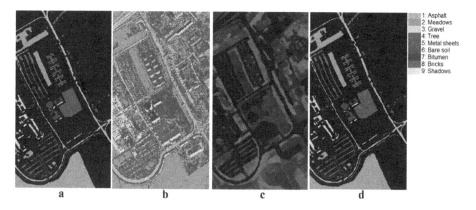

Fig. 2. a: Ground truth of the Pavia scene with the 9 classes. b: SVM based pixel classification applied to the to the original hyperspectral image. c: 2D angular transformation of the segmented image using ECAS-II, where each pixel corresponds to the angle between the spectrum of the segmented pixel and a reference spectrum with all of its bands at the maximum value. d: SVM based pixel classification applied to the to the segmented image provided by the ECAS-II algorithm showing only the labeled areas.

Table 1. Accuracy results (%) for different algorithms

	κ	OA	AA
ECAS-II + SVM ($C = 256, \quad \gamma = 16$)	**98.64**	**99.01**	**96.87**
Pixel-wise SVM ($C = 256, \quad \gamma = 16$)	89.43	92.16	93.00
Pixel-wise SVM [9] ($C = 128, \quad \gamma = 0.125$)	75.86	81.01	88.25
W-RCMG [9]	81.30	85.42	91.31
EMP [15]	96.05	97.07	96.79
AEAP [16]	91.21	93.42	91.02
V-ELM-1 [17]	95.00	96.66	95.92

(Fig. 2). These segmentation results using the ECAS-II algorithm and the subsequent classification by means of a SVM based pixel-wise method are displayed on the right images of Fig. 2. In this case, the evolved CA used was the one labeled as CA#3. To compare the overall classification results obtained using this image five reference algorithms were selected. The first one is a purely spectral algorithm based on a pixel-wise SVM using the implementation found in the LIBSVM library. In order to apply the SVM, the hyperspectral image was normalized with respect to the maximum spectral intensity found in the whole data cube. Then the SVM was applied considering $C = 256$ and $\gamma = 16$ as the SVM parameters. These were selected by means of a fivefold cross validation procedure. For comparison purposes, the results for the same pixel-wise SVM algorithm (that is, without the previous CA based segmentation stage) are also presented following the implementation and results found in [9]. The second

Table 2. Class specific accuracy results (%) for the raw SVM and the ECAS-II + SVM algorithm

	Number of Training Samples	Number of Test Samples	Class-specific accuracies (%) ECAS-II + SVM $C = 256, \gamma = 16$	SVM $C = 256, \gamma = 16$
1: Asphalt	548	6083	**99.34**	89.42
2: Meadows	540	18109	**99.94**	93.12
3: Gravel	392	1707	**98.24**	85.12
4: Tree	524	2540	95.59	**97.63**
5: Metal sheets	265	1080	97.59	**99.16**
6: Base soil	532	4497	**99.53**	92.75
7: Bitumen	375	955	**98.84**	92.04
8: Bricks	514	3168	**99.55**	85.63
9: Shadows	231	716	83.24	**99.44**

algorithm is W-RCMG, a watershed transformation-based algorithm presented by Tarabalka et al in [9]. The third reference algorithm is the EMP (extended morphological profiles), presented in [15]. The fourth algorithm is called automatic extended attribute profiles (AEAPs) and was presented in [16]. Finally, the last reference technique is a modified V-ELM-1, presented in [17]. The success obtained in the classification is quantitatively measured for all the methods in terms of overall accuracy (OA), average accuracy (AA), and kappa coefficient (Table 1). It can be observed that the ECAS-II technique followed by a SVM pixel-wise classification stage is the algorithm that provides the best solution in terms of the three accuracy measures considered. Table 2 shows the class specific accuracies when the raw SVM and ECAS-II + SVM algorithm are applied to the Pavia image. It can be noticed that most of the classes experiment an improvement in terms of accuracy, achieving almost 100%, except for the trees, the metal sheets (which suffer a slight reduction of accuracy) and the shadow class. Shadows are a controversial class because they do not represent a unique material by themselves, just different materials but with a low intensity spectrum. In other words, there are many different spectra under the shadows class. The ECAS-II algorithm, which operates with the full spectral information, obviously tends to homogenize these areas with the surrounding ones that belong to the same material but exhibit higher intensity spectra. This leads to poorer results in terms of the classes present in the hand labeled ground truth than those obtained by the SVM alone, which just considers low intensity spectra as a class. In our opinion, this is a case of mislabeling in the images. In what follows, we analyze how the segmentation of the hyperspectral images through the ECAS-II algorithm improves on the SVM classification. For each hyperspectral scene, ECAS-II is applied and the SVM is run using different percentages of samples for each class. Table 3 presents the quantitative improvement in terms of OA, AA and Kappa, showing that the ECAS-II segmentation algorithm improves the SVM results in every case, but, most importantly, even when the number of samples considered in the training process of the classification algorithm is very limited, which is a great advantage of the proposed method.

Table 3. Improvement in accuracy measurements (%) when the CA is applied before the SVM classification algorithm

		% of training samples				
		2	5	20	50	80
Indiana scene	$\Delta OA(\%)$	**20.42**	17.72	10.62	7.68	9.24
	$\Delta AA(\%)$	**28.12**	18.71	9.25	8.66	10.35
	$\kappa(\%)$	**23.59**	20.32	12.12	8.77	10.56
Salinas Scene	$\Delta OA(\%)$	3.36	3.36	**4.16**	3.88	3.60
	$\Delta AA(\%)$	**2.03**	1.24	1.65	1.47	1.32
	$\kappa(\%)$	3.75	3.74	**4.63**	4.32	4.01
Pavia Scene	$\Delta OA(\%)$	**5.20**	4.97	4.37	3.52	3.00
	$\Delta AA(\%)$	2.87	**3.59**	3.20	2.82	2.29
	$\kappa(\%)$	**6.93**	6.61	5.79	4.67	3.98

4   Conclusions

A new automata based evolutionary approach for the segmentation of images of high dimensionality has been presented in this paper. The approach permits obtaining segmenting CAs that are adapted to the level of segmentation required by the user through the appropriate construction of a synthetic training image set. One of the advantages of this algorithm is that it can be trained using RGB images and then used over hyperspectral ones as the distance measure it uses is independent from dimensionality. When compared to other approaches in the literature the performance of the algorithm is quite competitive. In fact, it can be seen that the results are superior to other approaches even when the number of samples taken for the classification task is very small.

Acknowledgments. This work was partially funded by the Spanish MICINN through project TIN2011-28753-C02-01.

References

1. Darwish, A., Leukert, K., Reinhardt, W.: Image segmentation for the purpose of object-based classification. In: International Geoscience and Remote Sensing Symposium, vol. 3, pp. 2039–2041 (2003)
2. Tilton, J.C.: Analysis of hierarchically related image segmentations. In: IEEE Workshop on Advances in Techniques for Analysis of Remotely Sensed Data, pp. 60–69 (2003)

3. Pesaresi, M., Benediktsson, J.A.: A new approach for the morphological segmentation of high-resolution satellite imagery. IEEE Transactions on Geoscience and Remote Sensing 39(2), 309–320 (2001)
4. Farag, A.A., Mohamed, R.M., El-Baz, A.: A unified framework for MAP estimation in remote sensing image segmentation. IEEE Transactions on Geoscience and Remote Sensing 43(7), 1617–1634 (2005)
5. Eches, O., Dobigeon, N., Tourneret, J.Y.: Markov random fields for joint unmixing and segmentation of hyperspectral images. In: 2nd Workshop on Hyperspectral Image and Signal Processing: Evolution in Remote Sensing (WHISPERS), pp. 1–4 (2010)
6. Flouzat, G., Amram, O., Cherchali, S.: Spatial and spectral segmentation of satellite remote sensing imagery using processing graphs by mathematical morphology. In: IEEE International Geoscience and Remote Sensing Symposium Proceedings, IGARSS 1998, vol. 4, pp. 1–3 (1998)
7. Li, P.L.P., Xiao, X.X.X.: Evaluation of multiscale morphological segmentation of multispectral imagery for land cover classification. In: Proceedings of the 2004 IEEE International Geoscience and Remote Sensing Symposium, IGARSS 2004, 4, 0-3 (2004)
8. Quesada-Barriuso, P., Argello, F., Heras, D.B.: Efficient segmentation of hyperspectral images on commodity. Advances in Knowledge Based and Intelligent Information and Engineering Systems 243, 2130–2139 (2012)
9. Tarabalka, Y., Chanussot, J., Benediktsson, J.A.: Segmentation and classification of hyperspectral images using watershed transformation. Pattern Recognition 43(7), 2367–2379 (2010)
10. Li, J., Bioucas-Dias, J.M., Plaza, A.: Hyperspectral Image Segmentation Using a New Bayesian Approach with Active Learning. IEEE Transactions on Geoscience and Remote Sensing 49(10), 3947–3960 (2011)
11. Veracini, T., Matteoli, S., Diani, M., Corsini, G.: Robust Hyperspectral Image Segmentation Based on a Non-Gaussian Model. In: 2010 2nd International Workshop on Cognitive Information Processing (CIP), pp. 192–197 (2010)
12. Duro, R.J., Lopez-Pena, F., Crespo, J.L.: Using Gaussian Synapse ANNs for Hyperspectral Image Segmentation and Endmember Extraction. In: Graña, M., Duro, R.J. (eds.) Computational Intelligence for Remote Sensing. SCI, vol. 133, pp. 341–362. Springer, Heidelberg (2008)
13. Priego, B., Souto, D., Bellas, F., Duro, R.J.: Hyperspectral image segmentation through evolved cellular automata. Pattern Recognition Letters 34(14), 1648–1658 (2013)
14. Storn, R., Price, K.: Differential evolution a simple and efficient heuristic for global optimization over continuous spaces. Journal of Global Optimization 11(4), 341–359 (1997)
15. Benediktsson, J., Pesaresi, M., Amason, K.: Classification and feature extraction for remote sensing images from urban areas based on morphological transformations. IEEE Trans. Geosci. Remote Sens. 41(9), 1940–1949 (2003)
16. Marpu, P.R., Pedergnana, M., Mura, M.D., Benediktsson, J.A., Bruzzone, L.: Automatic generation of standard deviation attribute profiles for spectral-spatial classification of remote sensing data. IEEE Geosci. Remote Sens. Lett. 10(2), 293–297 (2013)
17. Lopez-Fandino, J., Quesada-Barriuso, P., Heras, D., Arguello, F.: Efficient ELM-Based Techniques for the Classification of Hyperspectral Remote Sensing Images on Commodity GPUs. IEEE Journal of Selected Topics in Applied Earth Observations and Remote Sensing PP(99), 1–10 (2015)

Multicriteria Inventory Routing by Cooperative Swarms and Evolutionary Algorithms

Zhiwei Yang[✉], Michael Emmerich, Thomas Bäck, and Joost Kok

LIACS, Leiden University, Niels Bohrweg 1, 2333-CA Leiden, The Netherlands
z.yang@liacs.leidenuniv.nl

Abstract. The Inventory Routing Problem is an important problem in logistics and known to belong to the class of NP hard problems. In the bicriteria inventory routing problem the goal is to simultaneously minimize distance cost and inventory costs. This paper is about the application of indicator-based evolutionary algorithms and swarm algorithms for finding an approximation to the Pareto front of this problem. We consider also robust vehicle routing as a tricriteria version of the problem.

1 Introduction

This is a study on the application of two population-based multi-objective optimization algorithms and their application to the inventory routing problem in logistics. The goal is to reduce two different cost factors simultaneously – distance cost and inventory cost. Whereas the first is related to the economical and ecological aspect of the problem (fuel consumption), the latter is related to the quality of service that needs to be optimized. The goal of our study is to compare two different, though closely related, strategies for computing the Pareto optimal front of this problem.

In multi-objective optimization, optimization problems are solved with two or more objective functions. In general they are defined as

$$f_1(\mathbf{x}) \to \min, \ldots, f_m(\mathbf{x}) \to \min, \text{subject to } g_1(\mathbf{x}) \leq 0, \ldots, g_k(\mathbf{x}) \leq 0, \mathbf{x} \in \mathcal{X}$$

Here with $\mathcal{X}$ the decision space of the problem is denoted, that is the space of all possible alternative solutions. In this paper we will consider both continuous and discrete search spaces.

In this paper we compare two algorithms that seek to maximize the hypervolume indicator over the set of all subsets of $\mathcal{X}$ of size n. The algorithms produce a sequence of so called approximation sets that gradually converges to a diverse approximation of the Pareto front with maximal hypervolume indicator.

Indicator-based multicriteria optimization (IMO) seeks to improve the performance indicator of a solution set and by doing so achieve a good approximation to a Pareto front.

The contribution of this paper is to study the performance of two algorithms for IMO for the inventory routing problem. One algorithm will be the SMS-EMOA, which is an established IMO technique. It is a steady-state evolutionary algorithm and uses the hypervolume indicator as a selection criterion [3,1].

© Springer International Publishing Switzerland 2015
J.M. Ferrández Vicente et al. (Eds.): IWINAC 2015, Part II, LNCS 9108, pp. 127–137, 2015.
DOI: 10.1007/978-3-319-18833-1_14

Secondly, we will study a cooperative swarm to find a hypervolume optimal Pareto front approximation. Besides comparing the results, it is also studied how well problem specific local search strategies can be used in order to improve the results. Last but not least, we consider stock-out a third objective function and present first results on a 3-D Pareto front approximation.

The paper is structured as follows: In Section 2 of this paper we describe the inventory routing problem and existing solvers from literature. Section 3 provides a detailed description of the algorithms, including a cooperative swarm-based approach and an indicator-based evolutionary algorithm. Section 4 deals with the performance studies for different variations of the algorithms and problem. Finally, Section 5 concludes the work with a summarizing discussion.

2   Problem Description and Related Work

The inventory routing problem (IRP) is a multi-period distribution problem which involves two components: the inventory management problem and the vehicle routing problem[5]. In this problem, products are repeatedly delivered from a single supplier to a set of n geographically dispersed customers over a given planning horizon T (in days). On different days, each customer consumes a certain amount. Moreover, customers can maintain a local inventory with a maximum inventory level. The supplier has to service all the customers with a fleet of homogeneous vehicles with a equal maximal capacity. The objective in [5] was to minimize the average inventory cost and to minimize the total routing cost during the planning period. Here, we will also consider this problem (with alternative algorithms) and also a tricriteria problem that also states stock-out risk as a objective function.

Formally, the problem is set up as described in [5]: There are n customers and at most one vehicle with capacity C per customer. Deliveries cannot be split in the model, that is a customer can be visited at most one time by a vehicle per day. Each customer $i \in \{1, \ldots, n\}$ has a maximum inventory level denoted with Q_i. For each customer $i \in \{1, \ldots, n\}$ and time $t \in \{1, \ldots, T\}$ (denoting an index for the days) $L_{i,t}$ denotes the inventory level, $q_{i,t}$ is the shipping quantity, $d_{i,t}$ is the demand to be satisfied, At day 1 the values of $L_{i,1}$ are set to some pre-defined initial inventory level. The inventory levels are then updated according to the equation given by Geiger and Sevaux [5].

The stockout-level can be computed as $S_{i,t} = \max\{0, d_{i,t} - L_{i,t-1} - q_{i,t}\}$. A positive $S_{i,t}$ means that not enough product is available at time t and for customer i. In [5] positive $S_{i,t}$ were avoided altogether by considering a strict constraint.

If demands are not known beforehand, stock-out risk can be defined. Here we assume that $d_{i,t}$ is only the average demand, and an upper bound for the demand is given by $\alpha d_{i,t}$ for some $\alpha > 1$ (here we chose 1.5, that is in the worst case the demand is 50% higher than the average case). Then we can define the worst case stockout-level as $\bar{S}_{i,t} = \max\{0, \alpha d_{i,t} - L_{i,t-1} - q_{i,t}\}$.

A solution candidate is represented by a tuple of delivery frequencies $(\pi_1, ..., \pi_n)$. For each customer it determines how often it is visited by a vehicle. For instance, $\pi_i = 1$ would mean day-to-day delivery for customer i, $\pi_i = 2$ means that on every second day a delivery takes place, and so forth. The required shipping quantities are then determined by

$$q_{i,t} = \min \left\{ \left(\sum_{\ell=t}^{t-1+\pi_i} d_{i,\ell} \right) - L_{i,t-1}, \; Q_i - L_{i,t-1}, \; C \right\}$$

The two objectives stated in [5]:

$$f_1 = \sum_{t=1}^{T} \sum_{i=1}^{n} L_{i,t} \to \min \tag{1}$$

$$f_2 = \sum_{t=1}^{T} VRP_t(q_{1,t}, \ldots, q_{n,t}) \to \min \tag{2}$$

where Equation 1 describes the total inventory cost and Equation 2 describes the total cost for the routing. The latter is determined by solving for each day a vehicle routing problem with the given shipping quantities for that day (they are determined by the frequencies).

Finally, we also consider stock-out risk, by computing the total stock-out in the worst-case scenario as

$$f_3 = \sum_{t=1}^{T} \sum_{i=1}^{n} \bar{S}_{i,t} \to \min \tag{3}$$

thereby assuming that all other quantities in the problem are computed according to the average case scenario.

3   Multiobjective Optimization Algorithms

Two algorithms are used for optimization. Both algorithms aim to maximize the hypervolume indicator of a population of solutions and thereby create a diversified set on the Pareto front. The first algorithm is the SMS-EMOA, an evolutionary algorithms with steady state selection scheme, and the second one is a swarm based algorithms where each point in the population is viewed as a search agent that seeks to improve its individual contribution to the hypervolume indicator.

3.1   Multi-objective Particle Swarm Optimization Algorithm

In the interpretation we use in this paper a particle swarm optimization (PSO) algorithm is a randomized search heuristics where a swarm of particles moves gradually towards an optimal solution driven by randomized modification operators and interaction between the particles.

In conventional PSO algorithms, the swarm is driven by a leader, who is the currently best individual in a population, and by local memories of particles on their so-far best positions. In single-objective optimization such processes will typically converge to local, or sometimes even to global optima. In multiobjective optimization such an approach could be easily used to find a single point on the Pareto front, but is not well suited to distribute points across the Pareto front, because the particles all strive to resemble the leader which is counter-productive when searching for a diverse set of solutions. To a certain extent this can be compensated by assigning local leaders, but this makes the algorithm quite complicated and adds parameters to the algorithm (i.e., number of leaders).

The use of traditional PSO for multi-objective optimization problems has been addressed already in the literature, both in the context of general multiobjective optimization [2], and for finding Pareto fronts that maximize the hypervolume indicator [10]. Both approaches let to algorithms that can produce good approximations to Pareto fronts.

In this paper, however, we consider another approach that we will term *cooperative particle swarm*. This algorithm will have the following properties that distinguishes it from previous swarm-based approaches:

- Leader-free: The particles in the population cooperate in covering the Pareto front, instead of competing with each other. There is no leader in the swarm; each particle strives to contribute to the global performance of the swarm.
- Indicator-based: The algorithms seeks to maximize an unary performance indicator. Here the hypervolume indicator is used.

The new approach is deliberately kept very simple. This is for two reasons: Firstly we want to demonstrate that only a few essential components are needed to steer a swarm towards a Pareto front. Secondly, simplicity will make the algorithms easier accessible to a rigorous theoretical analysis. It will also be easier to compare it on a conceptual level to the later discussed set-gradient based algorithm.

We will term the approach Multiobjective Optimization by Cooperative Particle Swarms (MOCOPS).

The pseudo-code for the proposed MOCOPS algorithm is given in Algorithm 2. It starts with randomly initializing a set of particles. Then, in each iteration of the algorithm, a particle is randomly selected and a small random variation of this particle is generated by adding a random permutation.

If the fitness contribution of the mutated particle relative to the population is better than for the original position then the particle will move to the new position: Firstly, it will be tested which one of the two positions leads to a better hypervolume indicator of the population. Secondly, if both positions are equally good (which will typically occur for dominated solutions), the point that has a better value in the aggregated linear objective function with equal weights is considered. Note that if one solution is dominated by the other solution it will also be considered better in the latter comparison (because of positive equal weighting). Therefore, eventually all solutions will strive towards the non-dominated front and then their hypervolume contribution will be considered.

The cycle continues with picking a random particle again. The MOCOPS algorithm is displayed in pseudo code in algorithm 2. Care must be taken to ensure $\mathbf{x}_{new} \in \mathbb{S}$ (e.g. by rejecting infeasible vectors).

Algorithm 1. Multiobjective Optimization by Cooperative Swarms (MOCOPS)

Input initial population P_0
while termination criterion is not reached **do**
    $t \leftarrow t + 1$
    $s \sim u(\{1, 2, \ldots, n - 1, n\})$
    $\mathbf{x}_{old} = \mathbf{x}^{(s)}$
    $P \rightarrow P_t \setminus \{\mathbf{x}^{(s)}\}$
    {Try to improve position of particle $\mathbf{x}^{(s)}$}
    $\mathbf{z} \sim N(\mathbf{0}, \mathbf{I})$
    $\mathbf{x}_{new} = \mathbf{x}_{new} + \sigma \mathbf{z}$
    if $HV(P \cup \{\mathbf{x}_{new}\}) > HV(P \cup \{\mathbf{x}_{new}\})$ **then**
        $P_t = P \cup \{\mathbf{x}_{new}\}$
    else if $HV(P \cup \{\mathbf{x}_{new}\}) < HV(P \cup \{\mathbf{x}_{old}\})$ **then**
        $P_t = P \cup \{\mathbf{x}_{old}\}$
    else if $f_1(\mathbf{x}_{new}) + f_2(\mathbf{x}_{new}) < f_1(\mathbf{x}_{old}) + f_2(\mathbf{x}_{old})$ **then**
        $P_t = P \cup \{\mathbf{x}_{new}\}$
    else
        $P_t = P \cup \{\mathbf{x}_{old}\}$
    end if
end while
Return P_t

One iteration of the bicriteria MOCOPS algorithm can be performed with a time complexity in $\mathcal{O}(\mu \log \mu)$ and its complexity is related to the problem of computing the hypervolume contribution of a point which is discussed in [4]. However, by implementing the algorithm as an on-line algorithm, that is using incremental update steps, we can compute a single iteration with time complexity in $O(\log \mu)$ (amortized over the number of iterations) [7]. Fast - linear time - hypervolume update schemes are also known for three objective functions [6].

3.2 Evolutionary Algorithm: SMS-EMOA

As opposed to the swarm algorithm, in the evolutionary algorithms points can be removed and new points might appear in the course of evolution. Viewing both as stochastic systems, the swarm algorithms could be described as a multi-trajectory stochastic flow whereas the latter is a branching process, with 'birth events' creating a new branch, and 'death events' terminating a branch.

The SMS-EMOA is otherwise very similar to the swarm-based algorithm, as it also bases the decisions on hypervolume contributions of points. Given a

---

Algorithm 2. S-Metric Selection Evolutionary Algorithm (SMS-EMOA)*

Input initial population P_0
while termination criterion is not reached **do**
　$t \leftarrow t + 1$
　$s \sim u(\{1, 2, \ldots, n-1, n\})$
　$\mathbf{x}_{old} = \mathbf{x}^{(s)}$
　$P \leftarrow P_t \cup \{\mathbf{x}^{(s)}\}$
　$\mathbf{z} \sim \mathbf{N}(\mathbf{0}, \mathbf{I})$
　$\mathbf{x}_{new} = \mathbf{x}_{new} + \sigma \mathbf{z}$
　{Determine the (set of) least hypervolume contributors}
　$\mathcal{L} \leftarrow \arg\min_{\mathbf{x} \in P} \Delta H(\mathbf{x})$
　Chose randomly a solution in $\mathcal{L}$
　$P \leftarrow P_t \setminus \{\mathbf{x}^{(s)}\}$
end while
(*Simple version with random selection among dominated solutions.)

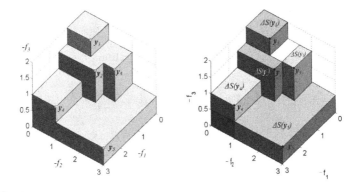

Fig. 1. Left: Hypervolume dominated by a set of 3-D objective vectors $\mathbf{y}_i = \mathbf{f}(\mathbf{x}^{(i)})$. Right: Exclusively dominated volume $\Delta\mathcal{S}$ of these vectors.

population P, the hypervolume contribution $\Delta H(\mathbf{x})$ is defined as:

$$\Delta H(\mathbf{x}) = HV(P) - HV(P \setminus \{\mathbf{x}\})$$

For a visualization for a population of 3-D objective vectors, see Figure 1. An asymptotically optimal algorithm for computing all hypervolume contributions in a population of 3-D vectors has been discovered by Emmerich and Fonseca [4]. The running time complexity is $\Theta(|P| \log |P|)$ and this step is therefore up to a constant factor as fast as computing the hypervolume of a population.

The simple version of SMS-EMOA that was used in our experiments is outlined in Algorithm 2. After initializing a population, in each iteration first a new solution is created by mutating an existing solution. Then it is added to the population. Subsequently, the hypervolume contributions of all population

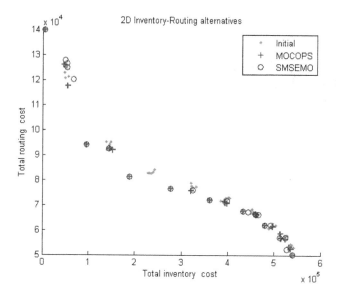

Fig. 2. Pareto front approximations of the bicriteria problem obtained with MOCOPS and SMS-EMOA

members are computed and those with the least contribution are determined. A randomly chosen 'least hypervolume contributor' is then discarded from the population. Due to the last step the population size is kept constant and there is a selection pressure towards sets that cover more hypervolume.

4   Benchmark and Results

We tested MOCOPS and SMS-EMOA on the bicriteria and tricriteria inventory routing problem. For the hypervolume indicator the reference point $r = (1864800, 576563.43, 183813)$ was used. For determining a reference point an upper bound for all objective function values was required: For the first objective (inventory cost) we assume all inventories are always at their maximum allowed level. For the second objective we assume that every day we serve each customer with one vehicle. Finally, for the third objective we assume the stock-out that would occur if no deliveries would take place.

The algorithm setup was as follows: The population size is always 50. For the mutation we use integer mutation with geometrical distribution [8] and if interval boundaries are exceeded we set to boundary. The vehicle routing was done with the parameter-free savings heuristics. The populations were initialized as in [5]. All other problem data was chosen according to [5]. The runs were conducted for 1000 iterations. The matlab code is made available via http://natcomp.liacs.nl/index.php?page=code.

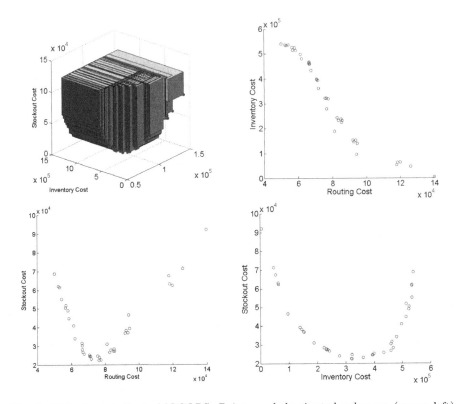

Fig. 3. Plots for tricriteria MOCOPS: Points and dominated subspace (upper left), routing cost vs. inventory cost (upper right), routing cost vs. stock-out cost (lower left),routing cost vs. stock-out cost (lower right)

The first comparison is on the bicriteria problem. Figure 2 shows the result of the comparison. Clearly the results of both algorithms look very similar and the Pareto front has a convex upper part and a concave lower part. The knee point appears as a isolated solution at $(1, 7.5)$. The input is that every two days all the customers are served. Moving further down with the inventory cost would cause a steep increase in the routing cost. Reducing the routing cost further would cause first a strong deterioration of the inventory cost to a point near $(3, 7.4)$, but a further decease from this point on has a more balanced trade-off and there is much room for improvement of the routing cost. According to our results, the knee point looks quite attractive as a solution, or otherwise a solution with much lower routing costs (more energy efficient) that will also allow a much higher inventory cost (e.g. a solution around $(4.7, 5.4)$). Note that the results are slightly better than those in [5].

Figure 3 and figure 4 show results obtained after 1000 iterations of the tricriteria MOCOPS and, respectively, SMS-EMO algorithm. We plotted a 3-D visualization of the obtained Pareto front and the attain-surface that separates the dominated subspace from the non-dominated subspace. In order to more

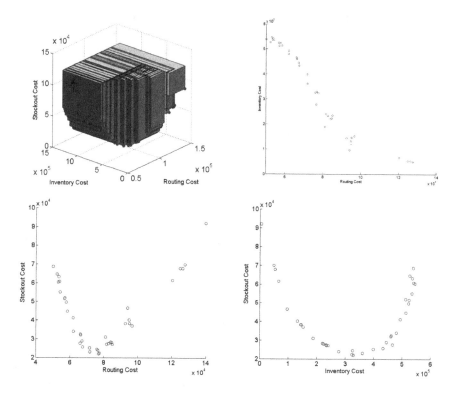

Fig. 4. Plots for tricriteria SMS-EMO: Points and dominated subspace (upper left), routing cost vs. inventory cost (upper right), routing cost vs. stock-out cost (lower left), routing cost vs. stock-out cost (lower right)

accurately assess the quality of single points we also provided the three projections as a scatter plot. Again both algorithms produced very similar results, which we interpret as an indication that a good approximation to the hypervolume maximal front was obtained. The interpretation of 3-D results is more involved, as three trade-offs need to be taken into account: Firstly, from the projection to Routing cost and Inventory cost we obtain a similar set of non-dominated solutions, in the 2-D projection, than for the 2-D study. The dominated points in the projection are almost all at low inventory costs, which means that a stock-out would be mainly caused by a too low maximal inventory level.

The relationship between stock-out cost, inventory level, and routing can be described as follows: When the frequency goes down, the stock-out will become bigger and the inventory level will become smaller, and the routing cost also. The isolated patch in the scatter plot of inventory cost and stock-out cost appears at a very low frequency. The isolated patch might be caused by the fact that at the lowest point the maximal inventory levels are reached. A further increase of the stock-out can benefit the routing cost objective while the inventory cost remains unchanged. The points who are performing well in the routing cost will not perform well in the stock-out cost, which explains the appearance of a parabolic (mirrored) shape in the pareto front. The flanks of the parabola reflect two different ways stock-out can be prevented – one is the increment of inventory cost and the other is the increment of routing cost. This is underpinned by an effect that cannot be seen in the plot. This is that along the parabola the frequency increases in the direction of lower investment costs.

5   Conclusion and Outlook

This paper presents results of two indicator-based multicriteria optimization algorithms on the bicriteria and tricriteria inventory routing problem. Both algorithms, a swarm-based approach and an evolutionary algorithm, produce similar results and we conclude that these results likely reflect the shape of the true Pareto fronts. The objective functions considered were routing cost and inventory cost (average case) and stock-out level (in a worst case scenario). It has been found that there are interesting trade-offs between the objectives and the Pareto front has a highly non-linear shape with both convex and concave parts. The algorithms could partly improve on previous results (routing cost) but also it provides first results on a tricriteria formulation (including the stock-out cost). This is a first step into uncertainty management for robust inventory routing. Clearly, either the inventory levels have to be increased or the routing costs have to be increased to achieve more robust solutions. The frequency of delivery can modulate this trade-off. However, there also exists a large number of dominated solutions in the 3-D case and also isolated patches on the Pareto front, the explanation of which will require further more detailed investigations. Future work will also have to be conducted on more detailed uncertainty models, for instance by replacing the possibilistic modeling by probabilistic modeling.

Acknowledgement. Z. Yang acknowledges financial support from China Scholarship Council (CSC), grant No. 201206110020.

References

1. Beume, N., Naujoks, B., Emmerich, M.: SMS-EMOA: Multiobjective selection based on dominated hypervolume. European Journal of Operational Research 181(3), 1653–1669 (2007)
2. Coello, C.A.C., Lechuga, M.S.: Mopso: A proposal for multiple objective particle swarm optimization. In: Proceedings of the 2002 Congress on Evolutionary Computation, CEC 2002, vol. 2, pp. 1051–1056. IEEE (2002)
3. Emmerich, M.T.M., Beume, N., Naujoks, B.: An emo algorithm using the hypervolume measure as selection criterion. In: Coello Coello, C.A., Hernández Aguirre, A., Zitzler, E. (eds.) EMO 2005. LNCS, vol. 3410, pp. 62–76. Springer, Heidelberg (2005)
4. Emmerich, M.T.M., Fonseca, C.M.: Computing hypervolume contributions in low dimensions: Asymptotically optimal algorithm and complexity results. In: Takahashi, R.H.C., Deb, K., Wanner, E.F., Greco, S. (eds.) EMO 2011. LNCS, vol. 6576, pp. 121–135. Springer, Heidelberg (2011)

5. Geiger, M.J., Sevaux, M.: The biobjective inventory routing problem–problem solution and decision support. In: Pahl, J., Reiners, T., Voß, S. (eds.) INOC 2011. LNCS, vol. 6701, pp. 365–378. Springer, Heidelberg (2011)
6. Guerreiro, A.P., Fonseca, C.M., Emmerich, M.T.M.: A fast dimension-sweep algorithm for the hypervolume indicator in four dimensions. In: CCCG, pp. 77–82 (2012)
7. Hupkens, I., Emmerich, M.: Logarithmic-time updates in sms-emoa and hypervolume-based archiving. In: Emmerich, M., et al. (eds.) EVOLVE - A Bridge between Probability, Set Oriented Numerics,and Evolutionary Computation IV. AISC, vol. 227, pp. 155–169. Springer, Heidelberg (2013)
8. Li, R., Emmerich, M., Eggermont, J., Bovenkamp, E.G.P.: Mixed-integer optimization of coronary vessel image analysis using evolution strategies. In: Proceedings of the 8th Annual Conference on Genetic and Evolutionary Computation, pp. 1645–1652. ACM (2006)
9. Miettinen, K.: Nonlinear multiobjective optimization, vol. 12. Springer Science & Business Media (1999)
10. Mostaghim, S., Branke, J., Schmeck, H.: Multi-objective particle swarm optimization on computer grids. In: Proceedings of the 9th Annual Conference on Genetic and Evolutionary Computation, GECCO 2007, pp. 869–875. ACM, New York (2007)

Embodied Evolution for Collective Indoor Surveillance and Location

Pedro Trueba, Abraham Prieto, Francisco Bellas$^{(\boxtimes)}$, and Richard J. Duro

Integrated Group for Engineering Research,
Universidade da Coruña, 15403, Ferrol, Spain
{pedro.trueba,abprieto,fran,richard}@udc.es
http://www.gii.udc.es

Abstract. This work is devoted with the application of a canonical Embodied Evolution algorithm in a collective task in which a fleet of Micro Aerial Vehicles (MAVs) have to survey an indoor scenario. The MAVs need to locate themselves to keep track of their trajectories and to share this information with other robots. This localization is performed using the IMU, artificial landmarks that can be sensed using the onboard camera and the position of other MAVs in sight. The accuracy in the decentralized location of each MAV has been included as a part of the problem to solve. Therefore, the collective control system is in charge of organizing the MAVs in the scenario in order to increase the accuracy of the fleet location, and consequently, the speed at which a new point of interest is reached.

Keywords: Embodied Evolution · Indoor Navigation · Collective Tasks

1 Introduction

Navigation on indoor scenarios is a widely studied topic in autonomous robotics [1][2]. To navigate with success, the first issue to solve is locating the robot properly in the environment by means of its onboard sensors, typically, an inertial measurement unit (IMU), a camera, distance sensors (laser, ultrasonic..), and encoders for the wheels in the case of mobile robots [2]. However, it is a well known fact that accurate localization on indoor spaces is far from trivial, mainly because the odometry performed with the onboard sensors has a high drift and using natural beacons is a problem far from being solved nowadays [3][4][5]. When developing a navigation system for an autonomous robot in a real environment, the impact of the localization on the main task is as strong as to determine the control system that performs the navigation. In fact, we must consider that in real scenarios the implicit task of locating the robot is as significant as the navigation task to be solved, if not more. Therefore, it is fundamental that the localization of the robot is included as part of the problem. This aspect should also be considered when working on simulated scenarios, since otherwise, the result will not be translatable to a real problem.

J.M. Ferrández Vicente et al. (Eds.): IWINAC 2015, Part II, LNCS 9108, pp. 138–147, 2015.
DOI: 10.1007/978-3-319-18833-1_15

In this work, we deal with a simulated environment in which a fleet of Micro Aerial Vehicles (MAVs) have to survey an indoor scenario. As part of the problem, the MAVs need to locate themselves to keep track of their trajectories and to share this information with other robots. The localization will be performed using their IMU, artificial landmarks that can be sensed using the onboard camera and the position of other MAVs in sight. The control of each of the MAVs will be obtained by evolution using a distributed Embodied Evolution algorithm that is in charge of organizing the MAVs in the scenario in order to increase the accuracy of the fleet location, and consequently, the speed at which a new point of interest is reached.

The validity of evolutionary algorithms in the design of control systems for teams of autonomous robots that exploit the coordination between them in a completely decentralized fashion has been widely studied [6][7]. The main drawback of these approaches is that all of them have been implemented and analyzed in off-line operation mainly due to the high computational cost of their execution. With the aim of solving this problem arose a new evolutionary approach: Embodied Evolution (EE) [8][9]. It is an evolutionary paradigm inspired by natural evolution where the individuals that make up the population are embodied and situated in an environment where they interact in a local, decentralized and asynchronous fashion. Hence, evolution in EE is open-ended, leading to a paradigm that is intrinsically adaptive and highly suitable for real time learning in distributed dynamic problems, like the one we are setting up here. EE interest has grown remarkably in the last decade, with several papers dealing successfully with different collective tasks, both in simulation [10][11] and real robots [12][13]. But, to the authors best knowledge, this work is the first attempt of using EE for coordinating a fleet of MAVs to perform optimal collective navigation, and where localization is an intrinsic parameter to be optimized.

The remainder of the paper is structured as follows. Section 2 contains a brief description of the EE algorithm that will be used to solve the collective navigation task. Section 3 is devoted to the presentation of the collective gathering task setup, including the scenario, the formal objective statement, the details of the localization and control, and the presentation of the main results that have been obtained. Finally, in section 4 the conclusions of this work are extracted.

2 Canonical Embodied Evolution Algorithm

A few years ago we started the development of a canonical EE algorithm [14][15] with two main objectives. On one hand, with the aim of understanding in depth the behavior of this type of algorithm, we decided to simplify and extract the barebones specification of a generalized EE in terms of as few parameters as possible. This generalization has been achieved by substituting the activation of particular operators triggered by events produced in the simulation with probability distributions which are not task or scenario dependent. Some preliminary tests were also performed by substituting the performance of the individuals on a real or simulated arenas with predefined mathematical models. However, when the canonical algorithm is applied to a real scenario two operators have to be adapted to the constrains that this scenario imposes, and therefore, their dependence on the task is unavoidable: the mating operator, which is constrained by the communication limitations of scenario and individuals to exchange their genetic codes, and the evaluation operator, which relies on the actual behavior and on the state of the scenario.The second objective was to do away with the bonds the environment imposes on the structure and operation of the algorithm. The adaptation to the environment determines the type of tasks that are considered tractable. It also makes the algorithm and its behavior even more task dependent, and as a result, it becomes more complicated to extract general conclusions from experiments. Consequently, the circumstantial/spatial interactions has been replaced in the canonical algorithm by stochastic variables, which follow probability functions. The pseudo-code of the canonical EE is the following:

```
1    While simulation active
2    For each interaction
3    For each individual
4       Assign fitness (Interaction with the scenario)
5       If random < Pmating
6          Look for fertile partners
7          Select a partner for recombination (Pelegibility)
8          Generate offspring and store it (Pls)
9       If random < Preplacement
10         New individual = current offspring genotype
11         Eliminate this individual
12   End
```

It can be observed that, as in any evolutionary algorithm, the canonical EE performs three basic processes: evaluation (line 4 in the pseudo-code), mating (lines 5 to 8) and replacement (lines 9 to 11). In this particular evolutionary paradigm, the individuals must perform these processes in a completely decentralized, and therefore asynchronous, fashion, which affects their definition. Moreover, in order to make it independent from the environment and specific task, the circumstantial/spatial interactions that guide natural evolution have been replaced by stochastic variables, which follow specific probability functions. The intrinsic parameters that define their response are:

- Mating selection: it has been modeled as an event that is triggered by a uniform probability function that depends on a single parameter, the probability of mating, that is: $P_{mating} = S_{max}/T_{max}$, where S_{max} is the maximum window size of the tournament, and T_{max} is the maximum lifetime.
- Selection policy: the probability of being eligible as a candidate for mating ($P_{elegibility}$) is defined through a function that is based on the fitness value
- Genotypic recombination: a new intrinsic parameter is defined: the probability of using a local search strategy (P_{ls}), that is, a mutation operator. It is a measure of the exploration and exploitation balance through the ratio between crossover and mutation frequency.

– Replacement: the current canonical EE algorithm considers a fixed population size. The replacement process is modeled here as triggered by a replacement probability ($P_{replacement}$) and it is defined based on a more intuitive and manageable parameter, which is the life expectancy (Texp): $P_{replacement} = 1/Texp$. The life expectancy is defined for each individual in each time step based on its current fitness, which depends on its genotype and the genotypes of the others [15].

3 Collective Gathering Task

3.1 Experimental Setup

The experimental setup has been defined in simulation, based on a real indoor surveillance task performed by MAVs as a previous step to translate the real embodied evolution, or directly the controllers, to a fleet of real MAVs. The specific MAV that has been modeled is the Parrot ARDrone 2.0, a very popular general-purpose commercial quadcopter that is equipped with an IMU and two cameras. The tests performed using this specific model with positioning algorithms based on natural landmarks captured with the frontal camera of the MAV were unsatisfactory, so we decided to introduce visual fiducial markers in the setup in order to facilitate the location.

The most important aspect of the simulation is the response of the location sensors when a certain maneuver is carried out. Firstly, the IMU will provide some velocity signals for each degree of freedom, which have to be integrated to produce the estimated motion. The estimation of the velocity is modeled as subject to a normal distribution centered on the real input velocity with its corresponding variance matrix as is frequently assumed on real navigation. Regarding the visual fiducial markers, we have selected the AprilTags [16] developed in the University of Michigan, so a simulation model of them has been created. In fact, the location estimation provided by the markers is based on a real accuracy model which was produced in our laboratory using an ARDrone 2.0 and 40 cm long AprilTags, and that can be formulated as a function ϕ which relates the variance of the estimation ($Var(\overrightarrow{p_{drone}})$) to the relative distance ($\| \overrightarrow{p_{drone}} - \overrightarrow{p_{tag}} \|$) and orientation between camera and tag: $yaw_{drone:tag}$

$$Var(\overrightarrow{p_{drone}}) = \phi(\| \overrightarrow{p_{drone}} - \overrightarrow{p_{tag}} \|, yaw_{drone:tag}) \tag{1}$$

These tags (permanent tags) provide an absolute and potentially accurate position estimation, which does not degrade with time. In order to improve the performance of the navigation by improving the accuracy of the MAVs, the same type of tags were also attached to the body of the MAVs (mobile tags), which will make up a hybrid location sensor. Therefore, the detection provided by a mobile tag still constitutes a direct location estimation but, unlike in the case of permanent tags, their accuracy degrades as the accuracy of the carrier of the tag decreases. The use of this type of mobile tags leads to accuracy becoming a resource that MAVs can get and share to be able to slow down its degradation, and therefore, to accomplish their main task more efficiently. It could also allow some of the MAVs not needing to visit static tags, which are frequently non optimally located, being "nourished" by mobile tags.

The scenario used in the simulation was created as a 768x768 (square length units) non-toroidal square area, which contains four fixed tags placed randomly, and 40 gatherer drones. Fig. 1 shows an snapshot of a portion of the scenario, where the main elements are displayed. It is divided into cells whose color indicates their exploration probability from blue to green. The pink square represents a permanent tag, while the blue and red elements represent two MAVs. It can be observed that these elements have an outline near them, which represents its estimated location accuracy (the one the MAV knows). If it is high, the outline is over the MAV, meaning that it knows its location accurately. On the other hand, a large separation means a low accuracy level in its location for the MAV. The estimated location accuracy is used to update the surveillance map. In addition to these main elements, Fig. 1 displays the exploration trail left by two MAVs (red and blue) where it can be observed that the exploration level of the scenario decreases as they move away from the tag. Table 1 contains the specific parameters that define the scenario.

The final objective of the surveillance task is for the fleet of MAVs to continuously cover the maximum area possible. Each MAV keeps a record of the areas they have explored in an "exploration map" that can also be shared with others when they meet. The exchange of this information allows the task to be solved cooperatively since it enables distributing the exploration among the group of MAVs. However, since the estimation of one's position has a varying accuracy, the updating of the exploration map has to take that in account. The exploration of a cell is

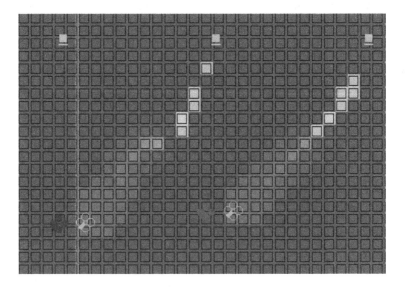

Fig. 1. Snapshot of a portion of the simulation scenario

modeled as an exploration probability (P_{ex}), which indicates the probability that a certain cell has of having been explored. This probability can be directly calculated as the ratio between the size of a cell (L_{cell}) and the location error range (E_{loc}) and is stored on the exploration map.

$$P_{ex} = \frac{L_{cell}^2}{\pi E_{loc}^2} \tag{2}$$

The individual fitness function of each individual is calculated based on two variables: the exploration performance and the accuracy provided to the rest of the fleet by means of its attached tags.

Regarding the MAVs, they are defined by their real spatial location $[x_r, y_r]$ (they are supposed to fly at a constant height $[z_r]$) and their estimated location $[x_e, y_e]$ (their orientation, given by the yaw angle, is defined by the direction of movement so it is not required explicitly). They also display an estimated accuracy (a_e), which is described by the standard deviation of their spatial coordinates. The MAV operation can be analyzed in terms of its sensors, actuators, and control unit.

- *Sensors*: three types of sensors are considered, namely, the exploitability (E), the available accuracy (A_a) and the current exploration level of the drone (D_{ex}). The first two sensors are associated to a certain portion of the areas (a group of neighboring cells), where the potential increase of exploration that the drone can provide if it visits it is calculated (E) together with an estimation of the available accuracy (A_a), as a combination of the distance and accuracy of the closest tag. Several groups of neighboring cells of different sizes are considered (1, 3x3, 6x6 and 12x12, up to 36 groups of cells in total) and their parameters are calculated. Those values together with the current exploration level of the drone (the exploration probability it will set if it visits a cell) will constitute the inputs to the control unit.
- *Actuators*: the MAVs can move throughout the scenario, detect fixed and mobile tags, compute their estimated positions based on the tags, and communicate exploration and location information to other MAVs.
- *Control unit*: the control unit performs the operation which converts the set of inputs coming from the sensors into one output which defines the current behavior that the drone will carry out. This conversion is defined by a set of

Table 1. Parameters that define the simulated scenario

Side of the arena (L)	768
Total area	L^2
Fixed tag detection range	$L/4$
Mobile tags detection range	$L/16$
Max velocity V_{max}	$L/50$
Standard deviation for the velocity	$V_{max}/50$
Max accuracy provided by an AprilTag	$L/10$

12 real values which correspond to the genetic parameters of the associated individual. In particular, the control unit operates in three stages. First, a discretization of the inputs is performed according to four thresholds which correspond to the first 4 genes to distinguish between close/distant neighborhood, highly and poorly explored neighborhoods, and great or reduced accuracy availability on the surrounding cells. Second, the discretized input space defines 8 different input sets, for each of them one behavior coefficient is assigned. And third, the behavior coefficient sets one of the five available behaviors: explore the near neighborhood, explore the distant neighborhood, search a tag, move apart from a tag and stay still to share the available accuracy with the rest of the drones acting as a temporal tag.

Finally, the canonical algorithm implementation used here follows the pseudocode presented in section 2. The specific values for the parameters used for the initial tests are those shown on Table 2. These values were selected according to the conclusions extracted from [15], where a theoretical characterization of the parameters was performed.

3.2   Results

The left plot of Fig. 2 displays a representation of the surveillance level of the scenario, which is a measure of the completion of the task, during 100000 time steps of evolution for a representative run. The maximum surveillance level in the scenario corresponds to a value of 16 which would imply that all the cells are being simultaneously surveilled with maximum accuracy. The surveillance level is, however, more closely bounded by a practical maximum which takes in consideration several factors, namely, the number of drones, the number of tags, the decay rate in location accuracy, etc, and which is complicated to compute exactly. The evolution was performed according to the configuration indicated in Table 2. As it can be observed, the global fitness increases quickly during the first 20000 iterations. Afterwards, the global fitness keeps growing but subject to significant fluctuations, which conceal the improvement. However, by performing a moving average the fitness value measured increases from 9.7 units for iteration 20000 to 10.2 units for iteration 100000.

Table 2. Parameters that define the canonical EE algorithm

Iterations	100.000
Population size	40
Maximum lifetime (T_{max})	1000
Selection criteria $(P_{elegibility})$	Higher fitness
Tournament max size (S_{max})	20
Local search probability (P_{ls})	0.99
Life expectancy (T_{exp})	[-100:100] (performance dependent)
Chromosome length	8 x [-1,1]

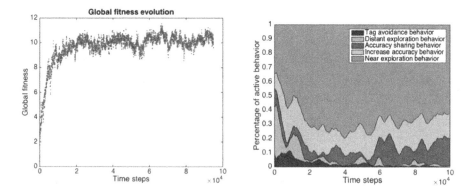

Fig. 2. Left: global fitness over time steps. Right: percentage area chart of the proportion of active behaviors within the population for each time step.

In order to provide a general idea of what happens in the population in terms of behavior as evolution takes place, the right plot of Fig. 2 shows a percentage area chart of the proportion of active behaviors within the population for each time step. During the first time steps the configuration is random, but the population quickly tends to increase the exploration and the tag search behavior together with decreasing the activity of behaviors which escape from the tags and move to distant unexplored regions. Therefore, during that first initial period the population performs a quick adjustment to perform the most direct and simple behavior to produce a satisfactory surveillance of the arena. After the first stage, these proportions are fine tuned to allow a slow increase of the performance. The most interesting change is the increase of the activity of the behavior which produces mobile tags that helps sharing accuracy among drones and therefore allows preserving an adequate accuracy level without moving towards the tags (red area in Fig. 2 right). It can be noted that after iteration 20000 its proportion gradually increases. It is also significant that the distant exploration and tag avoidance behaviors are also incrementally reduced to improve the overall behavior. Finally, to better understand but the structure of the population, not the general behavior of the group, in terms of individual behaviors, Fig. 3 shows a ternary graph of the composition of the population for three time intervals of the evolution. Each circle in the chart represents the ratio of the exploratory, tag search and accuracy sharing behaviors that an individual has performed during its lifetime up to a certain time step. The color assigned to each circle also represents the ratio of each basic behavior by proportionally combining the basic colors assigned to each behavior.

The chart on the left corresponds to the first one hundred time steps, and therefore to the composition of the population in terms of behavior shown during that interval. There is a clear uniform distribution of circles along the edges of the triangle, meaning that most of the individuals activate only two behaviors during their lifetime and that those behaviors are randomly selected.

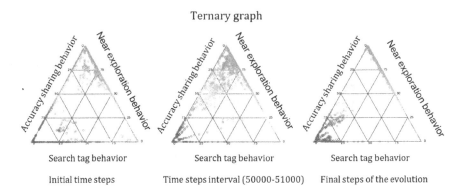

Fig. 3. Ternary graph of the composition of the population for three time intervals of the evolution

Although the presence of only two behaviors in most of the population may seem more arbitrary than random, it is produced due to the initial conditions of the scenario which generate very repetitive sensory inputs on the exploration sensors until the arena is explored sufficiently and the tag sensors indicate that the tag is either close or far. Therefore only two sets of sensory inputs are generated and consequently only two behaviors activated. The second graph corresponds to the behavior of the individuals during 1000 time steps in the middle of the evolution. A significant group of drones are now mainly exploring the scenario and combining the exploration with fitness sharing and also with searching for tags. Simply searching for tags seems not to be worth the effort and therefore that pure behavior seems to have been removed. Finally, the third graph indicates the distribution of the population for the last 1000 steps of the run. The population is clearly specialized. We can distinguish between individuals that mainly perform an accuracy sharing behavior and that every once in a while explore or look for a tag to preserve the accuracy. On the other hand there is a group of individuals which combine exploration and accuracy search in different proportions (from almost pure explorers, to very conservative individuals which keep their accuracy very high by visiting the tags very frequently and exploring less frequently) but which never share accuracy (right edge of the graph). Therefore, after the course of evolution, the population has specialized to be able to perform the task more efficiently, which implies not only the continuous coverage of the arena but also the preservation of an adequate location accuracy, which allows materializing the coverage into a reliable surveillance.

4   Conclusions

This work has shown the capability of the canonical Embodied Evolution algorithm to solve, on-line, a collective surveillance task with a fleet of autonomous MAVs. In this case, the optimization problem included the accuracy in the location of the MAVs as a new variable, leading to a coordinated control where the team must cover the scenario, while they are precisely positioned, by sharing their own location information. An emergent specialization has been observed in the final population obtained by the algorithm, with the individuals performing three main sub-tasks. The next step in this line is to transfer these results to the real MAVs and show the validity of the algorithm in real scenarios.

Acknowledgments. This project has received funding from the EU H2020 research and innovation programme under grant agreement No 640891.

References

1. Thrun, S., Burgard, W., Fox, D.: Probabilistic Robotics. MIT Press (2005)
2. Siegwart, R.: Nourbakhsh. I., Scaramuzza, D.: Introduction to Autonomous Mobile Robots. MIT Press (2011)
3. Chatterjee, A., Rakshit, A., Singh, N.: Vision Based Autonomous Robot Navigation: Algorithms and Implementations. SCI, vol. 455. Springer, Heidelberg (2013)
4. Shen, S.: Michael, N., Kumar, V.: Autonomous multi-floor indoor navigation with a computationally constrained MAV. In: Proceedings ICRA 2011, pp. 20–25 (2011)
5. Lippiello, V., Loianno, G., Siciliano, B.: MAV indoor navigation based on a closed-form solution for absolute scale velocity estimation using Optical Flow and inertial data. In: Proceedings CDC-ECC 2011, pp. 3566–3571 (2011)
6. Trianni, V., Nolfi, S.: Evolving collective control, cooperation and distributed cognition. In: Handbook of Collective Robotics, pp. 127–166. Springer (2012)
7. Nitschke, G.S.: Neuro-Evolution approaches to collective behavior. In: Proceedings CEC 2009, pp. 1554–1561 (2009)
8. Watson, R., Ficici, S., Pollack, J.: Embodied evolution: Distributing an evolutionary algorithm in a population of robots. Robotics and Autonomous Systems 39(1), 1–18 (2002)
9. Schut, M.C., Haasdijk, E., Prieto, A.: Is situated evolution an alternative for classical evolution?. In: Proceedings CEC 2009, pp. 2971–2976 (2009)
10. Haasdijk, E., Eiben, A.E., Karafotias, G.: On-line evolution of robot controllers by an encapsulated evolution strategy. In: Proceedings IEEE CEC 2010, pp. 1–7 (2010)
11. Elfwing, S., Uchibe, E., Doya, K., Christensen, H.: Darwinian embodied evolution of the learning ability for survival. Adaptive Behavior 19(2), 101–120 (2011)
12. Bredeche, N., Montanier, J.M., Liu, W., Winfield, A.: Environment-driven Distributed Evolutionary Adaptation in a Population of Autonomous Robotic Agents. Mathem. and Comput. Modelling of Dynamical Systems 18(1), 101–129 (2012)
13. Prieto, A., Becerra, J.A., Bellas, F., Duro, R.J.: Open-ended Evolution as a means to Self-Organize Heterogeneous Multi-Robot Systems in Real Time. Robotics and Autonomous Systems 58, 1282–1291 (2010)
14. Trueba, P., Prieto, A., Bellas, F., Caamao, P., Duro, R.J.: Specialization analysis of embodied evolution for robotic collective tasks. Robotics and Autonomous Systems 61(7), 682–693 (2012)
15. Trueba, P., Prieto, A., Bellas, F.: Distributed embodied evolution for collective tasks: parametric analysis of a canonical algorithm. In: Proc. GECCO 2013, pp. 37–38 (2013)
16. Olson, E.: AprilTag: A robust and flexible visual fiducial system. In: Proceedings ICRA 2011, pp. 3400–3407 (2011)

On the Spectral Clustering for Dynamic Data

D.H. Peluffo-Ordóñez[1](✉), J.C. Alvarado-Pérez[2], and A.E. Castro-Ospina[3]

[1] Universidad Cooperativa de Colombia – Pasto
diego.peluffo@campusucc.edu.com
[2] Universidad de Salamanca, Spain
jcalvarado@usal.es
[3] Research Center of the Instituto Tecnológico Metropolitano, Colombia
andrescastro@itm.edu.co

Abstract. Spectral clustering has shown to be a powerful technique for grouping and/or rank data as well as a proper alternative for unlabeled problems. Particularly, it is a suitable alternative when dealing with pattern recognition problems involving highly hardly separable classes. Due to its versatility, applicability and feasibility, this clustering technique results appealing for many applications. Nevertheless, conventional spectral clustering approaches lack the ability to process dynamic or time-varying data. Within a spectral framework, this work presents an overview of clustering techniques as well as their extensions to dynamic data analysis.

Keywords: Dynamic data · Kernels · Spectral clustering

1 Introduction

In the field of pattern recognition and classification, the clustering methods derived from graph theory and based on spectral matrix analysis are of great interest because of their usefulness for grouping highly non-linearly separable clusters. Some of their remarkable applications to be mentioned are human motion analysis and people identification [1], image segmentation [2,3] and video analysis [4], among others. The spectral clustering techniques carry out the grouping task without any prior knowledge –indication or hints about the structure of data to be grouped– and then partitions are built from the information obtained by the clustering process itself. Instead, they only require some initial parameters such as the number of groups and a similarity function. For spectral clustering, particularly, a global decision criterion is often assumed taking into account the estimated probability that two nodes or data points belong to the same cluster [5]. For this reason, this kind of clustering can be easily understood from a graph theory view point where such probability is to be associated to the similarity among nodes. Typically, clusters are formed in a lower dimensional space involving a dimensionality reduction process. This is done preserving the relationship

This work is supported by the Faculty of Engineering of Universidad Cooperativa de Colombia-Pasto, and the ESLINGA Research Group.

© Springer International Publishing Switzerland 2015
J.M. Ferrández Vicente et al. (Eds.): IWINAC 2015, Part II, LNCS 9108, pp. 148–155, 2015.
DOI: 10.1007/978-3-319-18833-1_16

among nodes as well as possible. Most approaches that deal with this matter are founded on linear combinations or latent variable models where the eigenvalue and eigenvector decomposition (here termed eigen-decomposition) of the normalized similarity matrix takes place. In other words, spectral clustering comprises all the clustering techniques using information of the eigen-decomposition from any standard similarity matrix obtained from the data to be clustered. In general, these methods are of interest in cases where, because of the complex structure, clusters are not readily separable and traditional grouping methods fail. In the state of the art on unsupervised data analysis, we can find several approaches for spectral clustering.

One of the biggest disadvantages of the spectral clustering methods is that most of them have been designed for analyzing only static data, that is to say, regardless of the temporary information. Therefore, when data are changing along time, clustering can be performed on single current data frame without analyzing the previous ones. Some works have addressed this important issue concerning applications such as human motion analysis [6]. Other approaches are focused on the design of dynamic kernels for clustering [7] as well as the use of dynamic KPCA [8]. In the literature, many approaches prioritize the use of kernel methods since they allow to incorporate prior knowledge into the clustering procedure. However, the design of a whole kernel-based clustering scheme able to group time-varying data achieving a high accuracy is still an open issue.

This work presents a short overview of recent spectral clustering approaches with a special focus on how they have been extended to dynamic data analysis. The rest of this paper is organized as follows: Section 2 outlines the conventional and recent methods based on both kernel and graph representations. Dynamic data extensions and adaptations for spectral clustering are presented in Section 3. Finally, Section 4 draws some final remarks.

2 Spectral Clustering

Following, a brief theoretical background with a bibliographic scope on spectral clustering based on both graph and kernel models is presented.

2.1 Graph-Based Approaches

Clustering based on spectral theory is a relatively new focus for unsupervised analysis; although it has been used in several studies that prove its efficiency on grouping tasks, especially in cases when the clusters are not linearly separable. This clustering technique has been widely used in a large amount of applications such as human motion analysis and people identification [1], image segmentation [2,3] and video analysis [4], among others. What makes spectral analysis applied to data clustering appealing is the use of the eigenvalue and eigenvector decomposition in order to obtain the local optima closest to the global continuous optima. The spectral clustering techniques take advantage of the topology of data from a non-directed and weighted graph-based representation. Such a topological approach is known as graph partitioning; in it an initial optimization problem is usually posed formulated under some constraints. Often this formulation is relaxed and then becomes a NP-complete problem [9,10,11]. In addition, the estimation of global optima in a relaxed continuous domain is done through the eigenvector decomposition (eigen-decomposition) [5], based on the theorem by Perron - Frobenius, which establishes that largest strictly real eigenvalues associated to a positive definite and irreducible matrix determine their own spectral ratio [12]. In this case, such matrix corresponds to the affinity or similarity matrix containing the similarities among data points. In other words, the space created by the eigenvectors in the affinity matrix is closely related to the clustering quality. Under this principle, and considering the possibility of obtaining a discreet solution through eigenvectors, there have been proposed multi-cluster approaches such as the k-way normalized cut method introduced in [5]. In such method, to obtain the discrete solution, another optimization problem must be solved, albeit in a lower dimensional domain. Besides, leading eigenvectors can be considered as a new data set that can be clustered by a conventional clustering algorithm such as K-means [9,13]. Alternatively, there are approaches that do not require the eigenvector calculation being recommendable when such a calculation is prohibitive (i.e. Big Data) [14,15].

2.2 Kernel-Based Approaches

More advanced approaches propose minimizing a feasibility measure between the solutions of the unconstrained problem and the allowed cluster indicator [16,17], finding peaks and valleys of a cost function that quantifies the cluster overlapping [18]. Then, a discretization process is carried out. The problem of continuos solution discretization is formulated and solved in [19]. There exist graph-based methods associated with normalized and non-normalized Laplacian, which provide relevant information contained in the Laplacian eigenvector is of great usefulness for the clustering task. As well, they often provide a straightforward interpretation about the clustering quality based on a random graph theory. Under the assumption that data is a random weigthed graph, it is demonstrated that spectral clustering can be seen as a maximal similarity-based clustering. In [17], some alternatives to solving open issues in spectral clustering are discussed, such as the selection of a proper analysis scale, extensions to multi-scale data management, the existence of irregular and partially known groups, and the automatic estimation of the number of groups. The authors propose a local scaling to calculate the affinity or similarity between each pair of data points. This scaled similarity improves the clustering algorithms in terms of both convergence and processing time. Additionally, taking advantage of the underlying information given by eigenvectors is suggested to automatically establish the number of groups. The output of this algorithm can be used as a starting point to initialize partitioning algorithms such as K-means. There are another approaches that have been pointed out to extend the clustering model to new data (testing data), i. e., out-of-sample extension. For instance, the methods presented in [20,21] allow for extending the clustering process to new data, approximating the eigen-function (a function based on the eigen-decomposition) using the

Nyström's method [22,23]. Therein, a clustering method and a searching criterion are typically chosen in a heuristic fashion. Another spectral clustering perspective is the Kernel Principal Component Analysis (KPCA) [24,25]. KPCA is an unsupervised technique for nonlinear feature extraction and dimensionality reduction [26]. Such method is a nonlinear version of PCA using positive definite matrices whose aim is to find the projected space onto an induced kernel feature space preserving the maximum explained variance [27]. The relationship between KPCA and spectral clustering is explained in [24,25]. Indeed, the work presented in [25] demonstrates that the classical spectral clustering, such as normalized cut [9], the NJW algorithm [10] and random walk model [28] can be seen as particular cases of weighted KPCA, just with modifications on the kernel matrix. In [27], a spectral clustering model is proposed which is based on a KPCA scheme on the basis of the approach proposed in [25] in order to extend it to multi-cluster clustering, wherein coding and decoding stages are involved. To that end, a formulation founded on least squares method-based support vector machines (LSSVM) considering weighted versions [29,30]. This formulation aims to a constrained optimization problem in a primal-dual framework that allows for extending the model to new isolated data without needing additional techniques (e. g. the Nyström method). In this connection, a modified similarity matrix is proposed whose eigenvectors are dual regarding a primal optimization problem formulated in a high dimensional space. Also, it is proven that such eigenvectors contain underlying relevant information useful for clustering purposes and display a special geometric structure when the resulting clusters are well formed -in terms of compactness and/or separability. The out-of-sample extensions are done by projecting cluster indicators onto the eigenvectors. As a model selection criterion, the method so-called Balanced Line Fit (BLF) criterion is proposed. This criterion explores the structure of the eigenvectors and their corresponding projections, which can be used to set the initial model parameters. Similarly, in [31], another kernel model for spectral clustering is presented. This method is based on the incomplete Cholesky decomposition. It is able to deal efficiently with large-scale clustering problems. The set-up is made on the basis of a WPCA scheme based on kernels (WKPCA) from a primal-dual formulation. In it, the similarity matrix is estimated by using the incomplete Cholesky decomposition. Also, it has recently introduced an improved method using optimal data projections [32].

3   Kernel-Based Clustering for Dynamic Data

Dynamic data analysis is of great interest, there are several remarkable applications such as video analysis [33] and motion identification [34], among others. By using spectral analysis and clustering, some works have been developed taking into account the temporal information for the clustering task, mainly in segmentation of human motion [35,6]. Other approaches include either the design of dynamic kernels for clustering [36,7] or a dynamic kernel principal component analysis (KPCA) based model [8,37]. Another study [38] modifies the primal functional of a KPCA formulation for spectral clustering to add the memory effect. There exists another alternative known as multiple kernel learning (MKL), which has emerged to deal with different issues in machine learning, mainly, regarding support vector machines (SVM) [39,40]. The intuitive idea of MKL is that learning can be enhanced when using different kernels instead of an unique kernel. Indeed, local analysis provided by each kernel is of benefit to examine the structure of the whole data when having local complex clusters. Following from this novel dynamic approaches have been proposed for both time-varying data tracking [41] and clustering [42,43].

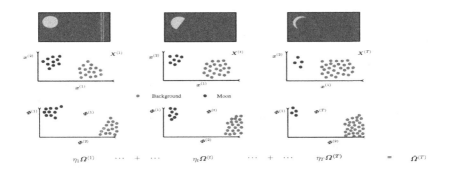

Fig. 1. Graphic explanation of MKL for clustering considering the example of changing moon

To understand the general idea of MKL applied to dynamic clustering, consider the following mathematical statements. Let us assume a sequence of N_f input data matrices such that $\{\boldsymbol{X}^{(1)}, \ldots, \boldsymbol{X}^{(N_f)}\}$, where $\boldsymbol{X}^{(t)} = [\boldsymbol{x}_1^{(t)\top}, \ldots, \boldsymbol{x}_N^{(t)\top}]^\top$ is the data matrix associated to time instance t. In order to take into consideration the time effect within the computation of kernel matrix, we can apply a multiple kernel learning approach, namely a linear combination of all the input kernel matrices until the current matrix. Then, at instance T, the cumulative kernel matrix can be computed as $\check{\boldsymbol{\Omega}}^{(T)} = \sum_{t=1}^{T} \eta_t \boldsymbol{\Omega}^{(t)}$, where $\boldsymbol{\eta} = [\eta_1, \ldots, \eta_T]$ are the weighting factors or coefficients and $\boldsymbol{\Omega}^{(t)}$ is the kernel matrix associated to $\boldsymbol{X}^{(t)}$ such that $\Omega_{ij}^{(t)} = \mathcal{K}(\boldsymbol{x}_i^{(t)}, \boldsymbol{x}_j^{(t)})$, where $\mathcal{K}$ denotes a kernel function. Since η_t is the weighting factor associated to the data matrix at time instance t within a sequence, $\boldsymbol{\eta}$ can be seen as a tracking vector. In Figure 1, the example of moving moon is recalled to graphically explained how the MKL works for clustering purposes. Each frame is represented by a feature space (data matrix) $\boldsymbol{X}^{(t)}$ which is mapped to a high-dimensional space $\boldsymbol{\Phi}^{(t)}$. Then, the corresponding kernels $\boldsymbol{\Omega}$ are calculated. Next, the cumulative kernel matrix $\check{\boldsymbol{\Omega}}$ is obtained by a linear combination of previously determined kernels. Coefficients or weighting factors for such linear combination are preset or calculated regarding each frame. Finally, some kernel-based clustering approach must be performed on the cumulative kernel matrix to get the cluster assignments.

4 Final Remarks

When analyzing a sequence of frames represented by a single data matrix, aiming the identification of underlying dynamic events, kernel-based approaches represent a suitable alternative. Certainly, kernel functions come from an estimation of inner product of high-dimensional representation spaces where clusters are assumed to be separable, and are often defined as a similarity measure between data points from the original space. Such similarity is designed for a local data analysis. In other words, kernels allow a piecewise data exploring by means of an estimation of a generalized variance. Therefore, we can infer that the evolutionary behavior of the sequence can be tracked by some ranking values derived from a kernel-based formulation.

Under the premise that a set of different kernels may represent more properly the input data than a single kernel, it has arisen the multiple kernel learning (MKL). In other words, a suitable alternative to analyze and explore data can be posed within a MKL framework.

This work gathered some cues and important aspects to understand spectral clustering techniques and their applicability in dynamic data analysis.

References

1. Truong Cong, D., Khoudour, L., Achard, C., Meurie, C., Lezoray, O.: People re-identification by spectral classification of silhouettes. Signal Processing 90(8), 2362–2374 (2010)
2. Wang, L., Dong, M.: Multi-level low-rank approximation-based spectral clustering for image segmentation. Pattern Recognition Letters (2012)
3. Molina-Giraldo, S., Álvarez-Meza, A.M., Peluffo-Ordoñez, D.H., Castellanos-Domínguez, G.: Image segmentation based on multi-kernel learning and feature relevance analysis. In: Pavón, J., Duque-Méndez, N.D., Fuentes-Fernández, R. (eds.) IBERAMIA 2012. LNCS, vol. 7637, pp. 501–510. Springer, Heidelberg (2012)
4. Zhang, D., Lin, C., Chang, S., Smith, J.: Semantic video clustering across sources using bipartite spectral clustering. In: 2004 IEEE International Conference on Multimedia and Expo, ICME 2004, vol. 1, pp. 117–120. IEEE (2004)
5. Stella, X.Y., Jianbo, S.: Multiclass spectral clustering. In: ICCV 2003: Proceedings of the Ninth IEEE International Conference on Computer Vision, p. 313. IEEE Computer Society, Washington, DC (2003)
6. Zhou, F., Torre, F., Hodgins, J.: Aligned cluster analysis for temporal segmentation of human motion. In: 8th IEEE International Conference on Automatic Face & Gesture Recognition, FG 2008, pp. 1–7. IEEE (2008)
7. Keshet, J., Bengio, S.: Automatic speech and speaker recognition: Large margin and kernel methods. Wiley (2009)
8. Maestri, M., Cassanello, M., Horowitz, G.: Kernel pca performance in processes with multiple operation modes. Chemical Product and Process Modeling 4(5), 7 (2009)
9. Shi, J., Malik, J.: Normalized cuts and image segmentation. IEEE Transactions on Pattern Analysis and Machine Intelligence 22(8), 888–905 (2000)

10. Ng, A.Y., Jordan, M.I., Weiss, Y.: On spectral clustering: Analysis and an algorithm. In: Advances in Neural Information Processing Systems 14, pp. 849–856. MIT Press (2001)
11. Shi, J., Malik, J.: A random walks view of spectral segmentation. In: AI and Statistics (AISTATS) (2001)
12. Anh, B.T., Thanh, D.D.X.: A perron-frobenius theorem for positive quasipolynomial matrices associated with homogeneous difference equations. Journal of Applied Mathematics (2007)
13. Ng, A.Y., Jordan, M.I., Weiss, Y.: On spectral clustering: Analysis and an algorithm. In: Advances in Neural Information Processing Systems 14, pp. 849–856. MIT Press (2001)
14. Peluffo-Ordóñez, D.H., Castro-Hoyos, C., Acosta-Medina, C.D., Castellanos-Domínguez, G.: Quadratic problem formulation with linear constraints for normalized cut clustering. In: Bayro-Corrochano, E., Hancock, E. (eds.) CIARP 2014. LNCS, vol. 8827, pp. 408–415. Springer, Heidelberg (2014)
15. Dhillon, I.S., Guan, Y., Kulis, B.: Weighted graph cuts without eigenvectors a multilevel approach. IEEE Trans. Pattern Anal. Mach. Intell. 29(11), 1944–1957 (2007)
16. Bach, F., Jordan, M.: Learning spectral clustering, with application to speech separation. The Journal of Machine Learning Research 7, 1963–2001 (2006)
17. Zelnik-manor, L., Perona, P.: Self-tuning spectral clustering. In: Advances in Neural Information Processing Systems 17, pp. 1601–1608. MIT Press (2004)
18. Ding, C., He, X.: Linearized cluster assignment via spectral ordering. In: Machine Learning-International Workshop Then Conference, vol. 21, p. 233. Citeseer (2004)
19. Higham, D., Kibble, M.: A unified view of spectral clustering. In: Mathematics Research Report 2. University of Strathclyde (2004)
20. Bengio, Y., Paiement, J., Vincent, P., Delalleau, O., Le Roux, N., Ouimet, M.: Out-of-sample extensions for lle, isomap, mds, eigenmaps, and spectral clustering. In: Advances in Neural Information Processing Systems 16: Proceedings of the 2003 Conference, p. 177. The MIT Press (2004)
21. Fowlkes, C., Belongie, S., Chung, F., Malik, J.: Spectral grouping using the Nyström method. IEEE Transactions on Pattern Analysis and Machine Intelligence, 214–225 (2004)
22. Baker, C.: The numerical treatment of integral equations, vol. 13. Clarendon press Oxford, UK (1977)
23. Williams, C., Seeger, M.: Using the Nyström method to speed up kernel machines. In: Advances in Neural Information Processing Systems 13. Citeseer (2001)
24. Ham, J., Lee, D., Mika, S., Schölkopf, B.: A kernel view of the dimensionality reduction of manifolds. In: Proceedings of the Twenty-first International Conference on Machine Learning, p. 47. ACM (2004)
25. Alzate, C., Suykens, J.: A weighted kernel PCA formulation with out-of-sample extensions for spectral clustering methods. In: International Joint Conference on Neural Networks, IJCNN 2006, pp. 138–144. IEEE (2006)
26. Schölkopf, B., Smola, A., Müller, K.: Nonlinear component analysis as a kernel eigenvalue problem. Neural Computation 10(5), 1299–1319 (1998)
27. Alzate, C., Suykens, J.: Multiway spectral clustering with out-of-sample extensions through weighted kernel PCA. IEEE Transactions on Pattern Analysis and Machine Intelligence, 335–347 (2008)

28. Chan, P.K., Schlag, M.D.F., Zien, J.T.: Spectral K-way ratio cut partitioning and clustering. IEEE Transactions on Computer-aided Design of Integrated Circuits and Systems 13, 1088–1096 (1994)
29. Suykens, J., Gestel, T.V., Brabanter, J.D., Moor, B.D., Vandewalle, J.: Least squares support vector machines. World Scientific Pub. Co. Inc. (2002)
30. Suykens, J., Van Gestel, T., Vandewalle, J., De Moor, B.: A support vector machine formulation to PCA analysis and its kernel version. IEEE Transactions on Neural Networks 14(2), 447–450 (2003)
31. Alzate, C., Suykens, J.: Sparse kernel models for spectral clustering using the incomplete cholesky decomposition. In: IEEE International Joint Conference on Neural Networks, IJCNN 2008 (IEEE World Congress on Computational Intelligence), pp. 3556–3563. IEEE (2008)
32. Peluffo-Ordez, D.H., Alzate, C., Suykens, J.A.K., Castellanos-Domnguez, G.: Optimal data projection for kernel spectral clustering. In: European Symposium on Artificial Neural Networks, Computational Intelligence and Machine Learning, pp. 553–558 (2014)
33. Shirazi, S., Harandi, M.T., Sanderson, C., Alavi, A., Lovell, B.C.: Clustering on grassmann manifolds via kernel embedding with application to action analysis. In: Proc. IEEE International Conference on Image Processing (2012)
34. Sudha, L., Bhavani, R.: Performance comparison of svm and knn in automatic classification of human gait patterns. Int. J. Comput. 6(1), 19–28 (2012)
35. Takács, B., Butler, S., Demiris, Y.: Multi-agent behaviour segmentation via spectral clustering. In: Proceedings of the AAAI 2007, PAIR Workshop, pp. 74–81 (2007)
36. Chan, A., Vasconcelos, N.: Probabilistic kernels for the classification of auto-regressive visual processes (2005)
37. Choi, S., Lee, I.: Nonlinear dynamic process monitoring based on dynamic kernel pca. Chemical Engineering Science 59(24), 5897–5908 (2004)
38. Langone, R., Alzate, C., Suykens, J.A.: Kernel spectral clustering with memory effect. Physica A: Statistical Mechanics and its Applications (2013)
39. González, F.A., Bermeo, D., Ramos, L., Nasraoui, O.: On the robustness of kernel-based clustering. In: Alvarez, L., Mejail, M., Gomez, L., Jacobo, J. (eds.) CIARP 2012. LNCS, vol. 7441, pp. 122–129. Springer, Heidelberg (2012)
40. Huang, H., Chuang, Y., Chen, C.: Multiple kernel fuzzy clustering. IEEE Transactions on Fuzzy Systems 20(1), 120–134 (2012)
41. Peluffo-Ordóñez, D., García-Vega, S., Castellanos-Domínguez, C.G.: Kernel spectral clustering for motion tracking: A first approach. In: Ferrández Vicente, J.M., Álvarez Sánchez, J.R., de la Paz López, F., Toledo Moreo, F. J. (eds.) IWINAC 2013, Part I. LNCS, vol. 7930, pp. 264–273. Springer, Heidelberg (2013)
42. Peluffo-Ordóñez, D.H., García-Vega, S., Álvarez-Meza, A.M., Castellanos-Domínguez, C.G.: Kernel spectral clustering for dynamic data. In: Ruiz-Shulcloper, J., Sanniti di Baja, G. (eds.) CIARP 2013, Part I. LNCS, vol. 8258, pp. 238–245. Springer, Heidelberg (2013)
43. Peluffo-Ordónez, D., Garcia-Vega, S., Langone, R., Suykens, J.A., Castellanos-Dominguez, G.: Kernel spectral clustering for dynamic data using multiple kernel learning. In: The 2013 International Joint Conference on Neural Networks, pp. 1–6. IEEE (2013)

A Multiple-Model Particle Filter Based Method for Slide Detection and Compensation in Road Vehicles

R. Toledo-Moreo[✉], Carlos Colodro-Conde, and F. Javier Toledo-Moreo

Technical University of Cartagena, DETCP, Edif. Antigones,
30202 Cartagena, Spain
rafael.toledo@upct.es, carlos.colodro@upct.es, javier.toledo@upct.es

Abstract. Continuous precise navigation is essential for many intelligent transportation systems (ITS) applications. It is a common understanding that for the challenges of advanced applications, global navigation satellite systems (GNSS) based solutions must be complemented with additional sensors. A very simple and still very competent option is the inclusion of one odometer and one gyroscope in the onboard sensing unit. This choice has the benefits that its errors can be easily characterized in normal conditions, and keeps the costs low. However, the intrinsic nature of the odometry system causes important diminutions of its performance in unusual friction conditions, such as slides and slips. This paper proposes a method to detect and compensate the errors made in the odometry in unusual friction conditions by means of a multiple model particle filter (MMPF) based hybridization. In concrete, we focus on slides since these appear more often in road vehicles. The theoretical contributions, along with the good results obtained in real trials are presented in the paper.

1   Introduction

Although GNSS (Global Navigation Satellite Systems) verify many of the requirements that some applications of road traffic demand, in areas with restricted visibility of the satellite constellation the lack of continuity in the simple GNSS positioning prevents from its direct application in many applications. For this reason, GNSS measurements are commonly fused with some other onboard sensors in order to supply a continuous positioning solution. In its minimum configuration, the set of sensors for assisting GNSS includes the odometry system that informs about the distance travelled by the vehicle. It is also quite usual to include a gyroscope for updating the heading angle by means of one integration process. On the one hand, this small set of aiding sensors presents the advantage of its low cost and relatively good performance. On the other hand, due to integration processes from rate of turn to obtain the heading angle and the errors in the odometry captors, the position supplied by them drifts with the time. Although there are a number of papers in the literature of different solutions to compensate from these drifts by means of electronic compasses or accelerometers

© Springer International Publishing Switzerland 2015
J.M. Ferrández Vicente et al. (Eds.): IWINAC 2015, Part II, LNCS 9108, pp. 156–165, 2015.
DOI: 10.1007/978-3-319-18833-1_17

[1,2,3,4], GNSS plus odometry plus gyroscope still is the most common configuration for assisted GNSS positioning. When applied, low cost accelerometers are normally included to support the odometry measurement, and not to replace it. Main problems associated to the use of odometry for car navigation are slips and slides, being in general slides more often than slips. The alteration of the friction between road and tires may introduce large errors in the odometry output at any time, and standard data-fusion filters are not prepared for that, causing unusual drifts and underestimated errors in the filter output. On the other hand, a filtering algorithm that would consider that the odometry errors are always of the order of those presented during slips and slides would normally overestimate the error, and consequently diminish the filter performance. An efficient data-fusion filter should be capable to represent these errors and modify its error considerations only when slides and slips appear.

This paper presents a fusion algorithm capable to adapt the error considerations of the filter according to the odometry errors and including slide situations. For concept proving, only slide errors are considered since they appear more often in road vehicles. However, the same approach can be applied to slips. Apart from the improvements in the positioning, the filter provides better error estimates. The proposed filter consists of a multiple model particle filter, in which different subsets of particles comply with the noise considerations represented by each model.

The most popular filtering methods for data-fusion are based on the Kalman filter (KF), and its version for non-linear systems, the extended Kalman filter (EKF) [5]. However, EKFs fail in case of severe non-linearities. For such cases, particle filters (PF) represent an excellent alternative [6,7]. PFs employ weighted sampling and resampling to generate Monte Carlo approximations of the intended posterior probability density functions.

Additionally, the use of multiple models to represent the different maneuvering state of road vehicles brings good results in terms of positioning, error considerations or even maneuvering prediction [8,9].

In the case under consideration, a set of multiple models allows applying the a priori information that we know about the unusual situations, and to use it in such a way that the filter is ready to cope with it. It can considered that when the vehicle slides, the odometry system supplies an underestimated value of the travelled distance.

The rest of the paper is organized as follows. Section 2 presents the proposed error models for the odometry in normal and slippery conditions. Next, Sections 3 and 4 introduce respectively the particle filter and the MMPF under consideration. In Section the results obtained by our method are discussed. Section 6 concludes the paper.

2   Error Model of the Odometry

2.1   In Normal Conditions

The error introduced due to the wear of the tires has a very low frequency, and it can be compensated by simply performing seldom calibrations of the odometry step value. It can be represented then as a calibration error with the form $\omega(k)\delta r$, where $\omega(k)$ is the angular rate of the wheels at instant k and δr the difference between the real wheel radius, and the nominal value r.

The errors due mechanical and electrical noises can be merged together in a common error term with a Gaussian characteristic and zero mean, $n(k) \sim N(0, \sigma_{odo})$. For our experiments, we have set the value of σ_{odo} to 0.26 m, that corresponds with the odometry step.

Therefore, under normal conditions the error in the odometry can be characterized as:

$$\varepsilon_{dt}(k) = \omega(k)\delta r + n(k) \tag{1}$$

2.2   Affected by Slides

We focus on slides in this paper, but analogous considerations could be done for slippage. A wheel slide appears when the wheel rotates slower than the value that would correspond to the current vehicle speed.

The equation for the characterization of the odometry error presented in (1) can be completed now to represent slides in the sensor model.

$$\varepsilon_{dt}^{slide}(k) = \delta^{slide}(k) + \omega(k)\delta r + n(k)$$

where $\delta^{slide}(k) \geq 0$ compensates for the sliding effect. We describe $\delta^{slide}(k)$ as:

$$\delta^{slide} = (1 - scv) \times c(k)$$

where $c(k) \sim N(0, \sigma_{odo})$, and scv (slide compensation value) may vary from 0 (maximum compensation) till 1 (minimum).

From now on, let us name the models for the odometry like goes: normal conditions model (NCM) and slide conditions model (SCM).

3   Particle Filter

The principle of the particle filter is to represent by means of a set of N samples $\{X^i(k)\}_{i=1}^{N}$ and their corresponding weights $\{w^i(k)\}_{i=1}^{N}$ the probability density function $p(X(k)/Y_{1:k})$ at instant k of the state vector $\mathbf{X}(k)$, given past observations, following next numerical expression:

$$p(\mathbf{X}(k)/Y_{1:k}) \simeq \sum_{i=1}^{N} X^i(k) \cdot w^i(k)$$

where $Y_{1:k}$ stands for the observations collected from the initialization till instant k. Each sample $X^i(k)$ can be described as a Dirac delta expression $\delta^i(k)(\mathbf{X}(k))$ in the form:

$$\delta^i(k)(\mathbf{X}(k)) = 0 \quad \text{if} \quad \mathbf{X}(k) \neq X^i(k) \tag{2}$$
$$\delta^i(k)(\mathbf{X}(k)) = 1 \quad \text{if} \quad \mathbf{X}(k) = X^i(k)$$

A detailed description of particle filtering can be found in [16,17].

4   Multiple-Model Particle Filter Based Method

The particle filter is an excellent alternative to Kalman filtering for non-linear systems, allowing non-gaussian noise considerations. Additionally, combined with multiple models bring new interesting possibilities to the characterization of the error of the sensors, such as abrupt changes from one instant to next one in the sensor error model.

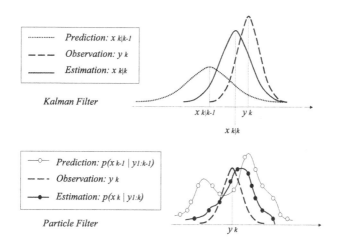

Fig. 1. Covariance of the prediction, observation and estimation phases in the Kalman filter (upper image) and the multiple-model particle filter (below)

The upper image of Fig. 1 shows an example of probability density functions (pdf) in the Kalman filter given by the prediction step (where the odometry is typically employed as a filter input), the observations (normally the GPS position), and the final filter estimate given by the Kalman update. The lower image of Fig. 1 represents also the pdf after prediction, observation and estimation steps, but this time for a PF. The distribution that represents now the pdf of the prediction is described by the combination of two sets of particles for the two different error models of the odometry (NCM and SCM). The observation will match better the prediction that describes more precisely the behavior of the odometry system. Therefore, when the update step is executed, the particles of the most suitable model are rewarded (their weights will increase as compared to the others). Therefore, the weights of the particles after the observation will indicate the probability that each model m represents properly the vehicle behavior, $\mu^m(k)$. This probability can be calculated following:

$$\mu^m(k) = \frac{\sum_i^{Nm} w(k)_i^m}{\sum_i^{Nm} w(k)_i^m + \sum_i^{Nm} w(k)_i^{\neq m}} \tag{3}$$

where $w(k)_i^m$ stands for the weight at instant k of the particle i that represents model m, and Nm is the number of particles of each model, that is assumed equal for all of them. The term $w(k)_i^{\neq m}$ stands for the weight at instant k of a particle i that is not driven by model m.

In our approach, a fixed number of particles are assigned to the each model: particles cannot switch models. Therefore, the probability values of each model vary depending on the particles weights, that represent the goodness of the models anytime.

Apart from the different assumptions in the odometry model, both filters will be identical. We employ the classical 2D kinematical model for a vehicle on a plane. The state vector is $[x(k)\ y(k)\ \psi(k)]$, representing east, north and heading (from north to east) at the center of the rear axle of the vehicle. The measurements of the odometry and the gyroscope are employed as inputs to the filter. The travelled distance measured by the odometer, $ds(k)$, is estimated when a gyroscope value, $\dot{\psi}(k)$, is processed. The equation for pose prediction in discrete time goes as follows:

$$x(k+1) = x(k) + ds(k)\ sinc(\dot{\psi}/2)\ cos(\psi(k) +$$
$$\dot{\psi}(k)T/2) - \dot{\psi}(k)T\big(Dx\ sin\psi(k) +$$
$$Dy\ cos\psi(k)\big)$$
$$y(k+1) = y(k) + ds(k)\ sinc(\dot{\psi}/2)\ sin(\psi(k) +$$
$$\omega(k)T/2) + \dot{\psi}(k)T\big(Dx\ cos\psi(k) -$$
$$Dy\ sin\psi(k)\big)$$
$$\psi(k+1) = \psi(k) + \dot{\psi}(k)T$$

being T the sampling period, Dx, Dy the coordinate of the GPS antenna in the body frame, and $sinc$ the cardinal sinus.

The observations $[x_{GPS}, y_{GPS}]$ are east and north values given by an EGNOS-capable GPS receiver after transformation into the appropriate reference frame.

The output of the MMPF will be that of the model with the highest probability.

5   Experiments and Results

The test vehicle was equipped with an EGNOS capable GPS receiver, a dual-frequency RTK receiver for evaluation, a yaw rate gyroscope and one odometer with resolution of 1 pulse per 26.15 cm placed on the gear box coupled to the rear wheels axle.

5.1   Results

In this Section we will analyze the goodness of the proposed filter to represent better the behavior of a road vehicle. In particular, we will a single-model PF (assuming not slides), with the multiple-model PF (MMPF) approach.

A test with two periods where the wheels slide to a considerable extend will be presented in this Section. As it will be seen, the slides were enough to impoverish the performance of a filter that does not take into account their effects. The decision of whether or not the slides are fair enough to launch the SCM filter is made based on the capability of the NCM to cope with the errors. As it will be seen, as long as the characterization of the odometry error is good, the probability of SCM will be higher than the one of the SCM. On the other hand, it will be lower when it cannot represent these errors. This way, the filter adapts itself to the nature of the odometry errors.

In the experiments, the output errors were estimated by using RTK measurements of high precision. The MMPF solution offers good values of 0.211, 0.147 and 0.667 for the mean, standard deviation and maximum horizontal positioning error. However, a comparison of the positioning errors under sliding conditions was found complicated. Due to the fact that the frequencies and time stamps

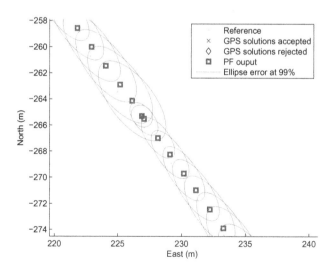

Fig. 2. NCM positioning output (red squares) and estimated 2σ ellipsis at 99% under normal conditions

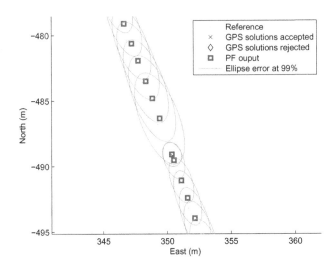

Fig. 3. NCM positioning output (red squares) and estimated 2 σ ellipsis at 99% affected by slides

of all the sensors and GPS receivers are different, we must estimate the error by means of an interpolation process, assuming that the error is the minimum distance between the RTK positions and the filter positioning outputs. Since in the experiments we focus on slides due to acceleration and deceleration maneuvers and the simplest case is when the vehicle follows a trajectory of very low curvature, a study of the errors in the horizontal position should focus only on the longitudinal errors of the body frame. To do it so, it would be needed that RTK and EGNOS measurements were collected exactly at the same instant. Nevertheless, it is possible to estimate the goodness of the filter by other means.

Figs. 2 and 3 show the results obtained by particle filter that describe the odometry by means of NCM under normal friction conditions and affected by slides. As it can be seen, the model can follow the nominal behavior of the vehicle, but fails to estimate correctly when slides appear.

The estimation of the covariance of the filter state at instant k is denoted as matrix $P^m(k)$, where m stands for the choice of odometry model. This value can be employed to compute and plot the ellipsis of positioning reliability. Its calculation can be carried out following:

$$P^m(k) = w(k)_i^m \times \left(x(k)_i^m - \bar{x}(k)^m\right) \times \left(x(k)_i^m - \bar{x}(k)^m\right)$$

where $\bar{x}(k)^m$ is the weighted mean value at instant k of the N particles i that belong to model m, calculated as

$$\bar{x}(k)^m = w(k)_i^m \times x(k)_i^m \quad i = 1 \ldots N$$

Time Average Autocorrelation Test. The time average autocorrelation test can be done to check the filter consistency is the time-average autocorrelation test [5]. This test can be executed in real time, and therefore, in single-runs trials. It is based on the ergodicity of the innovation sequence. If the number of time samples, K, is enough, this statistic is normally distributed and its variance is $1/K$. Its calculation for innovations 1 step apart can be done following:

$$\bar{\rho}(1) = \sum_{k=1}^{K} \nu(k)'\nu(k+1) \times \left[\sum_{k=1}^{K} \nu(k)'\nu(k) \times \right.$$
$$\left. \sum_{k=1}^{K} \nu(k+1)'\nu(k+1) \right]^{1/2} \tag{4}$$

where $\nu(k)$ stands for the innovation at instant k, that can be estimated as the difference between the predicted position and the observation.

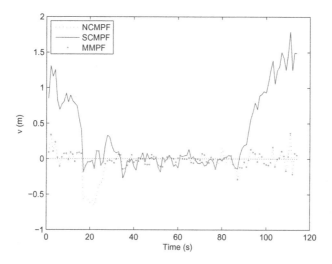

Fig. 4. Innovation values (ν) in meters for the different filter implementations during a test

Fig. 4 shows the values of the estimated innovation values during a test. The MMPF innovation will follow the values of both models, depending on the model probabilities that are calculated as shown in (3). Is it clearly visible the convenience of this switching process during the periods where slides appear, around seconds 20 and 80.

Table 1 shows the results obtained by the different filters in this test. The lower the value, the better the filter provides the aimed whiteness. As it can be seen, the MMPF truly outperforms the single model option. If we assume that the filter is correctly tuned, its variance would be $\sigma^2 = 1/K$, and for our experiment $\sigma = 0.0083$. Its 95% probability region can be obtained by applying a scaling factor of 1.95996, resulting the consistency interval at the 95% equal to

Table 1. Consistency results for NCMPF and MMPF

Filter	Time-average autocorrelation
NCMPF	0.5762
MMPF	0.0140

$[-0.01633 , 0.01633]$. As it can be seen, the single model PF is far from the threshold and the MMPF filter presents correct results, below the threshold. The proposed MMPF based method is therefore capable to correct the deficiencies of the individual filter under very varying conditions of dead-reckoning errors.

6   Conclusions

It is common nowadays to support GNSS based navigation systems with a dead-reckoning system consisting of an odometer and a gyro. To make the best out of the whole navigation system, it is important to characterize the errors made by every sensor. This paper focused on the problem of unusual odometry errors that cause significant and unexpected diminution of the dead-reckoning performance. Among all the errors, wheels slips and slides present the major changes respect to the normal system operation, appearing slides more often than slips in road vehicles.

This paper presented a solution to the problem of the characterization of unusual errors in the odometry system that is capable to detect and compensate these errors. The method is based on an particle filter running two models, one dedicated to normal conditions and the second to slide conditions.

The experiments performed in real trials show the convenience of the method under consideration. The designed multiple model particle filter deals with very different characteristics of the odometry errors, outperforming a solution that does not take into consideration slides. The particle filter allows the description of the non-linear system, resulting convenient to represent error distributions that are not Gaussian. The use of multiple models brings the possibility to characterize properly the vehicle model even with abrupt changes of the vehicle dynamics.

Future investigations in this line will focus on the characterization of more accurate road vehicle models and GNSS errors.

References

1. Sukkarieh, S., Nebot, E., Durrant-Whyte, H.: A High Integrity IMU/GPS Navigation Loop for Autonomous Land Vehicle Applications. IEEE Transactions on Robotics and Automation 15(3), 572–578 (1999)
2. Wang, J., Wilson, C.: Safety at the Wheel. Improving KGPS/INS Performance and Reliability, pp. 16–26. GPS World (May 2003)
3. Berdjag, D., Pomorski, D.: DGPS/INS data fusion for land navigation. Fusion 2004, Stockholm, Sweden (June 2004)

4. Toledo, R., Zamora, M.A., Úbeda, B., Gómez-Skarmeta, A.F.: An Integrity Navigation System based on GNSS/INS for Remote Services Implementation in Terrestrial
   Vehicles. In: IEEE Intelligent Transportation Systems Conference, pp. 477–480.
   Washington, D.C., USA (2004)
5. Bar-Shalom, Y.: Estimation and Tracking: principles, techniques, and sofware.
   Artech House Ed., Nonvood (1993)
6. Doucet, de Freitas, A., Gordon, N.: Sequential Monte Carlo Methods in Practice.
   Springer, New York (2001)
7. Djuric, P., Kotecha, J., Zhang, J., Huang, Y., Ghirmai, T., Bugallo, M., Miguez, J.:
   Particle filtering. IEEE Signal Proceessing Mag. 20(5), 19–38 (2003)
8. Toledo-Moreo, R., Zamora-Izquierdo, M.A., Ubeda-Miarro, B., Gomez-Skarmeta,
   A.F.: High-Integrity IMM-EKF-Based Road Vehicle Navigation With Low-Cost
   GPS/SBAS/INS. IEEE Transactions on Intelligent Transportation Systems 8(3),
   491–511 (2007)
9. Toledo-Moreo, R., Zamora-Izquierdo, M.A.: IMM-Based Lane-Change Prediction
   in Highways With Low-Cost GPS/INS. IEEE Transactions on Intelligent Trans-
   poration Systems (2009), Digital Object Identifier 10.1109/TITS.2008.2011691 (to
   be appeared)
10. Ekman, M., Sviestins, E.: Multiple Model Algorithm Based on Particle Filters for
    Ground Target Tracking. In: Proc. of the Information Fusion Conference, Quebec,
    Canada, pp. 1–8 (2006)
11. Hong, L., Cui, N., Bakich, M., Layne, J.R.: Multirate interacting multiple model
    particle filter for terrain-based ground target tracking. IEE Proc. Control Theory
    Appl. 153(6), 721–731 (2006)
12. Guo, R., Qin, Z., Li, X., Chen, J.: An IMMUPF Method for Ground Target Track-
    ing. In: Proc. of the International Conference on Systems, Mans and Cybernetics,
    ISIC, Montreal, Canada, pp. 96–101 (October 2007)
13. Foo, P.H., Ng, G.W.: Combining IMM Method with Particle Filters for 3D Maneu-
    vering Target Tracking. In: Proc. of the Information Fusion Conference, Quebec,
    Canada, pp. 1–8 (2007)
14. Wang, J., Zhao, D., Gao, W., Shan, S.: Interacting Multiple Model Particle Filter
    To Adaptive Visual Tracking. In: Proc. of the Third International Conference on
    Image and Graphics ICIG (2004)
15. Giremus, A., Tourneret, J.Y., Calmettes, V.: A Particle Filtering Approach for
    Joint Detection/Estimation of Multipath Effects on GPS Measurements. IEEE
    Trans. on Singal Processing 55(4), 1275–1285 (2007)
16. Gustafsson, F., Gunnarsson, F., Bergman, N., Forssell, U., Jansson, J., Karlsson,
    R., Nordlund, P.J.: Particle filters for positioning, navigation, and tracking. IEEE
    Trans. on Signal Processing 50, 425–437 (2002)
17. Toledo-Moreo, R., Betaille, D., Peyret, F.: Lane-level integrity provision for navi-
    gation and map matching with GNSS, dead reckoning, and enhanced maps. IEEE
    Trans. on Intelligent Transportation Systems 10(4), 100–112 (2010)

Pedestrian Tracking-by-Detection Using Image Density Projections and Particle Filters

B. Lacabex, A. Cuesta-Infante, A.S. Montemayor, and J.J. Pantrigo[✉]

Departamento de Ciencias de la Computación,
Universidad Rey Juan Carlos, Madria, Spain
b.lacabex@alumnos.urjc.es
{alfredo.cuesta,antonio.sanz,juanjose.pantrigo}@urjc.es

Abstract. Video-based people detection and tracking is an important task for a wide variety of applications concerning computer vision systems. In this work, we propose a pedestrian tracking-by-detection system focused on the role of computational performance. To this aim, we have developed a computationally efficient method for people detection, based on background subtraction and image density projections. Tracking is performed by a set of trackers based on particle filters that are properly associated with detections. We test our system on different well-known benchmark datasets. Experimental results reveal that the proposed method is efficient and effective. Specifically, it obtains a processing rate of 22 frames per second on average when tracking a maximum number of 9 people.

Keywords: People detection · People tracking · Tracking-by-detection · Image density projections · Particle filters

1 Introduction

Visual tracking is a research topic with an increasing demand in many fields; video surveillance, human-computer interaction, intelligent transportation, cell tracking, analysis of traffic, sport activities or human movement are only a few [15]. Such a variety usually imposes to begin with an appropriate object representation, according to what is intended to be tracked. Typically, we can distinguish two approaches to this task: feature-based models [8,9,2] and discriminative models [20,19,1]. The former relies on extracting features out of each frame by means of many efficient and well known algorithms. However, they are unable to determine whether such objects have to be tracked or not. Discriminative models are more complex and require more computational effort but, on the other hand, are able to categorize objects so one can focus only on those of interest. Once an object is found, the tracking system proceeds to search it in the subsequent frames according to a motion model. The search method can be deterministic, such as MeanShift [6] and Kalman filter [16], or stochastic like Particle Filters [11].

As tracking algorithms become commonplace, the demand for improvements rises; namely, initialization and termination of trajectories fully automated, tracking of multiple objects, robustness against camera movements, similar appearances, occlusions, cluttered and crowded scenes. Several works proposed to solve the tracking problem along a temporal window of length L videoframes. Thus, information from future

© Springer International Publishing Switzerland 2015
J.M. Ferrández Vicente et al. (Eds.): IWINAC 2015, Part II, LNCS 9108, pp. 166–174, 2015.
DOI: 10.1007/978-3-319-18833-1_18

frames up to time $t + L$ is used to locate the objects at time t. Results are quite accurate but obtained at least with a delay L, which may be not admissible for on-line applications [10]. On the other hand, particle filters (PF) are able to represent the uncertainty over the object location in the next frame conditioned only to the past observations. However, when there are multiple objects either the running time or the accuracy is compromised. As the number of particles increases the execution becomes computationally more intensive. On the contrary, for keeping the running time constant it is necessary to use the same number of particles, being easier to lose some targets.

To overcome difficulties that these improvements demand, in recent years a new paradigm known as *Tracking-by-detection* has become popular [1,4,2,12]. In this approach a detection algorithm is executed at every frame. Then, detections are associated to trackers that provide the dynamic of the target up to that time.

The performance of tracking-by-detection methods strongly depends on the correct association method, becoming a step in which most of the overall performance can be lost. The general approach is to have a measure of how close detections and trackers are, and then solve an assignment problem; but in practice there is a variety of implementations. Huang et al. present a three level method [14]. In the low-level only detections between consecutive frames are taken into account. Out of them, conflicting pairs are let aside and the remaining are associated in *tracklets* according a probability measure denoted as affinity. In a middle-level a motion model is constructed and the optimal tracklet association is obtained such that maximizes the probability of a sequence with given the set of tracklets. In the high-level entry, exit and scene occluder maps are used to construct a scene structure model and returns the complete trajectory set. Breitenstein et al. introduce the continuous detection confidence in [4]. To improve assignments they use an on-line person specific classifier, the distance to the tracker, a probability measure that takes into account size, motion direction and velocity, and inter-object occlusion reasoning. This work is extended in [5] for detections reappearing in the scene. Santhoshkumar et al. [17] compute the joint probability of size, position and appearance of each detection given each tracker; and then construct a score matrix with every pair. Association is the result of the Hungarian algorithm over the score matrix. If there are new detections, they are initialized with new trackers.

On-line and fully automatic tracking of multiple pedestrians is a challenge that remains open because frequently all the possible difficulties that prevent from a good performance are present in it. This paper presents a pedestrian tracking-by-detection system that is computationally efficient, up to the point that it is suitable for real-time applications, robust to occlusions, random trajectories and changing background.

The rest of the paper is organised as follows. The pedestrian detection method proposed is presented in section 2. Association and tracking procedures are described in section 3. The validation and performance evaluation of the overall procedure is discussed in section 4. Finally conclusions are pointed out.

2   Pedestrian Detection Method

A popular trend in tracking-by-detection systems consists of using feature-based algorithms in the detection stage. Their main shortcoming is the total bias for every single

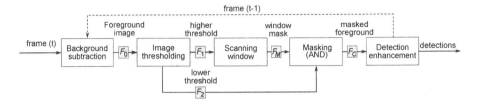

Fig. 1. An overview of our pedestrian detection method. The background subtractor extracts two different thresholded binary images. At that point, one passes through the scanning window to obtain its window mask. The other one gets masked with obtained window mask and proceeds to the detection enhancement section to obtain the final detections in the frame. To finish the execution, the current frame gets averaged in the background subtraction model for scenery renewal.

scene together with the tedious process of training [8,9]. In addition, due to the complexity of the pattern classification algorithms and high dimension of the features, the process of detection can lead to execution bottlenecks that prevent from real-time response. We overcome these difficulties with a detection method based on shape detections over the foreground image, depicted in Fig. 1. Although the method is general, we focus on pedestrian detection along this paper.

2.1 Background Subtraction

Following the background subtraction procedure, we propose a simple but cost-effective method. Firstly a subtractor uses the temporal median image created with input images over time and returns a background model. Then, the background model is used to subtract every input frame and reveal the foreground objects. This subtraction is done in each RGB channel and their outcome is summed up in a single image F_0. In the next step F_0 is binarized twice, with a high and a low threshold, resulting into two binary foreground images, F_1 and F_2 respectively. Once finished with frame t, it is used to update the background model for the next frame. The procedure has the following parameters: high and low threshold values, together with the number of frames used to model the background.

2.2 Scanning Window Detector and Image Masking

Usually the resulting foreground image from a background subtraction is used to extract the detection candidates with a connected components analysis. In this work, previous to that analysis, we introduce the *Scanning Window Detector*. This technique is a novel foreground image analysis to reduce unreliable detections or noise. The aim of this intermediate stage is to extract those labelled pixels that are surrounded by many others not labelled, and group them to form our pedestrian candidate. To this end, it consists of the following steps.

1. The high threshold ensures a low noise level but also introduces gaps in the foreground image F_1. In order to fill those gaps, we use a sliding window over F_1 seeking for large accumulations of foreground pixels inside their bounds.

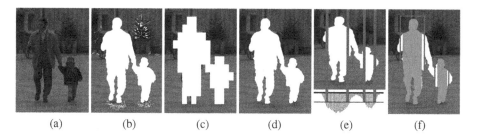

(a)   (b)   (c)   (d)   (e)   (f)

Fig. 2. Overview results at each stage of the detection algorithm. (a) Input frame. (b) White pixels are the image F_2 subtracted from the background model with low threshold. (c) White pixels are F_M, the result of the scanning window over F_1. (d) Image $F_C = F_2$ AND F_M (e) Apply detection enhacement via a vertical projection and its median threshold to separate nearby pedestrians. (f) Final resulting bounding boxes.

2. Whenever this condition is satisfied, we label the entire window as foreground information. The resulting binary image F_M consists of multiple overlayed windows labelled in white that describe separated objects as shown in Fig. 2c.
3. On the other hand, the low threshold ensures an accurate shape of the foreground image but introduces some noise. The resulting foreground image is F_2.
4. We use F_2 to obtain a *clean* foreground image $F_C = F_2$ AND F_M; i.e. masking F_2 with F_M in order to remove the undesired noise. Image F_C is shown in 2d.

The shape, size and the required number of foreground pixels of the window are defined according to two circumstances: the shape of the object that we want to detect and the scene where the object is projected. For this work we have focused on detecting pedestrians over a surveillance scene. Each pedestrian detection consists of the sum of multiple windows that contain a certain number of foreground pixels. For shape, we have followed the same configuration that [8] uses in its window scanner for detecting pedestrians. This shape is the one that best describes the simplified shape of a pedestrian almost from every single point of view. For size, we identify each pedestrian as a sum of positive window foreground detections. Setting the window size is subject to features such as the average size of pedestrians with respect to the whole frame, the position of the camera or its optic. So far it requires a training stage previous to be completely autonomous. For the datasets under test we have found the window size to be a $1/16$ of the averaged pedestrian size in the image. Regarding the number of pixels in the window, we consider at least a 50% of them must be foreground information in order to label the entire window as foreground. In addition, for improving the computational cost of the counting stage, we use the integral image [7] as the input format of the window scan process.

2.3 Detection Enhacement

The last step of our detection algorithm extracts pedestrians out of image F_C. We apply connected-component analysis [18] in order to separate each different pedestrian detection. For each connected-component we do a vertical projection by summing up

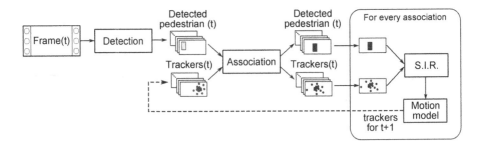

Fig. 3. Overview of the Tracking-by-detection algorithm proposed. The input is the detection response for the current frame. It includes a number of detected pedestrians that are associated to the current set of trackers. Then, for every association, a Sequential Importance Resampling of each particle of the tracker is carried out. Finally a motion model sets a prior to each particle for the next frame.

the number of binary components per column [13]. The data distribution that we obtain gets smoothed with a Gaussian kernel. Finally, the median of the resulting distribution is used as a threshold for the connected-component algorithm. This simple step allows the detections to asure that they are a desired object with high confidence. An example of these two last steps is shown in Fig. 2e and 2f respectively.

3   Tracking Method

Compared with conventional tracking systems, Tracking-by-detection have two main differencecs. The first in that it runs a detection algorithm in every frame. The second is that detection responses are matched with existing trackers in order to correctly associate one detection to one tracker. In the simplest implementation, trackers are just the sequence of detections for a given object, pedestrians in our case, up to the current time t. Hence matching is all about associating each detection at the current frame to the most likely sequence. Those detections and trackers not matched are then analysed in order to find out whether there is new people coming in or out of scene.

Particle filters (PF) are a sequential Monte Carlo method for approximating the probability distribution of a sequence of hidden variables $x_{1:t}$ up to time t, given a sequence of observations $y_{1:t}$ in terms of a finite number K of weighted samples (particles) drawn from the transition probability distribution; the weights being the likelihood of particles. The version of PF implemented in this paper is the known as Sequential Importance Resampling (SIR) that consists of resampling the particles and then weight all of them to $1/K$. Finally they are propagated according to a motion model for the next iteration.

1. Run the detection method for the current frame $F(t)$.
   Let $\mathcal{Y}_t = \{Y_i\}_{i=1:M}$ be the set of detection responses at time t, and $Y_i = y_{1:t}$ the sequence of detections along time for pedestrian i.
2. Run the association algorithm for the detections and trackers at time t.
   Let $\mathcal{X}_t = \{X_j\}_{j=1:N}$ be the set of trackers at time t; and let $X_j = \{x_{j,t}^{(k)}\}_{k=1:K}$ be the jth tracker, which consists of a set of K particles. The association consists of:

(a) Compute the score matrix S such that element s_{ij} is a measure of the affinity between detection Y_i and tracker X_j.

In this work affinity is a probability measure that takes into account position, size and appearance independently.

$$s_{ij} = p_{\text{pos}}(Y_i|X_j) \cdot p_{\text{size}}(Y_i|X_j) \cdot p_{\text{appr}}(Y_i|X_j);$$
$$\text{where} \begin{cases} p_{\text{pos}} \sim \mathcal{N}(\text{pos}(X_j), \sigma_{\text{pos}}^2), \\ p_{\text{size}} \sim \mathcal{N}(\text{size}(X_j), \sigma_{\text{size}}^2), \\ p_{\text{appr}} \sim \mathcal{N}(\text{appr}(X_j), \sigma_{\text{appr}}^2). \end{cases} \tag{1}$$

Notice that functions $\text{pos}(X_j)$, $\text{size}(X_j)$, $\text{appr}(X_j)$ receive K particles for each tracker but return a single value related to position, size and appearance respectively. In this work function $\text{pos}(X_j)$ returns the averaged position of the K particles from tracker X_j. Likewise with $\text{size}(X_j)$ and $\text{appr}(X_j)$ for size and mean RGB value of the detection.

(b) Run the Hungarian algorithm on the score matrix. The outcome is a binary matrix in which there should be only one 1 in every row and every column, indicating which detection is associated to which tracker. Otherwise:

 – rows without 1 are detections without tracker
   $\Rightarrow$ initialize a new tracker for them
 – whereas columns without 1 are trackers without detection
   $\Rightarrow$ remove trackers not associated.

(c) For every associated pair (X_i, Y_j)

  i. Estimate the likelihood of the detection $y_{i,t}$ given each particle $x_{j,t}^{(k)}$, which are the weights of each particle.
  ii. Resample K times the set of the particles from the multinomial distribution with weights obtained as its parameters.
  iii. Apply a motion model to each particle in order to approximate the particle to the areas where the next detection is expected.

4 Experimental Results

We have performed an experimental evaluation on the PETS 2009 dataset [1], a widely used benchmark for performance comparison for automatic video surveillance systems. The videos show an average of 6 to 8 people walking freely around a small area and feature a lot of occlusions and significant scale changes. Sequence S2L1 is taken from the VS-PETS 2009 benchmark; and only the first two viewpoints are used. The videos were recorded at 7 fps from an elevated viewpoint, are around 800 frames long and show a variable number of pedestrians. The hardware used for the experiment was an Intel Core i7-3740QM at 2.7 GHz, 1 Core with 8GB RAM. All the method has been coded in Python with Cython optimizations to minimize the iteration time over multiple dimension large arrays, which in Python often cause in execution bottlenecks .

---

[1] PETS 2009 dataset is available at http://ftp.pets.rdg.ac.uk/pub/PETS2009/

Table 1. Quantitative results of our Pedestrian Tracking-by-Detection method

Sequence	MOTA	MOTP	Processing rate (fps)
PETS 1	69.7%	79.2%	22
PETS 2	72.0%	84.3%	23

Fig. 4. Bounding boxes and tracking paths on 3 frames from each video sequence selected

We follow CLEAR-metrics introduced by [3]. This work introduces two metrics, Multiple Object Tracking Accuracy (MOTA) and Multiple Object Tracking Precision (MOTP), to evaluate the tracking performance of the system. Given that for every time frame t a multiple object tracker outputs a set of hypotheses $\{h_1...h_m\}$ for a set of visible objects $\{o_1...o_n\}$, the MOTP is simply the average distance d_t^i between true G_t^i and estimated targets D_t^i, and c_t be the number of object-hypothesis correspondences made for frame t:

$$MOTP = \frac{\sum_{i,t} d_t^i}{\sum_t c_t}, \quad \text{where } d_t^i = \frac{|G_T^i \bigcap D_T^i|}{|G_T^i \bigcup D_T^i|}. \tag{2}$$

The MOTA takes into account false positives fp_t, missed targets m_t and identity switches mme_t of the tracker regarding to the number of objects g_t:

$$MOTA = 1 - \frac{\sum(fp_t + m_t + mme_t)}{\sum g_t} \tag{3}$$

The results presented in Table 1 lead to several conclusions regarding the evaluation method and its outcome. First, due to the background substraction nature of the detector, the MOTP percentage is much higher than the one obtained with any feature based detector. This percentage could be even higher if a group management gets integrated in the system for pedestrian overlap situations. Regarding MOTA, the results appear promising as of the improvement that could be done in a future applying more complex association methods instead of just one based in distance. The acquired processing rate allows us to say that the method presented allows the execution to be done in real-time.

Figure 4 shows some selected frames from the two considered video sequences. We have noticed that the system performs a robust tracking in absence of occlusions, even when the number of pedestrian increases, without a significant loss of performance. In fact, our system is able to process frames at the same rate than they are acquired (see Table 1). Therefore it can be considered that the system works under real-time constraints. However, in the presence of people grouping and ungrouping, we observe that sometimes the identity of the involved pedestrians is not properly maintained. In order to diminish this shortcoming future work should improve the tracking method.

5   Conclusions

This work presents a novel pedestrian detection based on background subtraction and image density projections. Foreground images obtained from background subtraction are binarized twice. One of them is used for obtaining a coarse mask of the pedestrian that leaves noise pixels out. The other transforms the mask in more detailed image. Then image density projections are used to distinguish pedestrians when they are close one to each other. In addition we use this detection method into a Tracking-by-detection system. Each detection is associated to a tracker consisting of a particle filter. The method is tested with sequence S2L1 from the VS-PETS 2009. The overall performance has been evaluated with two metrics: Multiple Object Tracking Accuracy (MOTA) and Multiple Object Tracking Precision (MOTP) offering very accurate results and execution in real-time.

Acknowledgments. This research has been partially supported by the Spanish Government research funding refs. TIN 2011-28151 and TIN 2014-57633.

References

1. Andriluka, M., Roth, S., Schiele, B.: People-tracking-by-detection and people-detection-by-tracking. In: IEEE Conf. on Computer Vision and Pattern Recognition (CVPR 2008) (June 2008)
2. Azab, M.M., Shedeed, H.A., Hussein, A.S.: New technique for online object tracking-by-detection in video. IET Image Processing 8(12), 794–803 (2014)
3. Bernardin, K., Stiefelhagen, R.: Evaluating multiple object tracking performance: The clear mot metrics. J. Image Video Process., pp. 1–10 (February 2008)
4. Breitenstein, M.D., Reichlin, F., Leibe, B., Koller-Meier, E., Van Gool, L.: Robust tracking-by-detection using a detector confidence particle filter. In: 2009 IEEE 12th International Conference on Computer Vision, pp. 1515–1522 (September 2009)
5. Breitenstein, M.D., Reichlin, F., Leibe, B., Koller-Meier, E., Van Gool, L.: Online multiperson tracking-by-detection from a single, uncalibrated camera. IEEE Transactions on Pattern Analysis and Machine Intelligence 33(9), 1820–1833 (2011)
6. Comaniciu, D., Meer, P.: Mean shift: A robust approach toward feature space analysis. IEEE Transactions on Pattern Analysis and Machine Intelligence 24(5), 603–619 (2002)
7. Crow, F.C.: Summed-area tables for texture mapping. In: Proc. of the 11th Annual Conference on Computer Graphics and Interactive Techniques (1984)

8. Dalal, N., Triggs, B.: Histograms of oriented gradients for human detection. In: International Conference on Computer Vision & Pattern Recognition, vol. 2, pp. 886–893 (2005)

9. Felzenszwalb, P.F., Girshick, R.B., McAllerster, D., Ramanan, D.: Object detection with discriminatively trained part-based models. IEEE Trans. on Pattern Analysis and Machine Intelligence 32(9), 1627–1645 (2010)

10. Francois, J.B., Berclaz, J., Fleuret, F., Fua, P.: Robust people tracking with global trajectory optimization. In: Conference on Computer Vision and Pattern Recognition, pp. 744–750 (2006)

11. Gordon, N.J., Salmond, D.J., Smith, A.F.M.: Novel approach to nonlinear/non-gaussian bayesian state estimation. IEEE Proc. F, Radar and Signal Processing 140(2), 107–113 (1993)

12. Guan, Y., Chen, X., Yang, D., Wu, Y.: Multi-person tracking-by-detection with local particle filtering and global occlusion handling. In: 2014 IEEE International Conference on Multimedia and Expo (ICME), pp. 1–6 (2014) (July 2014)

13. Haritaoglu, I., Harwood, D., Davis, L.S.: W4: Real-time surveillance of people and their activities. IEEE Trans. on Pattern Analysis and Machine Intelligence 22(8), 809–830 (2000)

14. Huang, C., Wu, B., Nevatia, R.: Robust object tracking by hierarchical association of detection responses. In: Forsyth, D., Torr, P., Zisserman, A. (eds.) ECCV 2008, Part II. LNCS, vol. 5303, pp. 788–801. Springer, Heidelberg (2008)

15. Jalal, A.S., Singh, V.: The state-of-the-art in visual object. Informatica 36, 227–248 (2012)

16. Kalman, R.E.: A new approach to linear filtering and prediction problems. ASME Journal of Basic Engineering (1960)

17. Santhoshkumar, S., Karthikeyan, S., Manjunath, B.S.: Robust multiple object tracking by detection with interacting markov chain monte carlo. In: 2013 20th IEEE International Conference on Image Processing (ICIP), pp. 2953–2957 (September 2013)

18. Suzuki, S., Abe, K.: Topological structural analysis of digitized binary images by border following. CVGIP 30(1), 32–46 (1985)

19. Wu, B., Nevatia, R.: Detection and tracking of multiple, partially occluded humans by bayesian combination of edgelet based part detectors. Int. Journal of Computer Vision (2007)

20. Zhu, Q., Avidan, S., Yeh, M.C., Cheng, K.T.: Fast human detection using a cascade of histograms of oriented gradients. In: CVPR 2006, pp. 1491–1498 (2006)

Comparative Study of the Features Used by Algorithms Based on Viola and Jones Face Detection Algorithm

Alexandre Paz Mena[1(✉)], Margarita Bachiller Mayoral[1],
and Estela Díaz-López[1]

Department of Artificial Intelligence,
National University of Distance Education, Madrid, Spain
erzapito@gmail.com

Abstract The problem of face detection has been one of the main topics in computer vision investigation and lots of methods have been proposed to solve it. One of the most important is the algorithm proposed by Viola and Jones that offer good results. Many studies have used this algorithm but none have analysed the advantages or disadvantages of using a certain type of feature in either the detection or the computation time. In this article we analyse the Viola algorithm [12] and other derivatives from the point of view of input characteristics and computing time.

Keywords: Face detection · Boosting · Haar features · Real time

1 Introduction

The problem of face detection is a fundamental issue in applications such as video surveillance, driving assistance or facial recognition. Clearly, human face detection is a complex problem because of the variations in appearance. These variations are due to the position and orientation of the face in the image, the presence of glasses, a beard, a hat or other objects which partially hide the face and its expression. Moreover, in many of these applications it is essential to detect faces in the shortest possible time. Because of all this problems, face detection has been a challenge for researchers.

In the past, the problem of face detection in two dimensional images was analysed in many studies. However, the work presented by Viola and Jones [12] was a great success as it was the first algorithm with a high detection rate with a few computational time. Its principal contributions were a cascade of classifiers to achieve fast calculation time, use of a combination of simple classifiers to form a boosted classifier and simple rectangular Haar-like features that could be extracted and computed in a few operations thanks to an integral image. This work was so important that it became the starting point for much of other work.

One of the first things to be studied were the Haar-like rectangular features used by Viola and Jones. Although these features provided good performance for building frontal face detectors and were very efficient to compute due to

J.M. Ferrández Vicente et al. (Eds.): IWINAC 2015, Part II, LNCS 9108, pp. 175–183, 2015.
DOI: 10.1007/978-3-319-18833-1_19

the Integral Image technique, many researchers have proposed to extend them with many alternatives. For example, Viola et al [13] modified the features so they compared the same areas in different video frames. Lienhart and Maydt [7] proposed a new set of rotated Haar-like features that could also be calculated very efficiently. Levi et al [4] proposed that using local edge orientation histograms (EOH) as the source information for features could improve performance compared to the features used in contemporaneous systems. Dalal and Triggs [1] used features based on Histograms of Oriented Gradients (HOG). Ojala et al [9] proposed LBP features which were derived from a general definition of textures in local neighbourhood. Li and Zhang [6] proposed a novel learning procedure for training a boosted classifier that achieved a minimum error rate. Huang et al [3] presented sparse granular features, which represent the sum of pixel intensities in a square, instead of using rectangles. Moreover, they also proposed an efficient weak learning algorithm which uses a heuristic search method. Mita et al [8] proposed joint Haar-like features, which were based on the co-occurrence of multiple simple Haar-like features.

Another thing to be improved was AdaBoost, the learning algorithm used to train the boosted classifier. While AdaBoost uses binary features, many posterior works[5,3,8,6] use RealBoost, an alternative that feeds on features that output values between 0 and 1. Li and Zhang [5] proposed a novel learning procedure for training a boosted classifier that achieved a minimum error rate. A summary of the different proposals can be found in [14,10].

In this article we are going to focus on features improvement. Most of the commented proposals show good results, but they do not analyse which features were the most used or which of the new features really yield a performance improvement of the algorithm. That is, all papers lack an analysis of the obtained results regarding the features used and a demonstration that the new features really mean an improved face detection rate.

2   The Viola and Jones Face Detector

The algorithm presented by Viola and Jones [12] changed the style of work, going from a faces/non-faces classification problem to a rare event detection problem. This change was raised because in an image, there are few windows containing a face compared with the total possible detection windows.

To solve this problem the authors propose to use a classifier composed of a series of classifiers arranged in cascade. Each layer of the cascade is more complex than the previous one. This means that easy samples are eliminated soon letting the last layers, the most complex ones, deal with the difficult samples. Each layer is trained using a variant of the AdaBoost algorithm, looking for a target detection rate instead of just using all simple features available. These simple features, commonly known as "Haar" features for its similarity with Haar waves, check the difference of intensity of adjacent zones against a threshold.

Viola and Jones propose to use three types of feature: (1) A feature that checks the difference in an area divided into two rectangles of the same size; (2)

A feature that checks the central part of an area against the border parts; (3) A feature that divides an area into four rectangles and compares the difference between diagonal pairs. Figure 1 shows examples of areas divided as in these features. To calculate the intensity of an area they employ a data structure known as "integral image". This "integral image" stores the accumulation of intensities up to each point of the image. Thanks to these features, the use of windows of different sizes becomes trivial, the classifier only needs the new size of the search area to check features.

AdaBoost is a learning algorithm that combines a collection of binary weak classifiers to create a strong new one. Viola and Jones use this algorithm to create each layer of the cascade but, instead of using all weak classifiers. They stop when they reach a certain detection rate that depends on the previous layer. Briefly, AdaBoost is an iterative process that accumulates weighted weak classifiers. In each iteration, the algorithm selects the weak classifier that best improves the current accumulation detection.

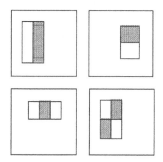

Fig. 1. "Haar" like features positioned in a detection window

One of the great advantages of the resulting cascade classifier is the speed, which is particularly interesting for real-time systems. Until the appearance of this paper, the techniques presented were heavy and inefficient. On its day, this new technique allowed analysing images of 384x288 pixels in 0.067 seconds with a 700MHz processor, away from the 40 seconds of Schneiderman [2]or the 1 second of Rowley et al [11].

3   Analysis of Detection Methods

Most of the research based on the Viola and Jones algorithm focuses on the modification of current features or the definition of new ones. Obviously, if we focus on the modification of features, especially adding new types, it is possible that the detector can codify a face with more precision due to it has major quantity of characteristics. However, there are some aspects to keep in mind and that have not yet been analysed.

Something very important to remember is that new features mean more training time. Assuming that in the original article the author says it took them a week to train the classifier, if the number of features are duplicated, the time will increase in the same proportion. For example, Levi et al [4] present features that use the gradient of intensity in addition to the intensity of the image. The calculation of the gradient needs much more time, which could be up to eight times normal time. That is to say, that the use of these this kind of features does not necessarily imply a faster classifier.

In short, when we use new features both the training time and the classification time tends to increase and, on the other hand, the benefits in terms of detection efficiency are not so evident. In our research, we analyse the Viola and Jones algorithm along with two modifications of it in order to investigate how adding new features affects the generated classifier, specially from the point of view of the detection rate and computation time.

3.1   First Technique: The Viola and Jones Algorithm

Before starting to examine the improvement achieved by some algorithms based on the Viola and Jones method, we will analyse different elements that will allow us to seek different objectives to improve the original algorithm:

☐ *Attention focus*: Searching for smaller elements would allow the generation of more complex classifiers while keeping speed. Therefore, one of the objectives is to find these elements which we will call attention focus.

☐ *Feature specialization:* The study of the use of each individual feature gives us information about its real utility in the detection process. Now, the goal will be to remove those features that do not provide information while increasing the number of those who do.

☐ *Search for new features:* The last objective will be analysing the current classifier to get new features that increase efficiency.

The focus attention of the classifier can be obtained by searching parts of the detection window that the classifier uses mainly. Figure 2 shows the areas most used by this classifier with bright color. The characteristics of the face that contain more information are the nose, the eyes and the cheeks. The mouth is hardly used although here does not have bright color. We can deduce that it would be interesting to design partial classifiers focused on detecting the eyes or the nose instead of designing a classifier that detects the whole face. These partial classifiers would be much faster thanks to their small size and should have the same detection rate.

The best way to eliminate non significant features is to study the utilisation of each type of feature in the final classifier. The results of this analysis are shown in table 1 where the most used features are double and crossed types. This contrasts with the alternative presented by Lienhart and Maydt [7], where they decided to eliminate the crossed features due to the fact that these can be represented with the rest of features. However, our results show that it would have been more useful to eliminate the triple features as they seem to provide less information.

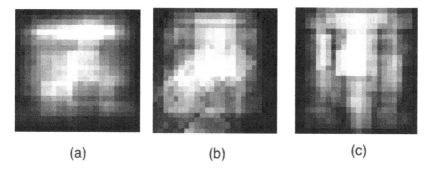

(a)   (b)   (c)

Fig. 2. Focus attention areas of different techniques: a) Viola and Jones b) Lienhart and Maydt c) Levi and Weiss

Table 1. Usage percentage of each type of feature in the classifier generated with the three techniques: 1) Viola and Jones, 2) Lienhart and Maydt and 3)Levi and Weiss

	Viola and Jones	Lienhart and Maydt		Levi and Weiss
		0°	45°	
Horizontal double features	45.771%	38.231%	8.707%	10.049%
Horizontal triple features	7.068%	8.980%	8.436%	7.084%
Vertical double features	4.751%	0.544%	9.660%	1.483%
Vertical triple features	7.764%	10.068%	10.885%	11.532%
Crossed	34.647%			12.191%
Central features		3.810%	0.680%	

We proposed to analyse the geometric properties to reduce the number of training features.. The aim is to eliminate, before the training, those features which have a geometrical value that is not used in the final classifier. Specially, we have checked the following geometric properties: position, area and ratio of width to height. Table 2 shows the maximum and minimum values of these properties in the final classifier. Some conclusion are explained below:

☐ *Position*: The results of the study indicate that we should use the whole image. All feature types use every point in the image.

☐ *Area*: Considering that each classifier is divided into different zones, two zones for double features, three zones for triple features and four zones for crossed features, the minimum size is one pixel for each zone. On the other hand, the maximum area of the whole feature is at least most 70% of the detection window.

☐ *Width to height ratio*: Horizontal features tend to have more width than height, while vertical features follow the other way round. However, crossed features do not show any kind of trend.

Table 2. Limit properties of each type of feature in the classifier generated with the Viola and Jones technique

	Horizontal		Vertical		Crossed
	Double	Triple	Double	Triple	
Left limit	0	0	0	0	0
Right limit	24	24	24	24	24
Top limit	0	0	0	0	0
Bottom limit	24	24	24	24	24
Minimum area	2	3	2	3	4
Maximum area	380	342	192	360	324
Maximum h/w ratio	10.5	7	8	24	10
Maximum w/h ratio	16	24	5	7	10

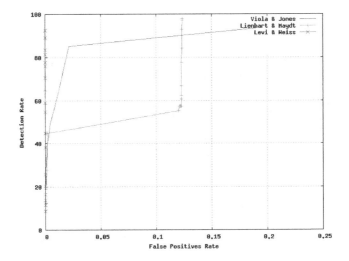

Fig. 3. ROC curve for all generated classifiers

The results of this classifier, analysing the CMU+MIT image database by Rowley et al [11], consisting of 40 black and white images. ROC curves are shown in figure 3 where it uses ROC curves . A ROC curve (Receiver Operating Characteristic) is a graphical representation of the detection rate against false positive rate that has a binary classifier of adjustable sensitivity. This figure will be used to compare the results of the three classifiers analysed using the Viola and Jones algorithm as a base.

An important point to rate is the classification speed. Table 3 shows the necessary time to analyse the whole CMU+MIT image database. This time will be very useful for comparing different classifiers in terms of speed.

Table 3. CMU+MIT database analysis time using the classifier generated with the three techniques: 1) Viola and Jones, 2) Lienhart and Maydt and 3)Levi and Weiss

	Viola and Jones	Lienhart and Maydt	Levi and Weiss
Integral image creation	0.89″	1.74″	7.02″
Image analysis	14.07″	9.31″	6.97″
Total	14.96″	11.05″	13.99″

3.2   Second Technique: 45° Rotation of the Viola and Jones Features

The following technique we have been analysed is a direct evolution of the algorithm of Viola and Jones. It was proposed by Lienhart and Maydt [7]. In that article, the authors proposed to add a new central feature type and to use a 45⁰ rotated version of the integral image along with a new set of features. Also, they eliminated the crossed features from the original article. Following the analysis made with the original technique, we analyse the attention focus. The use of the detection window by the classifier is shown in figure 2. As with the Viola and Jones algorithm, attention focus is mainly located on the eyes, nose and cheek.

However, if we analyse the usage percentage of the different types of features, visible in table 1, the conclusions are different from for the Viola and Jones classifier. On the one hand, central features are hardly used, they represent less than 5% of the total, even if we combine normal and rotated features. This data clearly contradicts the authors' decision to replace the crossed features with the central ones. On the other hand, rotated features seem very useful, being about 40% of total features. We can also see that vertical double features are almost not used in the final classifier. Finally, the horizontal double features are still the most commonly used ones, although their proportion is lower.

The results of this classifier obtained with the CMU+MIT database can be seen in figure 3. These results are slightly lower than those obtained by the Viola and Jones algorithm. This is because negative training samples are generated randomly so we are able to generate millions of different samples. This means that training is different each time it is performed.

Regarding the speed of the classifier, it can be seen in table 3 that the time to generate the integral images is greater than for the Viola and Jones technique. This is because we have to calculate two matrices, one for the original features and another for the rotated ones. However, the time of analysis decreases because it requires fewer features to encode the faces information.

3.3   Third Technique: Local Edge Orientations Histogram (EOH)

The next interesting modification of the Viola and Jones algorithm was presented by Levi et al [4]. They proposed to add new features, completely different to those used in the original article, which would employ the intensity gradient as the source of information. From this information, they suggested these new features: relation between gradients with different directions, dominant direction and gradient symmetry.

With this technique we also generated a classifier and analysed its results. The first difference of this classifier is the attention focus. In figure 2 it is shown that, in general, the attention focus is placed on the mouth, eyes and nose. This is due to the new features, based on gradient information, focusing on any part of the face where edges are sharper.

The use of the new features, shown in table 4, also provides interesting data. First, the usage percentage of the original feature types (table 1) is completely different from the other classifiers. Furthermore, the features based on gradient information are the most used, reaching 58% of the total. To continue, features that analyse the symmetry are minimal, only 3% of the total. If we want to accelerate the training process, it would be advisable to remove them.

Table 5 shows the use of different groups of angles. As the authors recommended, angles were standardised between 0° and 180° and grouped into four bins. Each bin is associated with the following orientations: horizontal, vertical, right diagonal and left diagonal. An important detail is that the use of the different bins in the final classifier is more or less matched.

The result of using this classifier against the CMU+MIT database shows that it has a higher detection rate than the rest of classifiers analysed, as shown in figure 3. However, the problem with this classifier lies on the time needed to generate all integral images. As it can be seen in table 3 this time is more than 50% of the total time. In comparison, Viola and Jones technique uses only 6% of the time to generate the integral images. Nevertheless the analysis time is considerably shorter than other techniques.

Table 4. Usage percentage of new type of feature in the classifier generated with the Levi and Weiss technique

	Levi and Weiss
Horizontal symmetry features	1.48%
Vertical symmetry features	1.32%
Dominant orientation features	24.38%
Orientation relation features	30.48%

Table 5. Usage percentage of each angle bin in the classifier generated with the Levi and Weiss technique

Angle	Levi and Weiss
0°	26.64%
45°	24.52%
90°	25.29%
135°	23.55%

4   Conclusions

The suitable selection of the feature set to use with the learning algorithm is fundamental for the success of a classifier. Also, in many cases it is necessary to take real time as a restriction. In this work, we have evaluated some methods based on the Viola and Jones algorithm in order to evaluate the use of features and classifier analysis speed. It is important to highlight that, until now, the authors of the different improvements only proposed new features to use in the classifier, but they did not analyse whether they had really been used in the final classifier. This study will allow to define a way to get a better set of features to use in a face high speed detection algorithm.

Acknowledgments. This work was partially supported by research project MCYT under grant number TIN2010-20845-C03-02.

References

1. Triggs, B., Dalal, N.: Histograms of oriented gradients for human detection. In: CVPR, pp. 886–893 (2005)
2. Schneiderman, H.: A statistical approach to 3d object detection applied to faces and cars. PhD thesis, Robotics Institute, Carnegie Mellon University, Pittsburgh, PA (2000)
3. Li, Y., Lao, S., Huang, C., Ai, H.: High-performance rotation invariant multiview face detection. IEEE Trans. Pattern Anal. Mach. Intell. 29(4), 671–686 (2007)
4. Levi, K., Weiss, Y.: Learning object detection from a small number of examples: the importance of good features (2004)
5. Zhang, Z., Li, S.: Floatboost learning and statistical face detection. IEEE Transactions Pattern Analysis and Machine Intelligence 26(9), 1112–1123 (2004)
6. Li, S.Z., Zhu, L., Zhang, Z., Blake, A., Zhang, H., Shum, H.-Y.: Statistical learning of multi-view face detection. In: Heyden, A., Sparr, G., Nielsen, M., Johansen, P. (eds.) ECCV 2002, Part IV. LNCS, vol. 2353, pp. 67–81. Springer, Heidelberg (2002)
7. Maydt, J., Lienhart, R.: An extended set of haar-like features for rapid object detection. In: IEEE ICIP, pp. 900–903 (2002)
8. Hori, O., Mita, T., Kaneko, T.: Joint haar-like features for face detection. In: Proceedings of the Tenth IEEE International Conference on Computer Vision, vol. 2, pp. 1619–1626 (2005)
9. Maenpaa, T., Ojala, T., Pietikainen, M.: Multiresolution gray-scale and rotation invariant texture classification with local binary patterns. IEEE Transactions Pattern Analysis and Machine Intelligence 24(7), 971–987 (2002)
10. Zhang, J., Paisitkriangkrai, S., Shen, C.: Face detection with effective feature extraction. CoRR (2010)
11. Kanade, T., Rowley, H., Baluja, S.: Rotation invariant neural network-based face detection. Technical report, Computer Science Department, Pittsburgh, PA (1997)
12. Jones, M., Viola, P.: Robust real-time face detection. International Journal of Computer Vision 5, 137–154 (2004)
13. Snow, D., Viola, P., Jones, M.: Detecting pedestrians using patterns of motion and appearance. International Journal of Computer Vision 63, 153–161 (2005)
14. Zhang, Z., Zhang, C.: A survey of recent advances in face detection. Technical report, Microsoft Research Microsoft Corporation (2010)

Experiments of Skin Detection in Hyperspectral Images

Manuel Graña[1,2(✉)] and Ion Marques[1]

[1] Computational Intelligence Group,
Universtiy of the Basque Country, UPV/EHU, Spain
[2] ENGINE project, Wrocław Unversity of Technology (WrUT), Wrocław, Poland
manuel.grana@ehu.es

Abstract. Skin detection in hyperspectral images has many potential applications to health monitoring and surveillance. In this paper we report on two different approaches that we have followed to tackle with this problem. First, the problem is treated as a classification problem using of active learning strategies to achieve a robust classifier in a short numver of interactions. Second, we approach the problem from the point of view of hyperspectral unmixing, looking for skin endmembers that would allow quick detection in large datasets. We test a new sparse lattice computing based algorithm. We provide experimental results over a dataset of human images in outdoors sunny environment.

1 Introduction

The target application in this paper is human skin detection in hyperspectral images, which is envisaged as a near future application for rescue missions [20], health monitoring, i.e. in border control, and others. Previous publications on this problem include identifying the optimal band set for skin detection [22], and a Normalized Difference Skin Index characterizing skin in near infrared regions [16]. In this paper, we report two approaches to deal with this problem. In the first approach, we solve it as a classification problem where the classifiers are trained in an active learning process. In the second, we assume that skin information is represented by a small number of endmembers, which can be identified by some endmember induction algorithm.

Classification by Active Learning. This supervised machine learning problem can be addressed using Active Learning. It is a useful tool when dealing with data containing scarce labeled samples. The Active Learning approach consists of two elements: a training algorithm and a query method. The training algorithm is used to build classifiers from the small set of labeled data. The query method is used to select unlabeled samples, which will be labeled by an oracle and added to the training set. This iterative process goes on until some stopping criteria are met. The oracle is a human expert, although for the computational experiments reported here the oracle is the provided ground truth. There are two main querying strategies to choose the samples for training[7]. If a single

© Springer International Publishing Switzerland 2015
J.M. Ferrández Vicente et al. (Eds.): IWINAC 2015, Part II, LNCS 9108, pp. 184–192, 2015.
DOI: 10.1007/978-3-319-18833-1_20

learner is used the choice depend on the selected measuring strategy. This is called *querying by a single model*. Some Active Learning schemes use ensembles of classifiers and they apply a *query by committee* method: Each member of the committee presents a labeling hypotheses. There is a voting of the label of the candidates. The sample whose labeling decision shows the biggest disagreement within the committee will be queried and labeled by the oracle. *Uncertainty reduction* methods chose the sample whose classification result shows the highest uncertainty. The problem with these methods is that they can easily include outliers in the learning process[19]. Avoiding this is the motivation behind other methods that take into account instance correlations. The *expected error reduction* approach tries to predict the future generalization error for all the samples, in order to choose the one that will lower the error the most. Techniques for combining the representativeness and informativeness of samples have also been recently explored[11]. In this paper, we use an ensemble classifier, hence the query method is uncertainty-based query-by-commitee.

Spectral Unmixing. Linear unmixing consists in the computation of the fractional abundances of the elementary materials in each pixel, whose spectra are called endmembers. The linear mixing model assumes that each pixel corresponds to an aggregation of materials in the scene due to reduced sensor spatial resolution. The true elementary materials are unknown most of the times, therefore the need of endmember induction from the hyperspectral image data. Endmember induction approaches include: (a) geometric methods searching for the vertexes of a convex polytope that covers the image data, such as N-FINDR [23], (b) projection methods, such as the orthogonal subspace projection based automatic target generation process (ATGP) [17], (c) heuristic methods, such as the Pixel Purity Index (PPI) algorithm [13], and (d) lattice computing based methods [8,9,18] linking lattice independence and affine independence. Unsupervised unmixing by sparse regression is reported in [15]. According to the theoretical model, fractional abundances must be non-negative and add to one. Therefore, fully constrained least squares (FCLS) [14] is used to compute them.

2 Classification Methods

Random Forest Classifiers. Random Forest (RF) algorithm is a classifier ensemble [4] that encompasses bagging [3] and random decision forests [1,10]. RF became popular due to its simplicity of training and tuning while offering a similar performance to boosting. They are built as a collection of decision tree predictors $\{h(\mathbf{x}, \psi_k); k = 1, ..., K\}$, where ψ_t are independent identically distributed random vectors whose nature depends on their use in the tree construction, and each tree casts a unit vote to find the most popular class of input $\mathbf{x}$. Given a dataset of N samples, a bootstrapped training dataset is used to grow a tree $h(\mathbf{x}; \psi_k)$.

Active Learning. Active Learning [6] focuses on the interaction between the user and the classifier. Let $X = \{\mathbf{x}_i, y_i\}_{i=1}^{l}$ be a training set consisting of labeled samples, with $\mathbf{x}_i \in \mathbb{R}^d$ and $y_i \in \{1, \ldots, N\}$. Let be $U = \{\mathbf{x}_i\}_{i=l+1}^{l+u} \in \mathbb{R}^d$ the *pool of candidates*, with $u \gg l$, corresponding to the set of unlabeled samples to be classified. Active Learning tries to find which input vector should be selected from the training set, in order to improve the learning capabilities of the classifier. At each learning iteration, the Active Learning algorithm selects from the pool U^t the q candidates that will, at the same time, maximize the gain in performance and reduce the uncertainty of the classification model when added to the current training set X^t. The selected samples are labeled by an oracle. Finally, the set of new labeled samples is added to the current training set and removed from the pool of candidates.

Classification Uncertainty in RF Classifiers. RF classifiers allow a committee approach for the estimation of unlabeled sample uncertainty [21]: assume that we have built a committee of k base classifiers, i.e. a RF with k trees. The output of the committee members provide k labels for each candidate sample $\mathbf{x}_i \in U$. The data sample class label is provided by the majority voting. Our heuristic is that the standard deviation $\sigma(\mathbf{x}_i)$ of the class labels is the measure of the classification uncertainty of $\mathbf{x}_i$.

3 Spectral Unmixing by Sparse Lattice Computing Approach

Given a hyperspectral image $\mathbf{H}$, whose pixels are vectors in L-dimensional space, $\mathbf{x} \in \mathbb{R}^L$ the Linear Mixing Model (LMM) states that each pixel spectral signature $\mathbf{x}$ is decomposed as $\mathbf{x} = \mathbf{E}\alpha + \mathbf{n}$, where $\mathbf{E} = \{\mathbf{e}_1, \mathbf{e}_2, ..., \mathbf{e}_q\}$, $\mathbf{e}_i \in \mathbb{R}^L$ is the set of endmembers, $\alpha \in \mathbb{R}^q$ is a vector of fractional abundances, and $\mathbf{n}$ represents the noise affecting each band. Spectral unmixing is the problem of finding the q dimensional abundance vector α. Sparse Regression (SR) has been applied to hyperspectral image unmixing [12,15] assuming that the set of constituent endmembers changes for each image pixel. Given a library of spectral signatures $\Phi \in \mathbb{R}^{q \times L}$ called the dictionary, whose q columns are referred as atoms, SR looks for the minimum size spectral unmixing coefficients $\min_{\alpha} \|\alpha\|_0$ subject to $\|\Phi\alpha - \mathbf{x}\|^2 < \delta$. To achieve this sparsification we use Conjugate Gradient Pursuit [2]. To assess unmixing quality we measure the reconstruction error, i.e. the difference between the reconstructed hyperspectral image $\hat{\mathbf{X}} = \mathbf{E}\alpha$ and the original $\mathbf{X}$, by the Mean Angular Distance (MAD).

Sparse Lattice Computing Unmixing. The proposed spectral unmixing method is composed of the following stages, first we use the WM algorithm[1] [18] to obtain an overcomplete set of endmembers, secondly, we select a reduced dictionary by clustering over the set of endmembers, and, finally, we perform sparse spectral unmixing to obtain the local accurate unmixing. We call this process Sparse WM (sWM).

---

[1] The name WM is not acronym, it refers to the two LAAM matrices employed by the algorithm.

The WM algorithm starts by computing the minimal hyperbox covering the data, $\mathcal{B}(\mathbf{v}, \mathbf{u})$, where $\mathbf{v}$ and $\mathbf{u}$ are the *minimal* and *maximal corners*, respectively, whose components are computed as follows: $v_k = \min_\xi x_k^\xi$ and $u_k = \max_\xi x_k^\xi$ with $k = 1, \ldots, L$ and $\xi = 1, \ldots, N$. Next, the WM algorithm computes the dual erosive and dilative Lattice Auto-Associative Memories (LAAMs), $\mathbf{W}_{XX}$ and $\mathbf{M}_{XX}$. The columns of $\mathbf{W}_{XX}$ and $\mathbf{M}_{XX}$ are scaled by $\mathbf{v}$ and $\mathbf{u}$, forming the additive scaled sets $W = \left\{ \mathbf{w}^k \right\}_{k=1}^L$ and $M = \left\{ \mathbf{m}^k \right\}_{k=1}^L$:

$$\mathbf{w}^k = u_k + \mathbf{W}^k; \ \mathbf{m}^k = v_k + \mathbf{M}^k, \ \forall k = 1, \ldots, L, \tag{1}$$

where $\mathbf{W}^k$ and $\mathbf{M}^k$ denote the k-th column of $\mathbf{W}_{XX}$ and $\mathbf{M}_{XX}$, respectively. Finally, the set $\mathbf{E} = W \cup M \cup \{\mathbf{v}, \mathbf{u}\}$ contains the candidate endmembers covering all the image pixel spectra [18], upon which we perform SR after clustering the endmembers.

The endmember dictionary is selected by correlation clustering: Highly correlated endmembers are variations of the same material spectrum. Loosely correlated endmembers correspond to different materials present in the scene. The number of endmember clusters is found by an iterative procedure consisting in the approximation of $\mathbf{E}$ by Gaussian Mixture Models (GMM) increasing number of Gaussian components $k = 2, 3, \ldots, 20$. GMM maximum likelihood parameter estimates are calculated using an Expectation Maximization (EM) algorithm. The GMM parameters provide the clustering of endmembers in $\mathbf{E}$. Finally, the WM endmembers closest to the k^* GMM centroids form the final endmember dictionary $\tilde{\mathbf{E}}$ for SR unmixing

4 Materials

The hyperspectral data was obtained using a SOC710 camera. It has a spectral coverage of 400 to 1000 μm, a spectral resolution of 4.6875 μm and a dynamic range of 12 bits. It delivers 696 pixel per line images, and 128 bands. We set an integration time of 20 milliseconds and an electronic gain of 2 (twice the gain). Obtaining one 696x520 image cube takes around 23.2 seconds. The only light source was the sun. This stable lighting conditions and the short image acquisition times minimize spectral variations due to illumination changes. We took images of three subjects with diverse color skins. The subjects were located in places with both artificial structures like concrete or rocks and natural backgrounds like grass or flowers. We have provided a hand made reference ground truth for validation.

For reflectance normalization, we use a single white target object as reference assuming surfaces of zero reflectance will produce zero radiance. The images collected for this experiments were set up with a "white standard" surface as reference. We averaged the pixels pertaining to the surface to obtain a reference radiance value for each image. In order to reduce noise, we applied a 1D smoothing technique to all pixel spectra in the images, the well known Robust Locally

Weighted Regression smoothing [5]. It is a moving average local method where each smoothed value is determined by neighboring data points. The method uses a regression weight function modeled by a linear polynomial.

5   Results of Active Leaning Classification

The goal is to classify image pixels into two classes, the target region and the background. In our experiment, we first manually label all the images. We use this ground truth to simulate the manual input of the user. For each RF, we begun with 5 training samples. The RF consisted on $k = 100$ trees, sampling $\sqrt{d}$ variables as candidates at each split. We selected on each run $q = 20$ uncertain pixels candidates using the criterion defined above. Active learning process on each image consists of 40 iterations, as can be seen in figure 1 . We have performed 100 repetitions of each experiment and reported the mean and standard deviation values of CR, precision and sensitivity.

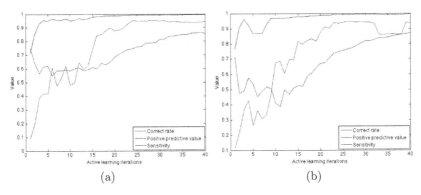

Fig. 1. Performance indicator on increasing active learning iterations for a) radiance and b) reflectance values, averaged over the 7 images

Correct Rate (CR) is shown in table 1. Variances, shown in parenthesis, are low for all images. Image A1 gives the worst results. There is no strong difference between reflectance and radiance. Image A3 shows the biggest standard deviation, which indicates that for this image the initial random selection of training pixels is crucial. Overall, the smoothing process has two effects: effectively enhancing the segmentation performance of the method and reducing the variance between experiment runs. Regarding precision, it halves the standard deviation while enhancing the average smoothing capabilities by 9.44%.

Figure 2 illustrate the segmentation results. The ground truth is shown in black (background) and white (skin). Red pixels indicate true positives, while blue pixels denote false positives. We show the image corresponding to the method that dropped the best precision for each hyperspectral cube. It is illustrated in figure2(a) where it can be appreciated that there are noticeable

Table 1. Correct rate obtained by active learning for each image (standard deviation in parenthesis)

		No smoothing		Smoothing	
		Radiance	Reflectance	Radiance	Reflectance
Cartesian coord.	A1	0.9890 (0.0007)	0.9888 (0.0010)	0.9888 (0.0009)	0.9882 (0.0011)
	B2	0.9891 (0.0015)	0.9890 (0.0027)	0.9919 (0.0008)	0.9922 (0.0004)
	A3	0.9949 (0.0013)	0.9953 (0.0011)	0.9965 (0.0004)	0.9963 (0.0005)
	C4	0.9977 (0.0024)	0.9985 (0.0001)	0.9984 (0.0003)	0.9983 (0.0004)
	C5	0.9935 (0.0025)	0.9948 (0.0005)	0.9961 (0.0014)	0.9965 (0.0005)
	C5b	0.9945 (0.0014)	0.9939 (0.0018)	0.9962 (0.0011)	0.9961 (0.0010)
	A5	0.9985 (0.0002)	0.9974 (0.0027)	0.9984 (0.0001)	0.9984 (0.0001)
	mean	0.9939 (0.0014)	0.9940 (0.0014)	0.9952 (0.0007)	0.9951 (0.0005)

white areas. The scene was windy, which can move not only the vegetation but also the subjects clothes and hair. It can be observed that big areas of the arms are not correctly segmented. Other sample segmentation results are much more precise. Notice that there are several blue pixels in areas that are expected to be labelled as skin. This might be consequence of the human error in the manual segmentation process. Every human segmentation step involves the possibility of miss-labeling a sample, therefore dragging that error across the Active Learning iterations. Moreover, slight movements of the subjects during the image acquisition can also introduce spatial noise that leads to these erroneous labels.

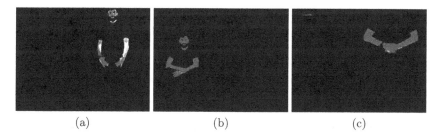

(a)   (b)   (c)

Fig. 2. Some active learning segmentation results

6   Results of Sparse Spectral Unmixing

To identify a viable skin endmember -i.e. an endmember that best represents the human skin pixels we compute the correlation between the abundance images obtained from each endmember with the hand delineated ground truth. We select the maximum correlation as the candidate skin endmember. Figure 3 shows the abundance image with overall maximum correlation to its corresponding ground truth image.

Fig. 3. Abundance image most correlated to ground truth, which corresponds to FIPPI algorithm's 19th endmember for image A3

7   Conclusions

We present a collection of hyperspectral images collected in outdoor conditions, which include windy moments introducing additional noise, due to the long capture times by the available hyperspectral camera.

Active Learning Classification. The experimental framework reported here enabled the exploration of diverse skin detection computational aspects. Firstly, it was shown that it is possible to segment skin in hyperspectral images, even in noisy situations. Secondly, an Active Learning methodology was proposed to label and segment the images. Finally, the accuracy of the proposed system was assessed under varying image preprocessing steps. Overall, results can be summarized as follows: (a) The use of reflectance normalization relative to white standards present on the scene does not improve over bare radiance signal. (b) Spectral smoothing enhances segmentation results. (c) The segmentation process is not robust to motion noise, such wind appearing in image A1. Results show that many errors are located in skin areas bordering non skin regions. This phenomenon can be partially caused by the manual segmentation step, where due to chromatic similarities it is difficult to asses whether one pixel is skin or not. Nevertheless, results are encouraging and show promise for an intereractive efficient system for human skin detection.

Spectral Unmixing. We report unmixing performances by state-of-the-art algorithms and a new proposed sparse method which achieves quite good

performance with highly parsimonious endmember set. We proceed to skin endmbe-
mber identification on the basis of a manually delineated ground truth correlation
with abundance images, obtaining high variability in the results, which suggest
that more extensive data capture and experiementation is needed. The process
is unsupervised, in the sense that no gorund truth is used to guide endmember
selection. Future works will be addressed to propose and experiment with supervised approaches, where endmember induction algorithms may be specifically
applied to the skin image regions.

Acknowledgments. I. Marques acknowledges the grant received from the Basque
Government through the Predoc Program of the Education, University and Research
Department. Thanks to Prof. Miguel Velez-Reyes University of Texas at El Paso for
providing the environment to capture the dataset. This research has been partially
funded by grant TIN2011-28753-C02-02 of the Ministerio de Ciencia e Innovación of
the Spanish Government (MINECO), and the Basque Government grant IT874-13 for
the GIC research group. ENGINE project is funded by the European Commission grant
316097.

References

1. Amit, Y., Geman, D.: Shape quantization and recognition with randomized trees.
   Neural Computation 9(7), 1545–1588 (1997)
2. Blumensath, T., Davies, M.E.: Gradient pursuits. IEEE Transactions on Signal
   Processing 56(6), 2370–2382 (2008)
3. Breiman, L.: Bagging predictors. Machine Learning 24(2), 123–140 (1996)
4. Breiman, L.: Random forests. Machine Learning 45(1), 5–32 (2001)
5. Cleveland, W.S.: Robust locally weighted regression and smoothing scatterplots.
   Journal of the American Statistical Association 74(368), 829–836 (1979)
6. Cohn, D., Atlas, L., Ladner, R.: Improving generalization with active learning.
   Machine Learning 15, 201–221 (1994), 10.1007/BF00993277
7. Fu, Y., Zhu, X., Li, B.: A survey on instance selection for active learning. Knowledge
   and Information Systems 35(2), 249–283 (2013)
8. Graña, M., Savio, A.M., Garcia-Sebastian, M., Fernandez, E.: A lattice computing
   approach for on-line FMRI analysis. Image and Vision Computing 28(7), 1155–1161
   (2010)
9. Graña, M., Villaverde, I., Maldonado, J.O., Hernandez, C.: Two lattice computing
   approaches for the unsupervised segmentation of hyperspectral images. Neurocomputing 72(10-12), 2111–2120 (2009)
10. Ho, T.K.: The random subspace method for constructing decision forests. IEEE
    Transactions on Pattern Analysis and Machine Intelligence 20(8), 832–844 (1998)
11. Huang, S.-J., Jin, R., Zhou, Z.-H.: Active learning by querying informative and
    representative examples. IEEE Transactions on Pattern Analysis and Machine In-
    telligence 36(10), 1936–1949 (2014)
12. Iordache, M.-D., Bioucas-Dias, J.M., Plaza, A.: Sparse unmixing of hyperspectral
    data. IEEE Transactions on Geoscience and Remote Sensing 49(6), 2014–2039
    (2011)
13. Green, R., Boardman, J., Kruse, F.: Mapping target signatures via partial unmixing
    of aviris data. Technical report, Jet Propulsion Laboratory (JPL) (1995)

14. Lee, D.D., Seung, H.S.: Learning the parts of objects by non-negative matrix factorization. Nature 401, 788–791 (1999)
15. Marques, I., Graña, M.: Hybrid sparse linear and lattice method for hyperspectral image unmixing. In: Polycarpou, M., de Carvalho, A.C.P.L.F., Pan, J.-S., Woźniak, M., Quintian, H., Corchado, E. (eds.) HAIS 2014. LNCS, vol. 8480, pp. 266–273. Springer, Heidelberg (2014)
16. Nunez, A.S., Mendenhall, M.J.: Detection of human skin in near infrared hyperspectral imagery. In: IEEE International on Geoscience and Remote Sensing Symposium, IGARSS 2008, vol. 2, pp. II–621–II–624 (2008)
17. Plaza, A., Chang, C.-I.: Impact of initialization on design of endmember extraction algorithms. IEEE Transactions on Geoscience and Remote Sensing 44, 3397–3407 (2006)
18. Ritter, G.X., Urcid, G.: A lattice matrix method for hyperspectral image unmixing. Information Sciences 181(10), 1787–1803 (2011)
19. Roy, N., Mccallum, A.: Toward optimal active learning through sampling estimation of error reduction. In: Proc. 18th International Conf. on Machine Learning, pp. 441–448. Morgan Kaufmann (2001)
20. Trierscheid, M., Pellenz, J., Paulus, D., Balthasar, D.: Hyperspectral imaging or victim detection with rescue robots. In: IEEE International Workshop on Safety, Security and Rescue Robotics, SSRR 2008, pp. 7–12 (2008)
21. Tuia, D., Pasolli, E., Emery, W.J.: Using active learning to adapt remote sensing image classifiers. Remote Sensing of Environment (2011)
22. Uto, K., Kosugi, Y., Murase, T., Takagishi, S.: Hyperspectral band selection for human detection. In: 2012 IEEE 7th Sensor Array and Multichannel Signal Processing Workshop (SAM), pp. 501–504 (2012)
23. Winter, M.E.: N-FINDR: An algorithm for fast autonomous spectral end-member determination in hyperspectral data. In: Imaging Spectrometry V. SPIE Proceedings, vol. 3753, pp. 266–275. SPIE (1999)

Scene Recognition for Robot Localization in Difficult Environments

D. Santos-Saavedra[1]([✉]), A. Canedo-Rodriguez[1], X.M. Pardo[1],
R. Iglesias[1], and C.V. Regueiro[2]

[1] CITIUS (Centro Singular de Investigación en Tecnoloxías da Información),
Universidade de Santiago de Compostela, Santiago de Compostela, Spain
[2] Department of Electronics and Systems, Universidade da Coruña, A Coruña, Spain
{david.santos,xose.pardo,roberto.iglesias}@usc.es

Abstract. Scene understanding is still an important challenge in robotics. In this paper we analyze the utility of scene recognition to determine the localization of a robot. We assume that multi-sensor localization systems may be very useful in crowded environments where there will be many people around the robot but not many changes of the furniture. In our localization system we categorize the sensors in two groups: accurate sensor models able to determine the pose of the robot accurately but which are sensible to noise or the presence of people. Robust sensor modalities able to provide rough information about the pose of the robot in almost any condition. The performance of our localization strategy was analyzed through two experiments realized in the Centro Singular de Investigacion en Tecnoloxias da Informacion (CITIUS), at the University of Santiago de Compostela.

Keywords: Scene recognition · Mobile robots · Multi-sensor fusion · Localization system

1 Introduction

One of the limitations of today's robots is scene understanding. Knowing "where am I" has always being an important research topic in robotics and computer vision [1]. There are many approaches that use different sensor modalities (sonar, laser, compass, Wi-Fi, etc) [2][3][4] to determine the topological localization of the robot, i.e., its position in a map. Nevertheless, robots are still unable to understand their environments, they are not aware if they are moving in a room that is similar to another one where they have been moving previously. The automatic detection of representative situations -a room without people that can be tidied up, people sitting in a sofa or children playing, people that have just entered home, amongst others- would represent an important qualitative leap forward, as robots would stop from being passive and transform into robots with "initiative". On the other hand, the identification of similarities amongst different working spaces would allow the retrieval of controllers. These are some simple examples of the benefits of scene understanding to mention but a few.

J.M. Ferrández Vicente et al. (Eds.): IWINAC 2015, Part II, LNCS 9108, pp. 193–202, 2015.
DOI: 10.1007/978-3-319-18833-1_21

Another field for which scene recognition might be useful is *robot localization*. In the past we have developed a multi-sensor fusion algorithm based on particle filters for mobile robot localization in crowded environments. The idea behind merging different sensor modalities is the achievement of robust solutions able to provide a reliable information about where the robot is, despite of difficult conditions such as change of illumination, occlusions due to the presence of many people around the robot, etc). In this sense identifying the scene where the robot is could help to determine the robot position faster and in a more reliable way. This is the objective of the research described in this paper.

2   Scene Recognition

We assume that the robot will take one observation at some location and it will use a classifier to identify this observation. In the past we have carried an exhaustive analysis of the performance of different SVM classifiers which use different image descriptors to solve the task of scene recognition [11]. As a result of this analysis we found out that working with a representation that combines global and local features is the best option [11].

The global descriptor we have chosen is Local Difference Binary Pattern. This representation is composed by the *Local Difference Sign Binary Pattern* (LSBP) and *Local Difference Magnitude Binary Pattern* (LMBP). The LSBP is a non-parametric local transform based on the comparison amongst the intensity value of each pixel of the image with its eight neighboring pixels, as illustrated in Fig. 1. As we can see in this figure, if the center pixel is bigger than (or equal to) one of its neighbors, a bit 1 is set in the corresponding location. Otherwise a bit 0 is set. The eight bits generated after all the comparisons have to be put together following always the same order, and then they are converted to a base-10 number in the interval $[0, 255]$. This process maps a 3×3 image patch to one of 256 cases, each corresponding to a special type of local structure, and it is repeated for every pixel of the original image. The *Local Difference Sign Binary Pattern* (LSBP) is also referred to as the *Census Transform* (CT) [5]. For the other component, LMBP is computed as the intensity difference between the center pixel and its neighboring pixels. If the difference in intensity amongst the center pixel and one of its neighbors is higher than a threshold T, a bit 1 is set, otherwise a bit 0 is set. Like in the case of the Census Transform, after conversion, we will obtain a base-10 number as result.

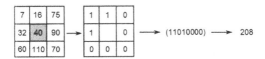

Fig. 1. Illustration of the LSBP Process on a 3 3 image patch

Thus, for every image we can compute the holistic representation given by the combination of the *LMBP* and the *LSBP*. Both the LSBP and the LMBP histograms are 256 dimensions (the bins of the histograms are each one of the values that the LMBP and LSBP can take), therefore, the new feature representation is 512 dimensional (256×2). It is a common practice to suppress the first and the last bins of these histograms, due to noise cancellation and the removal of not significant information, obtaining a final size of 508. We will call *LDBP* (local difference binary pattern) to the combination of the LSBP and LMBP histograms. We also used *spatial pyramids* to get an holistic representation at different abstraction levels, and principal component analysis (PCA) to reduce the dimension of the final descriptor.

Regarding the use of local descriptors, i.e., the discovering of salient points in the image, we used a bag-of-visual-words model. This model works on three stages: (i) Obtaining of the local descriptors; (ii) Quantization of the descriptors into a codebook; (iii) description of the image as a collection of words. We have used SURF [6] to detect the interest points and their description. Regarding the second stage, quantization of the descriptors, we have used k-means to cluster them. In this case the cluster centers act as our dictionary of visual words. Therefore, an image is represented by the histogram of the visual words, i.e. this histogram is treated as the feature vector of the image that is going to be used by the classifier.

Finally, our classifier, a *Support Vector Machine*, works with the combination of descriptors previously described, i.e, the concatenation of the representation based on the LDBP and the bag-of-visual-words.

3   Localization Algorithm

Robot localization is a problem that is considered mostly solved. Plenty of research has been published in the past where SLAM techniques using different sensors modalities are used to localize the robot and, at the same time, build a map of the environment. Nevertheless, if we consider robots that need to operate in environments that might be crowded with people or challenging conditions (such as high illumintion fluctuations), but that at the same time are static (i.e., the layout, furniture, does not change often), localization solutions that merge different sensor modalities are often preferable. On the other hand, in this kind of environments, continuous map updates, due to SLAM strategies, add little value while they consume important computational resources. Furthermore, in this kind of environments there will be many people moving around the robot, so continuous SLAM may eventually integrate them as part of the map, leading to failures. Because of this, we assume that a more functional solution might consist on dividing the problem in two stages: deployment and operation. During the deployment stage, we will construct the map of the environment and calibrate all the sensors. To create the map, we will use Simultaneous Localization and Mapping (SLAM) techniques [13]. Once created, the map will not be modified. On the operation stage, the robot will use the map and the sensor perceptions to localize itself.

3.1 Pose Probability Estimation

The localization algorithm used in this work [8] estimates at any time t, the robot pose s_t from perceptual information Z_t (sensor measurements) and robot movement u_t (provided by odometry encoders). The "pose probability estimation" step computes the likelihood $bel(s_t)$ of each pose from u_t and Z_t. After that, the "pose estimation" step calculates s_t, a state vector with the most probable robot pose.

Following a Bayesian Filtering approach, the likelihood assigned to each robot pose $bel(s_t)$ will be the posterior probability over the robot state space conditioned on the control data u_t and the sensor measurements Z_t [9]:

$$bel(\boldsymbol{s_t}) = p(\boldsymbol{s_t}|\boldsymbol{Z_t}, u_t, \boldsymbol{Z_{t-1}}, u_t, ..., \boldsymbol{Z_0}, u_0) \tag{1}$$

Assuming that the current state s_t suffices to explain the current u_t and Z_t (Markov assumption) and that the sensor measurements are conditionally independent given the state of the robot, we can estimate $bel(s_t)$ recursively [9]:

$$bel(\boldsymbol{s_t}) \propto \left[\int p(\boldsymbol{s_t}|\boldsymbol{s_{t-1}}, u_t)bel(\boldsymbol{s_{t-1}})d\boldsymbol{s_{t-1}} \right] \prod_{k=1}^{N_s(t)} p(z_t^k|\boldsymbol{s_t}) \tag{2}$$

Implementation with Particle Filters. Equation 2 can be approximated very efficiently using particle filters [9]. These filters approximate $bel(s_t)$ as a set of M random weighted samples or particles:

$$bel(\boldsymbol{s_t}) \approx \boldsymbol{P_t} = \{\boldsymbol{p_t^1}, \boldsymbol{p_t^2}, ..., \boldsymbol{p_t^M}\} = \{\{s_t^1, \omega_t^1\}, \{s_t^2, \omega_t^2\}, ..., \{s_t^M, \omega_t^M\}\} \tag{3}$$

where each particle p_t^i consists, at time t, of a possible robot pose s_t^i and a weight ω_t^i (likelihood) assigned to it. Equation 2 is calculated by means of the Augmented Montecarlo Localization algorithm with Low Variance Resampling [9].

The algorithm consist essentially on an initial phase and a repeatable cycle of 3 steps. In the initial phase all the particles are distributed randomly over the state space with weight $(\frac{1}{M})$. Then, the algorithm perform the 3 main steps:

1. Propagation. Each particle evolves to a new state according to robot movement.
2. Re-evaluation. The weight of each particle is re-computed taking into account the measurements of the available sensors.
3. Resample. The algorithm constructs a new set of particles P_t from P_{t-1} and new random particles.

Pose Estimation. In this stage, the pose estimation from particle set P_t most likely is chosen. The localization system performs a clustering on the particle set P_t. Each cluster represents a hypothesis of the robot pose. Then the most likely hypothesis is selected, by calculating the likelihood of each hypothesis. The likelihood of each hypothesis will be the sum of the weights of its particles.

4   Sensor Fusion

In this paper we analyze a localization system that merges the information provided by a laser scanner, Wi-Fi, a Magnetic compass, a external network of cameras, and the images taken from a Kinect and which are used to identify the scene where the robot is. Next we give a brief explanation of the sensor-models that are used to achieve an estimation of the robot pose.

4.1   Sensor Models

2D Laser Range Finder. At any time instant t, a 2D laser range finder provides a vector $l_t = \{l_t^1, l_t^2, ..., l_t^{N_L}\}$ of N_L range measurements between the robot and the nearby objects. This vector is usually called the laser signature. Using an occupancy map [9] of the environment, we could pre-compute the laser signature $l_e(s)$ expected for any possible pose in this map. Thus, we can approximate the laser "sensor model" by the similarity between $l_e(s)$ and l_t :

$$p(z^l|s) = \left[\sqrt{1 - \frac{\sqrt{\sum_{i=1}^{N_L} l_t^i \cdot l_e^i(s)}}{N\sqrt{\sum_{i=1}^{N_L} l_e^i(s) \sum_{i=1}^{N_L} l_t^i}}} \right] \left[\frac{1}{N_L} \sum_{i=1}^{N_L} max\left(1 - \frac{|l_e^i(s) - l_t^i|}{max_{LD}}, 0\right) \right] \quad (4)$$

Wi-Fi. The Wi-Fi positioning system provides an estimate of the robot position using the signals received from Wi-Fi landmarks such as Wi-Fi Access Points (APs). To train the WiFi observation model, we have used Gaussian processes regression (GP regression) [14]. GP regression is a *supervised learning* technique, therefore it requires a training data set. To build it we moved the robot around the environment. Our robot is equipped with a Wi-Fi card, and therefore it receives the signal power from the Wi-Fi APs in the environment. This information can be represented as $z_t^w = \{w_t^1, ..., w_t^{n_t^w}\}$, where n_t^w is the number of APs available at time t, and w_t^i is the power in dBms of the ith AP at time t. Using the data captured at the "data collection", we build a training set for each AP $D_i^w = \{(x_t, y_t) ; w_t^i\}$. Each sample of the training set consists on an output w_t^i (power of the ith AP received from the scan at time t), associated with an input (x_t, y_t) (position where the scan took place). With this training set, the GP regression computes for each AP the functions $\mu_i^w(x, y)$ and $\sigma_i^w(x, y)$, which represent the average and the typical deviation of the signal strength of the ith AP across the environment.

These functions will be the base of the observation model of the Wi-Fi. First of all, we will compute the observation model of each AP independently:

$$p(w_t^i|s) \propto \frac{1}{\sigma_i^w(x, y)\sqrt{2\pi}} exp\left[-\frac{1}{2}\left(\frac{w_t^i - \mu_i^w(x, y)}{\sigma_i^w(x, y)}\right)^2 \right] \quad (5)$$

we have observed that the combination of sensor models with a simple voting scheme, which averages the predictions of each model, gives us a robust solution:

$$p(z_t^w|s) \propto \sum_{i=1}^{n_t^w} p(w_t^i|s) \qquad (6)$$

Magnetic Compass. A magnetic compass provides the orientation $z_t^c = \theta_t^c$ of the robot with respect to the Earth's magnetic north. First of all, we collect a set of k true robot orientations $\theta^T = \{\theta_1^T, ..., \theta_k^T\}$, and we store the corresponding set of compass readings $\theta^c = \{\theta_1^c, ..., \theta_k^c\}$. Then, we compute the differences among both sets:

$$e^c = \{e_1^c, ..., e_k^c\} = \{|\theta_1^c - \theta_1^T|, ..., |\theta_k^c - \theta_k^T|\} \qquad (7)$$

With this information, we build a training set $D^c = \{(x_t, y_t) ; e_t^c\}$. Applying Gaussian processes regression to this training set, we build $\mu^c(x, y)$ and $\sigma^c(x, y)$, the average and typical deviation of the error across the environment. These functions are the base of the observation model that we propose:

$$p(z_t^c|s) = \frac{1}{\sigma^{ec}(x, y)\sqrt{2\pi}} exp\left[-\frac{1}{2}\left(\frac{z_t^c - \theta}{\sigma^{ec}(x, y)}\right)^2\right] \qquad (8)$$

where θ is the orientation component of the robot pose s, and $\sigma^{ec}(x, y)$ is a noise parameter.

Cameras. These cameras use a robot detection algorithm based on active markers (LED lights) that are placed on the robots [12]. This algorithm allows the cameras to detect multiple robots very robustly in crowded environments and in real time. While a camera detects a robot, it sends the robot a periodic message (e.g. every 2 seconds) that contains the identity of the camera. Therefore, the robot can construct vectors $z_t^C = \{C_t^1, ..., C_t^{n^C}\}$, where n^C is the total number of cameras and C_t^i states whether or not each camera is detecting the robot. With this information, we can reconstruct the FOV coverage of each camera over the map. To this extent, we construct a training set for each camera $D_i^C = \{(x_t, y_t); C_t^i\}$. Each sample of the training set consists on a sample output C_t^i (whether or not the camera i detected the robot at time t), associated with a sample input (x_t, y_t) (position at time t). Once again we compute the FOV map of each camera using the GP regression . This gives us the functions $\mu_i^C(x, y)$ and $\sigma_i^C(x, y)$ for each camera i. Therefore the observation model for the cameras is:

$$p(z_t^C|s) = \prod_{j \in \{i | C_t^i = 1\}} \mu_j^C(x, y) \qquad (9)$$

Kinect. Our image classifier provides a probability grid for the robot position based on the image taken from the Kinect. Using an occupancy map, we assign probabilities to each area of the map a probability depending of its associated class or category. Both the number of classes and are defined previously. The probability of each category for a given image is defined by the confusion matrix obtained in the training process of the classifier. The classifier provides for a problem of n-classes a matrix $v(n \times n)$ where the i-value $v(i,i)$ represents the probability with which an image of class i is correctly classified. Each other value $v(i,j)$, $\forall j \neq i$ reflect the probabilities of getting wrong classifications of a image of class i. Once the classifier calculates the class for an image, we can assign to its category the probability corresponding to a correct classification in the confusion matrix. Other categories will take their probability value from the matrix according to the input image label and their own class value.

5   Experimental Results

We performed 2 experiments at the CITIUS research building (Centro Singular de Investigación en Tecnoloxías da Información da Universidade de Santiago de Compostela, Spain). The basement floor of the building was the environment for the first experiment and the ground floor for the second. We used a Pioneer P3DX robot equipped with a SICK-LMS100 laser, a Devantech CMPS10 magnetic compass, a kinect camera, a laptop with Wi-Fi connectivity (Intel Core i7-3610QM @ 2.3GHz, 6GB RAM), 5 WiFi APs, and a network of 5 Microsoft HD-5000 usb cameras running connected to other laptops. We established a period of 1000ms (control cycle) to execute our algorithm.

The methodology was the same for both experiments. First, we moved the robot along the environment following different trajectories and collecting all sensors information (training data sets). Then, we build an occupancy map of the environment using the GMapping SLAM algorithm [10]. After this, we run the localization algorithm over the trajectories with all the sensor-combinations. The analysis of the impact of each sensor-combination in the performance of the localization system is repeated over the same trajectories 10 times but with different random initialization conditions. For each execution, the robot starts with zero information, i.e. without any knowledge about its global localization. We analyzed how scene classification can improve the localization of the robot, in particular we considered the Error in position, Convergence time (tconv), which is the time interval until the algorithm provides a first estimation of the robot pose, and Stability (%Tloc), which is the percentage of time that the algorithm provides an estimation of the pose after the first convergence. For the explanation of the experiments we will use L to refer to Laser, W to Wi-Fi, C to cams, c to compass and K to Kinect with scene recognition.

5.1   Performance of the Localization System

Basement Floor Experiment. The first experiment took place at basement floor of CITIUS building. The occupancy map Fig.2.a was created using the data collected with the robot during the initial phase.

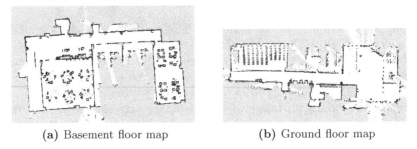

(a) Basement floor map   (b) Ground floor map

Fig. 2. Occupancy maps created using GMapping SLAM algorithm for the experiments

Fig. 3 shows the performances achieved with all the sensor combinations. We are mainly interested in combinations which include the Kinect sensor but we also show the performance of the localization system achieved with other sensors. We can see than in terms of position estimation, Kinect has similar behavior to another robust sensor, the Wi-Fi sensor.

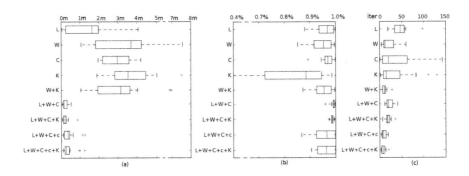

Fig. 3. Results of Error XY (a), tconv (b) and tfloc (c) for Basement floor

As we can see in the Fig. 3.a, Kinect performance achieves an error in XY comparable to Wi-Fi and Cameras. Regarding tfloc, Kinect improves results of the cameras but not Wi-Fi. Finally, comparing the convergence times we can see that kinect got worse results. This result is directly related to the accuracy of the image classifier. It is interesting to study how Kinect sensor works compared to Wi-Fi since both sensors provide a probability grid for the localization system. They do not estimate an exact position but they provides a probability for each position in the occupancy map. The combination W+K shows how they work together. The results reveals that they provide a similar information and, therefore, their combination does not achieve better accuracy. However, if we consider the performance grouping all the possible sensors like L+W+C+c+K or L+W+C+K, including Kinect as part of the system improves slightly the accuracy in pose estimation and time required to get the first robot pose.

Ground Floor Experiment. The second experiment took place at ground floor of CiTIUS building. Fig.2.b shows the occupancy map created for this case. Fig. 4 show like in the first experiment, the performances achieved with all the sensor combinations and new trajectories.

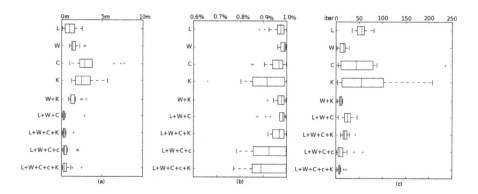

Fig. 4. Results of Error XY (a), tconv (b) and tfloc (c) for Ground floor

The new results confirm the similarity between Wi-Fi and Kinect. Although their performances are different, their combination does not achieve improvements in pose estimation. However, in terms of tfloc, their combination reduces the time that the algorithms needs to converge. The inclusion of Kinect worsens the tconv but it reduces the time needed for tfloc in both cases.

6 Conclusions and Future Work

In this work, we applied scene recognition to robot localization. We have developed an image classifier able to categorize the images taken from a Kinect sensor on top of the robot. Our goal in this paper was to analyze how scene recognition can help to determine the pose of the robot. In this paper we suggest the use of a multi-sensor localization system able to provide the pose of the robot in crowded and difficult environments. In this system we distinguish two types of sensors: on one hand there are sensors which provide an accurate estimation of the robot pose but which, on the other hand, can suffer from occlusion or convergence problems due to the presence of many people around the robot. On the other hand, there are modalities, such as WIFI positioning, able to provide information about the whereabouts of the robot in almost any condition. Scene recognition would be included in this second group. The results show that our image classifier can contribute to localization in most of the cases in terms of the percentage of times that the algorithm provides an estimation of the robot pose after the first convergence (%tloc). Nevertheless, the convergence time is not improved due to the use of scene recognition.

Acknowledgements. This work was supported by grants: GPC2013/040 (FEDER), TIN2012-32262.

References

1. Wu, J., Christensen, H.I., Rehg, J.M.: Visual Place Categorization. Problem, Dataset, and Algorithm. In: The 2009 IEEE/RSJ International Conference on Intelligent Robots and Systems, pp. 4763–4770 (2009)
2. Canedo-Rodriguez, A., Alvarez-Santos, V., Santos-Saavedra, D., Gamallo, C., Fernandez-Delgado, M., Iglesias, R., Regueiro, C.V.: Robust multi-sensor system for mobile robot localization. In: Ferrández Vicente, J.M., Álvarez Sánchez, J.R., de la Paz López, F., Toledo Moreo, F. J. (eds.) IWINAC 2013, Part II. LNCS, vol. 7931, pp. 92–101. Springer, Heidelberg (2013)
3. Drumheller, M.: Mobile robot localization using sonar. IEEE Transactions on Pattern Analysis and Machine Intelligence (2), 325–332 (1987)
4. Hahnel, D., Burgard, W., Fox, D., Fishkin, K., Philipose, M.: Mapping and localization with RFID technology. In: Proceedings of 2004 IEEE International Conference on Robotics and Automation, ICRA 2004, vol. 1, pp. 1015–1020. IEEE (April 2004)
5. Wu, J., Rehg, J.M.: CENTRIST: A Visual Descriptor for Scene Categorization. IEEE Trans. Pattern Analysis and Machine Intelligence 33(8), 1489–1501 (2011)
6. Bay, H., Ess, A., Tuytelaars, T., Van Gool, L.: Speeded-up robust features (SURF). Computer Vision and Image Understanding 110(3), 346–359 (2008)
7. Luo, J., Pronobis, A., Caputo, B., Jensfelt, P.: The KTH-IDOL2 database. Technical Report CVAP304, Kungliga Tekniska Hgskolan, CVAP/CAS (October 2006), http://cogvis.nada.kth.se/IDOL/
8. Canedo-Rodriguez, A., Alvarez-Santos, V., Santos-Saavedra, D., Gamallo, C., Fernandez-Delgado, M., Iglesias, R., Regueiro, C.V.: Robust multi-sensor system for mobile robot localization. In: Ferrández Vicente, J.M., Álvarez Sánchez, J.R., de la Paz López, F., Toledo Moreo, F. J. (eds.) IWINAC 2013, Part II. LNCS, vol. 7931, pp. 92–101. Springer, Heidelberg (2013)
9. Thrun, S., Burgard, W., Fox, D., et al.: Probabilistic robotics, vol. 1. MIT press, Cambridge (2005)
10. Grisetti, G., Stachniss, C., Burgard, W.: Improved techniques for grid mapping with rao-blackwellized particle filters. IEEE Transactions on Robotics 23(1), 34–46 (2007)
11. Santos-Saavedra, D., Pardo, X.M., Iglesias, R., Canedo-Rodríguez, A., Álvarez-Santos, V.: Scene recognition invariant to symmetrical reflections and illumination conditions in robotics. Accepted to ibPRIA 2015 (2015)
12. Canedo-Rodriguez, A., Iglesias, R., Regueiro, C.V., Alvarez-Santos, V., Pardo, X.M.: Self-organized multi-camera network for a fast and easy deployment of ubiquitous robots in unknown environments. Sensors 13(1), 426–454 (2013)
13. Thrun, S., Burgard, W., Fox, D.: Probabilistic robotics. MIT Press, Cambridge (2005)
14. Rasmussen, C.E., Williams, C.K.I.: Gaussian processes for machine learning, MIT Press (2006)

Pyramid Representations of the Set of Actions in Reinforcement Learning

R. Iglesias[1], V. Alvarez-Santos[2], M.A. Rodriguez[1], D. Santos-Saavedra[1(✉)], C.V. Regueiro[3], and X.M. Pardo[1]

[1] CITIUS, Universidade de Santiago de Compostela, Santiago de Compostela, Spain
[2] Situm Technologies S. L., Santiago de Compostela, Spain
[3] Department of Electronics and Systems, Universidade da Coruña, A Coruña, Spain
{david.santos,roberto.iglesias}@usc.es

Abstract. Future robot systems will perform increasingly complex tasks in decreasingly well-structured and known environments. Robots will need to adapt their hardware and software, first only to foreseen, but ultimately to more complex changes of the environment. In this paper we describe a learning strategy based on reinforcement which allows fast robot learning from scratch using only its interaction with the environment, even when the reward is provided by a human observer and therefore is highly non-deterministic and noisy. To get this our proposal uses a novel representation of the action space together with an ensemble of learners able to forecast the time interval before a robot failure

Keywords: Reinforcement learning · Robotics · Ensembles · Learning and adaptation

1 Introduction

If we consider the main scenarios where future robots are expected to move, or the tasks they are expected to carry out (assisting with the housework, security and vigilance, rehabilitation, collaborating in the care-entertainment, etc), we immediately realize that this new generation of robots must be able to learn on their own. They can not rely on an expert programmer, on the contrary, robots should be able to learn from what the user does, but also from their interaction with the physical and social environment.

One of the most suitable learning paradigms to get a robot learning from its interaction with the environment is reinforcement learning (RL). What makes this learning paradigm appealing is that the system learns on its own, through trial and error, relying only on its own experiences and a feedback reward signal – There are, however, known drawbacks to the application of RL in robotics. Thus the large numbers of random actions taken by the robot, especially during the early stages, or the very long convergence times showed by most of the traditional algorithms, make the application of this learning paradigm in real robots very difficult. In our case we have investigated a new proposal that combines some of the results we have already achieved in our research group with

© Springer International Publishing Switzerland 2015
J.M. Ferrández Vicente et al. (Eds.): IWINAC 2015, Part II, LNCS 9108, pp. 203–212, 2015.
DOI: 10.1007/978-3-319-18833-1_22

new developments, to achieve an algorithm able to reach fast learning process on a real robot interacting in the environment, even when the action space is multidimensional.

2 Value Functions: Evaluating a Control Policy

Let us say that there is a control policy π that determines what the robot does at every instant, i.e., this policy π is a mapping from relevant and distinguishable states to actions:

$$
\begin{aligned}
\pi : S \times A &\to [0,1] \\
(s,a) &\to \pi(s,a)
\end{aligned}
\tag{1}
$$

where S is the set of states that represent the environment around the robot, and A is the set of possible actions the robot can carry out, and $\pi(s,a)$ is the probability of performing action a in state s. The goal of reinforcement learning is to discover an optimal policy π^* that maps states to actions so as to maximize the expected return (sum of rewards) the robot will receive.

To evaluate a control policy, i.e. to quantify how long a policy will be able to move the robot before it makes something wrong and the robot receives negative reinforcement, we will use an algorithm that we have published in the past and which is called *increasing the time interval before a robot failure* [1]. Using this algorithm our system will not learn the expected discount reward the robot will receive – as it is habitual in reinforcement learning–, rather the expected time before failure (*Tbf*). This will make it easier to assess the evolution of the learning process as a high discrepancy between the time interval before failure predicted and what is actually observed on the real robot is a clear sign of an erroneous learning.

To assess our control policy we will build a utility function of the states the robot might encounter, termed *V-function*. Thus, $V^\pi(s)$ is a function of the expected time interval before a robot failure when the robot starts moving in s, performs the action determined by the policy for that state $\pi(s)$, and follows the same policy π thereafter:

$$
V^\pi(s) = E[-e^{(-Tbf^\pi(s_0 = s)/50T)}],
\tag{2}
$$

where $Tbf^\pi(s_0)$ represents the expected time interval (in seconds) before the robot does something wrong, when it performs action $\pi(s)$ in s, and then follows the control policy π. T is the control period of the robot (expressed in seconds). The term $-e^{-Tbf/50T}$ in Eq. 2 is a continuous function that takes values in the interval $[-1, 0]$, and varies smoothly as the expected time before failure increases.

Therefore, according to Eq. 3, if the robot performs an action every 250 milliseconds (value of T in Eq. 3), a V-value equal to -0.8 (for example) will clearly mean that the robot will probably receive a negative reinforcement after 2.8 seconds.

$$
Tbf^\pi(s) = -50 * T * Ln(-V^\pi(s)),
\tag{3}
$$

The definition of $V^{\pi}(s)$, Tbf^{π}, determine the relationship between consecutive states:

$$Tbf^{\pi}(s_t) = \begin{cases} T & \text{if } r_t < 0 \\ T + Tbf^{\pi}(s_{t+1}) & \text{otherwise} \end{cases} \tag{4}$$

r_t is the reinforcement the robot receives when it follows π in state s_t. If we combine Eq. 2, 3, and 4, it is true to say that in the general case:

$$V^{\pi}(s_t) = -e^{-1/50} \prod_a \prod_{s'} (-V^{\pi}(s'))^{\pi(s_t,a)P(s_t,a,s')}. \tag{5}$$

where $P(s_t, a, s')$ represents the probability of moving from state s_t to state s' when the robot performs action a in s.

Instead of the value function $V^{\pi}(s)$ many algorithms rely on the *state-action value function* $Q^{\pi}(s, a)$ instead, which has advantages for determining the optimal policy. This new function is defined as the expected time before a robot failure when the robot starts from s, by executing action a, and then following policy π:

$$Q^{\pi}(s, a) = E[-e^{(-Tbf^{\pi}(s_0=s,a_0=a)/50T)}], \tag{6}$$

hence:

$$Tbf^{\pi}(s, a) = -50 * T * Ln(-Q^{\pi}(s, a)),$$

Like we did for the V-function, in the case of the Q-values it is possible to say that in general:

$$Q^{\pi}(s_t, a) = -e^{-1/50} \prod_{s'} (-V^{\pi}(s'))^{P(s_t,a,s')}. \tag{7}$$

3   Pyramid Representation of the Set of Actions

According to the explanation provided in the previous sections, our robot will learn an utility function of states and actions $Q(s, a)$. Nevertheless this involves the existence discrete sets of states and actions. Regarding the discrete representation of actions, there are some issues that must taken into account: if the set of actions is obtained as a thin partition of the action space, the robot could determine the best control policy with a high accuracy. Nevertheless, due to the fact that the number of actions is high, the learning process could take too long and, in the worst case, might not converge at all. On the other end, if the set of actions is obtained as a rough partition of the action space, the learning process will take shorter, but there could be convergence problems due to the fact that the partition of actions might be too coarse and thus the robot cannot attain the right and specific action for some of the states. Due to this, we opted for an intermediate alternative that will allow a gradual and incremental learning of which is the best action for each state, in particular we will inspire in what is called *Spatial Pyramid Representation* [2,3] in computer vision.

We inspired in this kind of strategy to divide the action space into a set of increasingly finer intervals. In particular we used a connected tree where each level of the tree divides the action space into a set of Voronoi regions (Fig. 1).

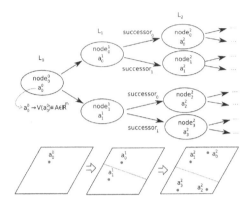

Fig. 1. Pyramid representation of the action space. This representation is obtained at the beginning of the learning process. Each node of the tree represents a Voronoi region, the centre of which is represented in the node (a_j^i). Each level of the tree divides the action space in regions that are smaller as we move further in the tree.

$best\_Tbf$ is computed for the nodes in the last level of the tree,
for *for i=0* **to** *num_nodes_level(L)* **do**
  $best\_Tbf(s, node_i^L) =$
  $-50.T.ln(-max\{Q(s, Vor(succ_0)), Q(s, Vor(succ_1))\})$
end
$best\_Tbf$ is computed for the rest of nodes in the tree,
The computation is done backwards, from the penultimate level to the first one
for *for l=L-1* **to** *0* **do**
  for *for i=0* **to** *num_nodes_level(l)* **do**
    the $best\_Tbf$ for the branch 0 is computed
    $best\_Tbf_0 = \frac{-50.T.ln(-Q(s, Vor(succ_0)))+(L-l)*best\_Tbf(s, succ_0)}{1+(L-l)}$;
    the $best\_Tbf$ for the branch 1 is computed
    $best\_Tbf_1 = \frac{-50.T.ln(Q(s, Vor(succ_1)))+(L-l)*best\_Tbf(s, succ_1)}{1+(L-l)}$;
    the $best\_Tbf$ for the node being analysed is estimated
    $best\_Tbf(s, node_i^l) = \ldots$
    $max\{best\_Tbf_0, best\_Tbf_1\}$
  end
end

Algorithm 1. Estimation of the best_Tbf values for the pyramid representation

Basically, as we can see in this figure, every level of the tree divides the action space into a set of regions that are smaller as we move further in the tree, i.e. as we move down in the tree the number of Voronoi regions is higher and their size is smaller. Finally, this tree is binary, i.e, each node has only two successors. Considering that a_j^i represents the centroid of the jth Voronoi region in which level i partitions the action space, it is true to say that the successors of this node verify that their Voronoi centres are within the Voronoi region of the parent node. Thus, in the example shown in Fig. 1, $a_0^2 \in Vor(a_0^1)$, $a_1^2 \in Vor(a_0^1)$, $a_2^2 \in Vor(a_1^1)$, $a_3^2 \in Vor(a_1^1)$, and so on.

Now, instead of having a Q value for every possible action, our algorithm will keep a Q value for every Voronoi region, these Q values will evolve during the learning procedure. The value $Q(s, Vor(a_j^i))$ represents the *expected time interval* before a robot failure when the robot performs any action included in $Vor(a_j^i)$ in state s.

$best\_Q$ is computed for the nodes in the last level of the tree,
for *for i=0* **to** *num_nodes_level(L)* **do**
$\quad | \quad best\_Q(s, node_i^L) = max\{Q(s, Vor(succ_0)), Q(s, Vor(succ_1))\}$
end
$best\_Q$ is computed for the rest of nodes in the tree,
The computation is done backwards, from the penultimate level to the first one
for *for l=L-1* **to** *0* **do**
$\quad$ **for** *for i=0* **to** *num_nodes_level(l)* **do**
$\quad\quad$ the $best\_Q$ for the branch 0 is computed
$\quad\quad best\_Q_0 = -(-Q(s, Vor(succ_0)).(-best\_Q(s, succ_0))^{(L-l)})^{1+(L-l)};$
$\quad\quad$ the $best\_Q$ for the branch 1 is computed
$\quad\quad best\_Q_1 = -(-Q(s, Vor(succ_1)).(-best\_Q(s, succ_1))^{(L-l)})^{1+(L-l)};$
$\quad\quad$ the $best\_Q$ for the node being analysed is estimated
$\quad\quad best\_Q(s, node_i^l) = \ldots$
$\quad\quad max\{best\_Q_0, best\_Q_1\}$
$\quad$ **end**
end

Algorithm 2. Estimation of the best_Q values for the pyramid representation

---

3.1 Use of the Pyramid Representation to Determine the Greedy Policy

As the learning progresses the robot should tend to take more and more actions that are greedy, i.e. which maximise the time interval before a robot failure. Let us remember that $Q(s, Vor(a_j^i))$ represents the *expected time interval* before a robot failure when the robot performs any action included in $Vor(a_j^i)$ in state s. Nevertheless these values let us guarantee a good convergence of the learning procedure. To understand this it is enough to consider that when there is a Voronoi region which is too big, and therefore contains too many actions that might be inappropriate for a particular state, its corresponding Q value will tend to be low, no matter that there might be a very good action inside this region. This could significantly slow the learning process or even cause a convergence towards suboptimal actions. To prevent this problem, we will use another value, which we will call *best time before a robot failure*. By the $best\_Tbf(Vor(a_j^i))$ we represent the highest time before a robot failure which can be obtained with an action included in $Vor(a_j^i)$, i.e., if we compute the *expected time interval* before a robot failure for every action included in $Vor(a_j^i)$, then $best\_Tbf(Vor(a_j^i))$ would be the highest of these times (Algorithm 1).

As in the case of the expected times before a robot failure (Tbf), it is possible to project the $best\_Tbf()$ in the interval $[-1, 0]$:

$$best\_Q(Vor(a_j^i)) = -e^{(-best\_Tbf(Vor(a_j^i))/50T)}$$

Therefore, instead of working in the time domain, which might involve working with very big numbers when the learning converges to a good control policy, we prefer to work only with the Q-values (Algorithm 2).

This pyramid representation (binary tree) can be used to decide the action the robot should carry out at every instant (Algorithm 3). Nevertheless, in this case the decision process will be a sequential procedure (an L-step procedure, being L the number of levels of the pyramid). Obviously, it is possible to incorporate these new values $best\_Q()$ during the sequential action decision process.

According to algorithm 3 $best\_set\_of\_actions$, is the Voronoi region that contains all the possible actions that can be performed by the robot. Nevertheless, if we consider that the intersection of Voronoi regions is a Voronoi region, is straightforward to deduced that the Voronoi region represented by $best\_set\_of\_actions$ is reduced progressively during the sequential procedure described in this algorithm.

$best\_node = node_0^0$
$best\_set\_of\_actions = Vor(a_0^0)$
for *level=1* **to** L **do**
  the successor with the $best\_Q$ is chosen
  $best\_sucessor = arg\_max_{i=0,1}\{best\_Q(s, sucessor_i(best\_node))\};$
  the best set of actions is updated according to the chosen successor
  $best\_set\_of\_actions = best\_set\_of\_actions \cap Vor(best\_sucessor);$
  $best\_node = best\_successor$
end

Algorithm 3. Sequential process to determine the action to be taken

4 Experimental Results

In this section we describe several experiments carried out at the 100sqm robotics laboratory of our research centre (CITIUS). In all the experiments described

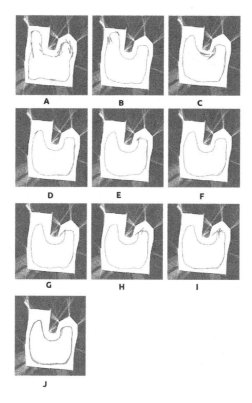

Fig. 2. Trajectories followed by our robot to learn the wall following task. Both linear and angular velocities are learnt by the algorithm. Lap A,B and C concentrates most reinforcements, while laps D to I receive few of them. Figure J shows the final three autonomous laps performed by the robot.

in this section we built an ensemble of 100 learners [4]. Every member of the ensemble used its own representation of states and actions. To represent the environment around the robot every learner used a Fuzzy ART neural network [5], the vigilance parameter (ρ) is a random number between 0.92 and 0.97. Every learner used a pyramid representation of the action space, but the partitioning of this space into Voronoi regions was random. The number of levels in the pyramid is 3. We have also made some tests with values up to 6 but it the results are similar. Finally, to update the Q-values every learner used a Monte Carlo method, in particular the *every-visit* Monte Carlo method [6]. Regarding the reinforcement, we decided that it should come from a human observer that is seeing what the robot does. This observer will be able to punish the robot by simply pressing a button in a wireless joystick. The robot's angular velocity must be between -0.7859 and 0.7859 radians per second. The experiments are divided in two parts: first, we started testing our proposal to learn a wall following controller, and then, we moved to the learning of a human following controller.

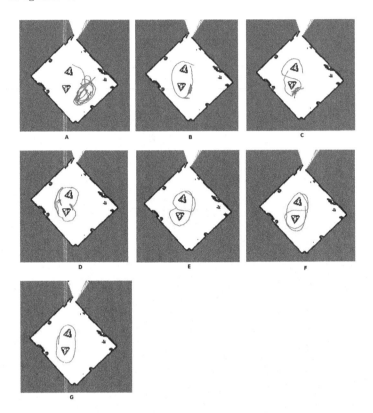

Fig. 3. Trajectories performed by our robot while learning the task of following a human while avoiding obstacles. Figure A shows the part where the robot received many reinforcements to teach him how to follow a human without obstacles. Figures B to D show several laps around the obstacles where the robot learnt how to avoid them and keep following the human. Finally, figures E to G show how the robot followed a human among the obstacles with very few or no reinforcements at all.

In the wall following controller task the robot had to learn both angular and linear velocities, but we had to include the following restrictions due to the robot's characteristics: linear velocity between 0.0 and 0.6 m/s and linear velocity $\leq (-0.5|angular\_vel|)/0.78 + 0.6$. The experiments that we have carried out confirmed the validity of our algorithm. It was able to learn the task in the first two or three laps, and then it needed from 2 to 6 more laps with a few reinforcements to autonomously complete three consecutive laps without any reinforcement (Fig. 2).

Finally, we wanted to test our algorithm with a second behaviour: human following while avoiding obstacles. For this reason, we set up a new environment in which we put a few obstacles. This time the human will walk around the obstacles while the robot learns to follow him. In this new set of experiments we seek that the robot learned how to avoid obstacles even if that meant to lose sight of the human for a few seconds, that is, the robot should learn how to recover the sight of the human once he has avoided the obstacle. The inputs were the distance to the robot's target and his angle with respect to the robot's front direction, and also information about the obstacles: we calculated a repulsive force using the data from the laser scanner. In these experiments we learnt that the best strategy was to start teaching the robot how to follow a human without obstacles, which is achieved in few minutes. Then, the human could start walking around the obstacles to allow the robot to learn how to avoid them (Fig. 3).

5   Conclusions

Most service robots should be constant learners: it should acquire new skills in an active, open-ended way, and develop as a result of constant interaction and co-operation with humans. Future service robots will be required to run autonomously over really long periods of time in environments that change over time.These robots will be required to live together with people, and to adapt to the changes that people make to the world. In this paper we have described a learning strategy based on reinforcement which allows fast robot learning from scratch using only its interaction with the environment. The strategy we have developed has the following characteristics: a) *Use of pyramid representations of the action space* (binary tree in this case). We call it pyramid because in the general case it would not have to be a binary tree (i.e. the number of successors could be higher than two). On the other hand, and like in computer vision, the joint consideration of the Q-values associated to all levels of the tree influence the decision of the action to be taken by the robot. The use of this pyramid representation allows a progressive learning of the task, i.e., the robot will discover first the big interval which contains the right actions for every state, but this knowledge will be progressively refined so that the robot finally knows the correct action or narrow interval of actions that are suitable for every state. b) Use of *an ensemble of learners able to forecast the time interval before a robot failure*. The use of an ensemble helps to get better generalization. We have tested our strategy in real world experiments. It is important to realize that in all these experiments a human user provided the reinforcement with a joystick. This is another challenge since sometimes the user can change his mind (regarding when something is right or wrong) during the learning procedure, etc. Our proposal was robust enough to cope with this.

Acknowledgements. This work was supported by grants: GPC2013/040 (FEDER), TIN2012-32262.

References

1. Quintia, P., Iglesias, R., Rodriguez, M.A., Regueiro, C.V.: "Simultaneus learning of perception and action in mobile robots2. Robotics and Autonomous Systems 58(12), 1306–1315 (2010)

2. Lazebnik, S., Schmid, C., Ponce, J.: Beyond Bags of Features: Spatial Pyramid Matching for Recognizing Natural Scene Categories. In: Proceedings of the 2006 IEEE Computer Society Conference on Computer Vision and Pattern Recognition (CVPR 2006), pp. 2169–2178 (2006)
3. Kristo, K., Chua, C.S.: Image representation for object recognition: utilizing overlapping windows in Spatial Pyramid Matching. In: 20th IEEE International Conference on Image Processing (ICIP) (2013)
4. Quintia Vidal, P., Iglesias Rodriguez, R., Rodriguez Gonzalez, M.A., Vazquez Regueiro, C.: Learning on real robots from experience and simple user feedback. Journal of Physical Agents 7(1) (2013)
5. Carpenter, C.A., Grossberg, S., Rosen, D.B.: Fuzzy art: Fast stable learning and categorization of analog pattern by an adaptive resonance system. Neural Network 4(6), 759–771 (1991)
6. Barto, A.G.: Reinforcement learning: An introduction. MIT press (1998)

Related Tasks Selection
to Multitask Learning Schemes

Andrés Bueno-Crespo[1(✉)], Rosa-María Menchón-Lara[2],
and José Luis Sancho-Gómez[2]

[1] Departamento Informática de Sistemas,
Universidad Católica San Antonio, Murcia, Spain
abueno@ucam.edu
[2] Departamento Tecnologías de la Información y las Comunicaciones,
Universidad Politécnica de Cartagena, Cartagena(Murcia), Spain

Abstract. In Multitask Learning (MTL), a task is learned together with other related tasks, producing a transfer of information between them which can be advantageous for learning of the first one. However, rarely can solve a problem under an MTL scheme since no data are available that satisfying the conditions that need a MTL scheme. This paper presents a method to detect related tasks with the main one that allow to implement a multitask learning scheme. The method use the advantages of the Extreme Learning Machine and selects the secondary tasks without testing/error methodologies that increase the computational complexity.

Keywords: Neural networks · Multitask learning · Extreme learning machine · MultiLayer perceptron

1 Introduction

Multitask learning (MTL) is a machine learning approach to modeling a problem (considered as main task) together with other related problems (considered as secondary tasks), using a shared representation [1, 2, 3, 4]. This makes the existence of an inductive information transfer in most cases improve the generalization of the main task. However, in real world applications, it is difficult to find related problems which is to be learned, and this makes this type of scheme to be abandoned. This drawback raises need for a method that is able to find related tasks within the same domain and even to determine which part of the information can be transferred for a task is really important to the learning of another.

This paper presents a new method to select secondary tasks for MTL architectures, taking advantage of the benefits of Extreme Learning Machine algorithm (ELM) [10], specifically the Optimal Pruned ELM (OP-ELM) [5]. In addition, the method find connection with the main subtasks.

The rest of the paper is organized as follows. Section 2 describes the multilayer perceptron (MLP) design method called ASELM (Architecture Selection based on ELM) and which the proposed method makes use. Section 3 explains what is an MTL scheme. The proposed method is described in Section 4. Section 5 shows the results and finally conclusions and future work shown in Section 6.

J.M. Ferrández Vicente et al. (Eds.): IWINAC 2015, Part II, LNCS 9108, pp. 213–221, 2015.
DOI: 10.1007/978-3-319-18833-1_23

2 Architecture Selection Using Extreme Learning Machine

The Extreme Learning Machine (ELM) is based on the concept that if the MLP input weights are fixed to random values, the MLP can be considered as a linear system and the output weights can be easily obtained using the pseudo-inverse hidden neurons outputs matrix $\mathbf{H}$ for a given training set. Although that idea was previously analyzed in other works [6, 7]. It was Huang who formalized demonstrating the capability that the ELM has to be an universal approxima-tor for a wide range of random computational nodes, and all the hidden node parameters can randomly be generated according to any continuous probability distribution without any prior knowledge [8, 9]. Thus, given a set of N input vectors, an MLP can approximate N cases with zero error, $\sum_{i=1}^{N} \|\mathbf{y}_i - \mathbf{t}_i\| = 0$, being $\mathbf{y}_i$ the output network for the input vector $\mathbf{x}_i$ with target vector $\mathbf{t}_i$. Thus, there exist β_j, $\mathbf{w}_j$ and b_j such that,

$$\mathbf{y}_i = \sum_{j=1}^{M} \beta_j f(\mathbf{w}_j \cdot \mathbf{x}_i + b_j) = \mathbf{t}_i, \quad i = 1, ..., N. \tag{1}$$

where $\beta_j = [\beta_{j1}, \beta_{j2}, ..., \beta_{jm}]^T$ is the weight vector connecting the jth hidden node and the output nodes, $\mathbf{w}_j = [w_{j1}, w_{j2}, ..., w_{jn}]^T$ is the weight vector con-necting the jth hidden node and the input nodes, and b_j is the bias of the jth hidden node.

The previous N equations can be expressed by:

$$\mathbf{HB} = \mathbf{T}, \tag{2}$$

where

$$\mathbf{H}(\mathbf{w}_1, \ldots, \mathbf{w}_M, b_1, \ldots, b_M, \mathbf{x}_1, \ldots, \mathbf{x}_N) =$$

$$= \begin{bmatrix} f(\mathbf{w}_1 \cdot \mathbf{x}_1 + b_1) & \cdots & f(\mathbf{w}_M \cdot \mathbf{x}_1 + b_M) \\ \vdots & \cdots & \vdots \\ f(\mathbf{w}_1 \cdot \mathbf{x}_N + b_1) & \cdots & f(\mathbf{w}_M \cdot \mathbf{x}_N + b_M) \end{bmatrix}_{N \times M} \tag{3}$$

$$\mathbf{B} = \begin{bmatrix} \beta_1^T \\ \vdots \\ \beta_M^T \end{bmatrix}_{M \times m} \quad \text{and} \quad \mathbf{T} = \begin{bmatrix} \mathbf{t}_1^T \\ \vdots \\ \mathbf{t}_N^T \end{bmatrix}_{N \times m} \tag{4}$$

where $\mathbf{H} \in \Re^{N \times M}$ is the matrix of hidden neurons output layer of the MLP, $\mathbf{B} \in \Re^{M \times m}$ is the output weight matrix, and $\mathbf{T} \in \Re^{N \times m}$ is the target matrix of the N training cases. Thus, as $\mathbf{w}_j$ and b_j with $j = 1, ..., N$, are randomly selected, the MLP training is given by the solution of the least square problem of (2), i.e., the optimal output weight layer is $\hat{\mathbf{B}} = \mathbf{H}^\dagger \mathbf{T}$, where $\mathbf{H}^\dagger$ is the Moore-Penrose pseudo-inverse [11].

Finally, ELM for training MLPs can be summarized as follows:

Algorithm 1. Extreme Learning Machine (ELM)

Require: Given a training set $\mathcal{D} = \{(\mathbf{x}_i, \mathbf{t}_i) | \mathbf{x}_i \in \mathbb{R}^n, \mathbf{t}_i \in \mathbb{R}^m, i = 1, \ldots, N\}$, an activation function f and an hidden neuron number M,
1: Assign arbitrary input weights $\mathbf{w}_j$ and biases b_j, $j = 1, \ldots, M$.
2: Compute the hidden layer output matrix $\mathbf{H}$ using (3).
3: Calculate the output weight matrix $\mathbf{B} = \mathbf{H}^\dagger \mathbf{T}$, where $\mathbf{B}$ and $\mathbf{T}$ are both defined in (4).

ELM provides a fast and efficient MLP training [10], but it needs to fix the number of hidden neurons. In order to avoid the exhaustive search for the optimal value of M, several pruned methods have been proposed [13, 14, 15, 18, 16, 17], among them, the most commonly used is is the ELM Optimal Pruned (OP-ELM) [17]. The OP-ELM sets a very high initial number of hidden neurons ($M \gg N$) and, by using Least Angle Regression algorithm (LARS) [19], sorts the neurons according to their importance to solve the problem (2). The pruning of neurons is done using Leave-One-Out Cross-Validation (LOO-CV) by choosing that combination of neurons (which have been previously sorted by the LARS algorithm) that provides lower LOO error. The LOO-CV error is efficiently computed using the Allen's formula [17].

ASELM ("Architecture Selection using Extreme Learning Machine") [20] use OP-ELM, once the initial MLP architecture is defined, the OP-ELM optimally discards those hidden neurons whose combination of input variables is not relevant to the target task. Note that, because of the binary value of the input weights, the selection of hidden nodes implies also the selection of the relevant connections between the input and hidden layers. Thus, only input connections corresponding to selected hidden neurons and with weights values equal to 1 will be part of the final architecture. A summary of the ASELM algorithm is shown next.

Algorithm 2. Architecture Selection ELM (ASELM)

Require: Given a training set $\mathcal{D} = \{(\mathbf{x}_i, \mathbf{t}_i) | \mathbf{x}_i \in \mathbb{R}^n, \mathbf{t}_i \in \mathbb{R}^m, i = 1, \ldots, N\}$, activation function f, an hidden neuron number $2^n - 1$, where n is the number of input features, proceed as follows:
1: The weights of the input layer are initialized with binary values by considering all possible combinations of inputs. The case of all weights set to zero is discarded.
2: MLP network is trained by the OP-ELM and, then, useless hidden neurons are discarded according to the ranking given by LARS and LOO-CV procedure.
3: The final MLP architecture is given by the selected hidden neurons with its corresponding input(s) weight(s) equal to one.

---

3   Multitask Learning Architecture

The MTL architecture for a neural network is similar to the classical scheme STL (Single Task Learning), except MTL scheme has one output for each task to learn, while in the STL scheme has a separate network for each task (Figure 1).

Thus, when we speak about MTL, we are referring to a type of learning where a main task and other tasks (that are considered secondary), are learning all at once in order to help the learning of the main task.

In a MTL scheme, there is a common part shared by all tasks and a specific part, own of each task. The common part is formed by the weights connections from the input features to the hidden layer, allowing common internal representation for all tasks, [21]. Thanks to this internal representation, learning can be transferred from one task to another, [3]. The specific part, formed by the weights that connect the hidden layer to the output layer, specifically allow to model each task from the common representation. The main problem with this type of learning is to find tasks related to the main one. Even finding them, it may be that we do not know the kind of relationship they have, because it can be a positive or negative influence to learn the main task.

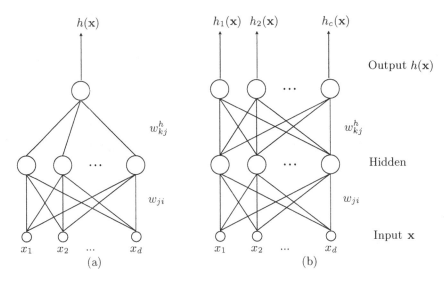

Fig. 1. Different learning schemes. In (a), an scheme is shown where a single task is learned alone (STL), while in (b), a set of tasks are learned simultaneously (MTL), there a common part (from the input to the hidden layer) and other specific (from the hidden layer to the output) of each task.

4 Proposed Method

ASELM [20] method allows to find an architecture that facilitates the learning of a task, it is due to the method eliminates neurons and connections those hinder learning or are irrelevant. The ASELM favors the positive inductive bias because it removes those elements that hinder the neural network learning.

This paper proposes the use of the ASELM methodology to select relevant secondary tasks. To achieve this, the targets of secondary tasks will be used as new input features (removing them from the outputs of the classic MTL scheme), therefore, there is only a single output corresponding to the main task. This is going to allow that ASELM provide information about common connections with respect to the main task. The selection of relevant secondary tasks is performed since they are now part of the input vector.

This idea of exchanging outputs for inputs, is not new. Caruana proposed that some inputs may work better as outputs, as a new secondary tasks, [3]. This is very interesting in the case of missing data when classifying a new pattern features.

The following section explains in detail how these related tasks are selected.

5 Experiments

In order to show the goodness of the method to select the relevant secondary tasks, it has been used the "Logic Domain" dataset. This dataset is a toy problem specially designed for multitask learning, [22]. In this problem, the targets are represented by the combination of four real variables, considering the first task as the main task, while the others are considered as secondary ones.

Table 1 shows the logical expression of each tasks. The main task consists of the first four variables and although secondary tasks share with the main task, one or more variables, is the second secondary task that really shares a common logic subexpression ($x_3 > 0.5 \lor x_4 > 0.5$).

Table 1. Description of the "Logic Domain" tasks. Each task is a logical combination of four input features.

Task	Logical Expression
T_P	$(x_1 > 0.5 \lor x_2 > 0.5) \land (x_3 > 0.5 \lor x_4 > 0.5)$
T_{Sec_1}	$(x_2 > 0.5 \lor x_3 > 0.5) \land (x_4 > 0.5 \lor x_5 > 0.5)$
T_{Sec_2}	$(x_3 > 0.5 \lor x_4 > 0.5) \land (x_5 > 0.5 \lor x_6 > 0.5)$
T_{Sec_3}	$(x_4 > 0.5 \lor x_5 > 0.5) \land (x_6 > 0.5 \lor x_7 > 0.5)$

Figure 2 shows the influence of the T_{Sec_2} as input feature to learn T_p with the inputs x_3 and x_4, where it can be observed the T_{Sec_2} best separates the samples than other secondary tasks. The architecture used for ASELM can be observed

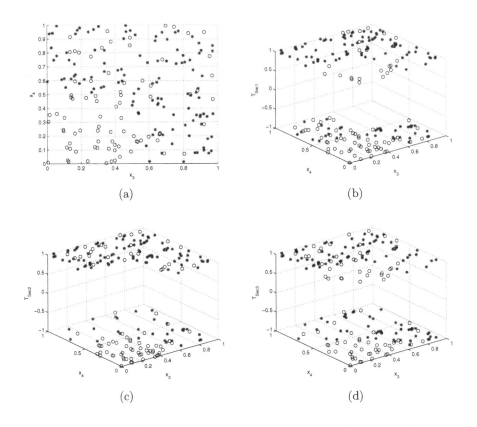

Fig. 2. Conjunto de datos "Logic Domain". Representation of the inputs features x_3 and x_4 (a). In the other subfigures were added secondary tasks (T_{Sec_1} (b), T_{Sec_2} (c) and T_{Sec_3} (d)) respectively, increasing the dimensionality. It can be observed the T_{Sec2} is the one that will most help T_p in their learning.

in Figure 3, where the only target is the main task (T_P), while the others three secondary tasks are used as extra inputs.

Initially, the neural network architecture uses 1023 hidden units. Once this model is trained with ASELM method, the result is quite significant, the model selects only two neurons, with hidden weights vector $w_1 = [0\ 0\ 1\ 1\ 0\ 0\ 0\ 0\ 1\ 0]$ and $w_2 = [1\ 1\ 0\ 0\ 0\ 0\ 0\ 0\ 0\ 0]$. It can be observed as the first selected hidden neuron is composed of input features x_3 and x_4, and the second secondary task (Figure 4). This means that the T_{Sec_2} is influencing in the learning of the T_p, through the neuron that learns the input features x_3 and x_4, which was the expected result (see Table 1).The second selected hidden neuron is only composed by the feature inputs x_1 and x_2, without any connection of secondary tasks.

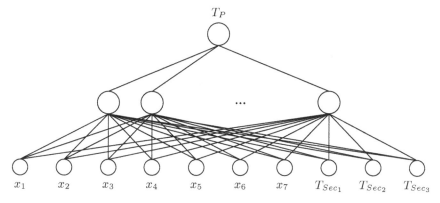

Fig. 3. Logic Domain. Scheme to learn the main task using secondary tasks as inputs.

Inputs x_1 .. x_7	Secondary Tasks		
0 0 1 1 0 0 0	0 1 0	First neuron	Hidden Layer
1 1 0 0 0 0 0	0 0 0	Second neuron	

Fig. 4. Logic Domain. Neurons selected by ASELM. The first neuron has two connections corresponding to the input features x_3 and x_4. The second neuron is represented by the input features x_1 and x_2.

6   Conclusions and Future Work

This paper presents a method to select tasks to be used in a MTL scheme providing information about weights connection, hidden nodes, input features and secondary tasks most helpful for learning of the main task. This method is based on the ASELM algorithm [20], which proves to be an efficient method and single solution for the complete design of a MLP (input features, weights connections and hidden nodes). Using secondary tasks as input features, the ASELM select the secondary tasks more relevant for learning of the main one. Thus, one of the main drawback of multitask learning is eliminated, that is, the negative influence of unrelated tasks. Is worth highlighting that the method provides a single solution. In this way, does not depend on random initializations or trial/error methods. In the experiments section, has been observed as efficiently, the method makes a selection of relevant secondary tasks, as well as providing additional information about the inputs features relation with the selected secondary tasks. As future work is to extend the method to other models of learning, such as Radial Basis Functions (RBF) and to solve regression problems since the ASELM is optimized for classification due to the elimination of bias [24].

References

[1]   Caruana, R.: Learning many related tasks at the same time with backpropagation. In: Advanced in Neural Information Processing Systems, pp. 656–664 (1995)

[2]   Baxter, J.: The evolution of learning algorithms for artificial neural networks, Complex Systems. In: Complex Systems. IOS Press (1995)

[3]   Caruana, R.: Multitask Learning. Phd Thesis, School of Computer Science, Carnegie Mellon University, Pittsburg, PA (1997)

[4]   Silver, D.L., Mercer, R.E.: Selective functional transfer: Inductive bias from related tasks. In: Proceedings of the IASTED International Conference on Artificial Intelligence and Soft Computing, pp. 182–191 (2001)

[5]   Miche, Y., Sorjamaa, A., Bas, P., Simula, O., Jutten, C., Lendasse, A.: OP-ELM: optimally pruned extreme learning machine. IEEE Transactions on Neural Networks 21(1), 158–162 (2010)

[6]   Pao, Y.H., Park, G.H., Sobajic, D.J.: Learning and Generalization Characteristics of the Random Vector Functional-Link Net. Neurocomputing 6(2), 163–180 (1994)

[7]   Igelnik, B., Pao, Y.H.: Stochastic Choice of Basis Functions in Adaptive Function Approximation and the Functional-Link Net. IEEE Transactions on Neural Networks 8(2), 452–454 (1997)

[8]   Huang, G.B., Chen, L.: Convex incremental Extreme Learning Machine. Neurocomputing 70, 3056–3062 (2007)

[9]   Huang, G.B., Wang, D.H., Lan, Y.: Extreme Learning Machines: A survey. International Journal of Machine Leaning and Cybernetics 2(2), 107–122 (2011)

[10] Huang, G.B., Zhu, Q., Siew, C.K.: Extreme learning machine: Theory and applications. Neurocomputing 70(1-3), 489–501 (2006)

[11] Serre, D.: Matrices: Theory and Applications. Springer (2002)

[12] Rao, C.R., Mitra, S.K.: Generalized Inverse of Matrices and its Applications. Wiley (1971)

[13] Rong, H.J., Ong, Y.S., Tan, A.H., Zhu, Z.: A fast pruned-extreme learning machine for classification problem. Neurocomputing 72(1-3), 359–366 (2008)

[14] Miche, Y., Bas, P., Jutten, C., Simula, O., Lendasse, A.: A Methodology for Building Regression Models using Extreme Learning Machine: OP-ELM. In: Proceedings of the European Symposium on Artificial Neural Networks (ESANN), pp. 247–252 (2008)

[15] Miche, Y., Sorjamaa, A., Lendasse, A.: OP-ELM: Theory, Experiments and a Toolbox. In: Kůrková, V., Neruda, R., Koutník, J. (eds.) ICANN 2008, Part I. LNCS, vol. 5163, pp. 145–154. Springer, Heidelberg (2008)

[16] Miche, Y., Lendasse, A.: A Faster Model Selection Criterion for OP-ELM and OP-KNN: Hannan-Quinn Criterion. In: Proceeding of the European Symposium on Artificial Neural Networks (ESANN), pp. 177–182 (2009)

[17] Miche, Y., Sorjamaa, A., Bas, P., Simula, O., Jutten, C., Lendasse, A.: OP-ELM: Optimally Pruned Extreme Learning Machine. IEEE Transactions on Neural Networks 21(1), 158–162 (2009)

[18] Mateo, F., Lendasse, A.: A variable selection approach based on the Delta Test for Extreme Learning Machine models. In: Proceedings of the European Symposium on Time Series Prediction (ESTP), pp. 57–66 (2008)

[19] Similä, T., Tikka, J.: Multiresponse Sparse Regression with Application to Multidimensional Scaling. In: Duch, W., Kacprzyk, J., Oja, E., Zadrożny, S. (eds.) ICANN 2005. LNCS, vol. 3697, pp. 97–102. Springer, Heidelberg (2005)

[20] Bueno-Crespo, A., Garca-Laencina, P.J., Sancho-Gmez, J.L.: Neural architecture design based on Extreme Learning Machine. Neural Networks 48, 19–24 (2013)

[21] Caruana, R.: Multitask Connectionist Learning. In: Proceedings of the 1993 Connectionist Models Summer School (1993)

[22] McCracken, P.J.: Selective Representational Transfer Using Stochastic Noise. Honour Thesis, Jodrey School of Computer Science, Acadia University, Wolfville, Nova Scotia, Canada (2003)

[23] Pal, S.K., Majumder, D.D.: Fuzzy sets and decision making approaches in vowel and speaker recognition. IEEE Transactions on Systems, Man, and Cybernetics 7, 625–629 (1977)

[24] Huang, G.B., Ding, X., Zhou, H.: Optimization method based Extreme Learning Machine for classification. Neurocomputing 74, 155–163 (2010)

[25] van Heeswijk, M., Miche, Y., Oja, E., Lendasse, A.: Gpu accelerated and parallelized ELM ensembles for large-scale regression. Neurocomputing 74(16), 2430–2437 (2011)

[26] Tapson, J., van Schaik, A.: Learning the pseudoinverse solution to network weights. Neural Networks (March 13, 2013)

Towards Robot Localization Using Bluetooth Low Energy Beacons RSSI Measures

J.M. Cuadra-Troncoso[1(✉)], A. Rivas-Casado[2], J.R. Álvarez-Sánchez[1],
F. de la Paz López[1], and D. Obregón-Castellanos[1]

Departamento de Inteligencia Artificial - UNED - Madrid, Spain
Guapu Technologies, Madrid, Spain
jmcuadra@dia.uned.es

Abstract. This article presents a preliminary study in order to explore the possibilities for robot localization using measured received signal strength indicator (RSSI) from Bluetooth low energy (BLE) beacons. BLE is a new brand technology focused on information transmission using very low energy consumption. It is being included in mobile devices from year 2011, nowadays almost every new mobile phone is shipped with this technology. Robot localization using particles filter has been developed in recent years using wireless technologies with a significant success. BLE beacons measures are rather noisier than measures from similar wireless devices. In this work we make an initial model of BLE measures and their noise. The model is used to generate data to be processed by a particle filter designed for localization using only ultra-wide band (UWB) beacons ranges. Data are generated with different noise level in order to explore localization errors behavior, these levels cover real noise levels founded in RSSI measure characterization.

Keywords: Robot localization · Mobile devices localization · Bluetooth low energy · Particle filter · RSSI based trilateration

1 Introduction

In recent years a number of localization or positioning techniques based on wireless technology has been developed, see cites for surveys [1,2,3,4,5]. One of the newest wireless technologies is BLE [6], aka Bluetooth Smart, and it is included in the Bluetooth protocol starting with version 4.0. BLE is focused on information transmission using very low energy consumption. Production and maintenance costs of BLE devices are low [4]. Commercial mobile devices started to be shipped with BLE from year 2011, nowadays almost every new smartphone, tablet and many wearable devices integrate this technology. Also cheap BLE USB modules enable this technology for personal computers and kits as Raspberry Pi or Arduino. Robots or vehicles localization has been the main field of study in localization techniques, but with the arrival of phenomena as Internet of People and Internet of Things, the field of study has acquired a new interest and research on localization using BLE is necessary.

© Springer International Publishing Switzerland 2015
J.M. Ferrández Vicente et al. (Eds.): IWINAC 2015, Part II, LNCS 9108, pp. 222–231, 2015.
DOI: 10.1007/978-3-319-18833-1_24

The simplest BLE device is the BLE beacon or antenna, see figure 1, it is available at market at prices ranging from 5\$ to 30\$, depending on configuration. BLE beacons are devices that basically only emits a short identification message periodically. Measuring some signal features should be possible to estimate distance between a receptor and a BLE beacon.

Fig. 1. BLE beacon. The simple circuitry is visible. It is powered by a usual 3 volts clock battery. The beacon diameter is about 5 cm.

The distance estimation could be done using the following characteristics: RSSI [7,8,9], Time of Arrival (ToA) [10,11], the Time Difference of Arrival (TDoA) [12,13], the Angle of Arrival (AoA) [14,15], and the Time of Flight (ToF) [16,17]. The only procedures available for BLE beacons are those based on RSSI. Time measure needs external synchronization or at least beacons have to be receptors, but BLE ones are only emitters. Angle measure is used to complement distance measures for robot heading estimation. It usually uses another sensors as compass o gyroscope available in robots and mobile devices, but in this preliminary study we have choose not to take AoA into account, and restrict it to position estimation.

Signal reception, specially in indoors environments, can be affected for obstacles between or surrounding emitter and receptor. If emitter and receptor are in a non-LOS (NLOS) situation, signal reception is affected by obstacles wave absorption. Surroundings obstacles cause multipath effects due reflections, affecting signal reception also. We will focus this work in the LOS situation. In section 2 a statistical study on RSSI is done using an Android smartphone as receptor and a commercial BLE beacon as emitter. An approximate characterization, under LOS situation, of RSSI measure and its noise is given, the relation between RSSI and reception distance is also approximately modeled.

Particle filters (PF) [18] are specially suited for high non linear environments and could solve global position problem [19]. Building with people are high non linear environments and global positioning is useful to track vehicles or humans carrying mobile devices. UWB beacons ranges has been successfully applied to robot localization using trilateration. Results from section 2 show that BLE measure noise is considerably higher than UWB noise, in section 3 we use a UWB range only robot localization procedure using PF [20] to test its performance depending on BLE RSSI noise levels. The paper ends with conclusions and possible future work in section 4.

2   BLE Beacons Measures Modeling

In this section we study the characteristics of BLE beacons signal strength and its relations with distance between receptor and beacon. The study in made for a particular pair of Android smartphone and commercial BLE beacon, other pairs can give different RSSI measures in the same situation, as we have tested. Results are comparable to those given in [21].

In all experiments in this section beacons broadcasting rate was 10Hz with power set to +4dBm, the highest values admitted by these beacons, in order to obtain more reliable measures in a short time (1.1s.). An Android application for measuring RSSI was implemented. This application uses an API [22] that filters measures and gives a 10% trimmed mean of the RSSI values sampled along a certain time period, we fix that period to 1.1 seconds.

2.1   Measuring RSSI at 1 Meter

We are going to use r as a symbol for RSSI and r_0 for RSSI at 1 m, these two quantities appear en eq. 1. In this experiment phone and beacon were in a medium sized room in LOS situation, multipath effects could appear due to walls proximity, no humans around. Three groups of 100 measures each one were taking. For the first group the angle between phone and beacon was approximately 0°, for the second one the angle was 90° and for the third one it was 180°. The whole set of measures was taking again interchanging phone and beacon positions, in order to take into account asymmetric reflections, so we got 600 measures at end. Means, medians and standard deviation of the six groups are given in table 1. Differences between means depending on angle are apparently significant, this indicates that an AoA procedure could help to distance estimation accuracy. In the other hand, differences between results corresponding to position interchange do not look so different, thus we will mix measures from both positions for our next statistic.

Table 1. RSSI measures at 1 meter. Statistics for r_0 depending on phone orientation respect to beacon. The second set of data is obtained interchanging phone and beacon positions.

	mean		median		stdev.	
0°	-39.23	-38.85	-40	-39	4.406	5.227
90°	-43.08	-42.12	-43	-42	2.561	3.112
180°	-32.60	-32.03	-35	-32	7.398	5.008

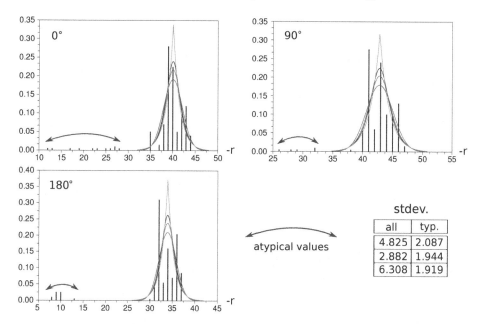

Fig. 2. RSSI measures at 1 meter. r is always lesser than zero, we have changed sign for convenience. Frequencies bar graph for different phone orientations respect to beacon. Some few atypical values could appear far from the main group of measures. The curves are density probability functions for several distributions, see text. The table show standard deviations for all values and for typical ones.

Figure 2 shows frequencies bar graph for different phone orientations respect to beacon. A small number of atypical values appears far from the main group of measures. These values increase sample variability, see table in figure, and can introduce larger errors in location procedures. In the experiment atypical values appear with a frequency of 3%. They should be filtered to obtain a more reliable location.

The figure also shows density probability functions for several distributions with increasing kurtosis. These distributions are, in ascending kurtosis order, Gaussian, logistic, hyperbolic secant and Laplace. Mean and standard deviation of distributions are estimated from samples after removing atypical values. Measured typical data seems to show a high kurtosis.

2.2 Distance Estimation

In order to obtain distance estimations from RSSI measures in LOS situation, we did two experiments measuring RSSI from known distances ranging from 0.2 m. to 20 m. The first experiment was done outdoors far away from other Bluetooth or Wi-Fi devices. The second one was done in a long corridor in our faculty, a building with problems for waves transmission. In the first experiment 20 RSSI measures were taken for each distance, in the second one 10 measures for each distance.

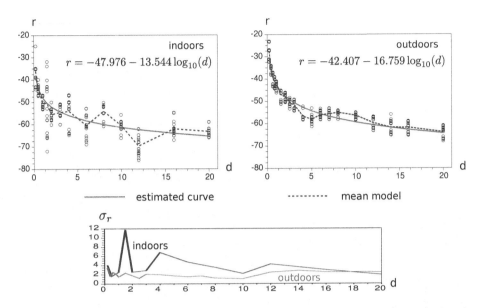

Fig. 3. Results for distance estimation. Upper figures show raw data (circles), fitted model and mean model (line passing through means of r for each d) graphs for experiments in indoors and outdoors situation. Estimated equations are shown. r is measured in dBm and d is measured in meters. Lower figure shows r standard deviation, σ_r, as function of d for both situations.

Using non linear regression we fit data to one usual expression for RF power loss as function of distance, as in [21]:

$$r = r_0 - 10n \log_{10}(d) \tag{1}$$

being d the distance and n the propagation constant which depends on environmental features.

Atypical measures were not discarded in this initial experiments. Given the small atypical measures frequency, experiments with larger number of samples are needed. Results of both experiments are summarized in fig. 3. Upper figures show raw data as circles and it is visible a larger dispersion of measures in indoor situation. This can also be seen in lower figure, it shows r standard deviation, σ_r, as function of d.

Upper figures show also the estimated curve and the mean model, a line passing through means of r for each d, the mean model has the best fit for a set of data. The fitted curve approximately represents the measures and we going to use outdoors estimation to simulate measures for experiments in section 3.

3   Range Only Robot Localization

In [20] a Rao-Blackwellised PF [23] is used for robot location using only ranges. They use UWB beacons with known position and TDoA for ranging. An implementation for simulations of localization using the PF proposed in [20] was given by the authors and it is available as a MRPT tool [24], this tool has been used for experiments in this section. Authors have provided another MRPT tool for beacons ranging simulation.

UWB and TDoA provide a much higher accuracy than BLE and RSSI. In these experiments we test several values for RSSI standard deviation and compute errors in robot pose estimation: distance between ground truth pose and estimated one.

The ranging simulation tool uses ranges directly and adds a Gaussian noise to the true measure with 0 mean and a selectable constant standard deviation. As we want to use RSSI, we get the true distance and convert it to RSSI using eq. 1, then we add to RSSI noise of 0 mean and a selectable constant standard deviation. The new RSSI value is converted into distance using inverse of eq. 1. RSSI standard deviation is more or less constant for outdoors situation, see fig. 3, but when RSSI is converted to distance, distance noise standard deviation increases with true distance, apparently in a linear way, and gets really large values, see table 2. Atypical values were not generated for these experiments, we did some of these experiment generating a %3 of atypical values finding an error increment especially large for small values of RSSI standard deviation, for larger ones the increment was roughly less than 50%.

Table 2. RSSI standard deviations, σ_r in dBm, used in experiments and the resultant intervals for range noise standard deviations, $\sigma_{\triangle d}$ in meters. True ranges from beacons were between 1.4 and 8.2 meters. Intervals endpoints were computed over the 100 smallest true ranges and over the 100 largest true ranges. Intervals endpoints are for Laplacian noise, in the case of Gaussian noise they are bigger, specially the upper limits.

σ_r	0.2	0.5	1	2	3	5
$\sigma_{\triangle d}$	0.02-0.13	0.05-0.32	0.11-0.59	0.23-1.16	0.34-1.58	0.7-2.87

Two maps were used in experiments. The first map had 4 beacon situated at the corners of an 8m. square and the second one had 9 beacons forming a 4m. cell grid over the 8m. square. The robot traced a 4.5 m. diameter circle inside the square and it runs for 100m. Gaussian, lesser kurtosis, and Laplacian, higher kurtosis, noise was added to RSSI with 0 mean and standard deviation ranging according to table 2. In experiments described in subsection 2.2, the mean for σ_r in outdoors situation is approximately 2 and in the indoors one it is approximately 5, so range used for σ_r covers real cases.

For every standard deviation in table 2, four robot runs were generated mixing maps and noises distributions. A set of five PF runs were computed for every

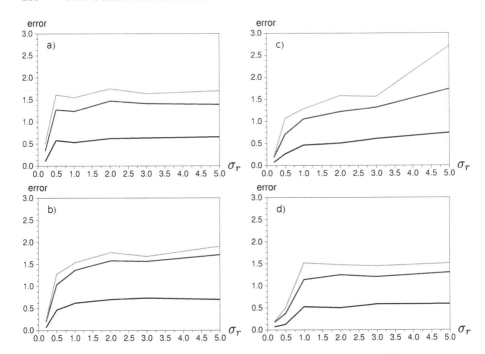

Fig. 4. Positioning error in meters depending on RSSI standard deviation in dBm. Lower curves are errors mean, medium curves are mean of errors maximums and upper ones are maximums of error maximums in 5 PF runs. In a) and b) we used 4 beacons situated at the corners of an 8m. square, in a) Gaussian noise was added to RSSI, in b) Laplacian one. In c) and d) we used 9 beacons forming a 4m. cell grid over the 8m. square, noise is Gaussian in c) and Laplacian in d).

robot runs. The results are summarized in fig. 4. For each PF runs we have computed the mean and the maximum for the position error. Curves in fig. 4 represent means of errors in PF runs sets, means of maximums of errors in PF runs sets and maximums of maximums of errors in PF runs sets.

The most remarkable fact is the stabilization, or at least slow increment, of position error as σ_r increases. The mean tends approximately to 0.6 in all cases, and maximums mean to 1.5. This mean error could be acceptable for a lot of applications.

The default value for range standard deviation given in the MRPT tool, used for UWB beacons in [20], is 0.01m. and ranges for $\sigma_{\triangle d}$ in table 2 are really larger especially for larger of σ_r, so the PF if working at a really high noise level.

Other facts, possible differences between cell grid size or noise distributions, are not so noticeable and larger samples in a variety of situations will be necessary to take a stance.

In fig. 5 the positioning error evolution for PF is compared with odometry positioning error evolution. Curves show the most common behavior observed in several experiments. PF error keeps its values bounded along with path length,

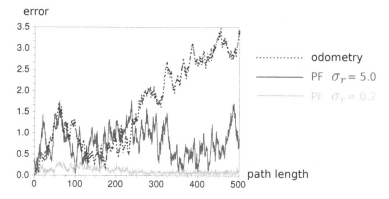

Fig. 5. Positioning error evolution along with path length. Typical behavior of instantaneous positioning error, in meters, for a circular robot path 500 m. long using four beacons. PF error for $\sigma_r = 0.2$dBm and $\sigma_r = 5.0$dBm compared with odometry error.

this is a known PF behavior, while odometry error grows without apparent limit, this is also a known fact. Even with $\sigma_r = 5.0$, the maximum value we used in experiments, the PF error is lower than odometry one when the path is long enough.

For heading error, not shown in figures, we only comment that it show in experiments a behavior similar to heading odometry error plus a constant that not seems to depend on σ_r. A complementary procedure is needed for heading estimation.

4   Conclusions and Future Work

In this paper we have done a initial study in order to characterize BLE beacons RSSI as it is received by a mobile phone in LOS situation. Mainly two features has been considered: RSSI noise and RSSI dependency on distance between emitter and receptor. Orientation between emitter and receptor seems to have influence on measures. Noise study has shown the possible existence of atypical values that increase localization error. The distribution of typical values is not clear and it seems to show high kurtosis. Special interest has to give an approximate estimate RSSI noise standard deviation, as noise level measure, and to convert it in ranges noise standard deviation, in order to appreciate how big noise is. For real values of RSSI noise standard deviation we measure, distance noise standard deviation can be of the order of meters, it increases with the measured range. RSSI dependency on distance can be approximately fitted to a known relation between wave power loss and distance.

After we obtained a RSSI measures characterization. We used it to generate simulated RSSI measures from a robot path with several noise levels. A Rao-Blackwellised PF, previously designed to be used with UWB beacons, has been applied to generated data in order to localize a robot using trilateration based on BLE beacons RSSI measures converted into distance. Distance measure using UWB beacons gives errors much smaller than using BLE ones, but even in these adverse circumstances, the PF could maintain mean position error in values suitable for some types of application, as human tracking.

Given the novelty of these technology a lot of research lines can be open. Some are roughed out in this work: to improve RSSI measures characterization and filtering and their relation with distances model, to test other PF algorithm in order to try to design a custom procedure for noise reduction and to add an AoA procedure for heading estimation. Other possible research lines are: to repeat this work under the NLOS situation in order to study more realistic cases, to design procedures for devices calibration, remember that this study is made for a particular phone and a particular beacon, and to test everything in real environments localizing real receptors carried by robots or people.

References

1. Liu, H., Darabi, H., Banerjee, P., Liu, J.: Survey of Wireless Indoor Positioning Techniques and Systems. IEEE Transactions on Systems, Man, and Cybernetics, Part C: Applications and Reviews 37(6), 1067–1080 (2007), doi:10.1109/TSMCC.2007.905750
2. Sana: A Survey of Indoor Localization Techniques. IOSR Journal of Electrical and Electronics Engineering 6(3), 69–76 (2013)
3. Pirzada, N., Yunus Nayan, M., Subhan, F., Fadzil Hassan, M., Amir Khan, M.: Comparative Analysis of Active and Passive Indoor Localization Systems. AASRI Procedia 5, 92–97 (2013), 2013 AASRI Conference on Parallel and Distributed Computing and Systems
4. Stojanović, D., Stojanović, N.: Indoor Localization and Tracking: Methods, Technologies and Research Challenges. Facta Universitatis, Series: Automatic Control and Robotics 13(1), 57–72 (2014)
5. Mainetti, L., Patrono, L., Sergi, I.: A survey on indoor positioning systems. In: 2014 22nd International Conference on Software, Telecommunications and Computer Networks (SoftCOM), pp. 111–120 (September 2014)
6. Heydon, R.: Bluetooth Low Energy - The Developer's HandBook. Prentice Hall (October 2012)
7. Papamanthou, C., Preparata, F.P., Tamassia, R.: Algorithms for Location Estimation Based on RSSI Sampling. In: Fekete, S.P. (ed.) ALGOSENSORS 2008. LNCS, vol. 5389, pp. 72–86. Springer, Heidelberg (2008)
8. Zanca, G., Zorzi, F., Zanella, A., Zorzi, M.: Experimental Comparison of RSSI-based Localization Algorithms for Indoor Wireless Sensor Networks. In: Proceedings of the Workshop on Real-world Wireless Sensor Networks, REALWSN, pp. 1–5. ACM, New York (2008)
9. Oliveira, L., Li, H., Almeida, L., Abrudan, T.E.: RSSI-based relative localisation for mobile robots. Ad Hoc Networks 13(pt.B(0)), 321–335 (2014)
10. Guvenc, I., Chong, C.-C.: A Survey on TOA Based Wireless Localization and NLOS Mitigation Techniques. IEEE Communications Surveys Tutorials 11(3), 107–124 (2009)
11. Shen, J., Molisch, A.F., Salmi, J.: Accurate Passive Location Estimation Using TOA Measurements. IEEE Transactions on Wireless Communications 11(6), 2182–2192 (2012)

12. Gustafsson, F., Gunnarsson, F.: Positioning using time-difference of arrival measurements. In: Proceedings of 2003 IEEE International Conference on Acoustics, Speech, and Signal Processing (ICASSP 2003), vol. 6, p. VI-553-6 (April 2003)
13. McGuire, M., Plataniotis, K.N., Venetsanopoulos, A.N.: Location of mobile terminals using time measurements and survey points. IEEE Transactions on Vehicular Technology 52(4), 999–1011 (2003)
14. Niculescu, D., Nath, B.: Ad hoc positioning system (APS) using AOA. In: Twenty-Second Annual Joint Conference of the IEEE Computer and Communications, INFOCOM 2003, vol. 3, pp. 1734–1743. IEEE Societies (March 2003)
15. Cong, L., Zhuang, W.: Nonline-of-sight error mitigation in mobile location. In: Twenty-third AnnualJoint Conference of the IEEE Computer and Communications Societies, INFOCOM 2004, vol. 1, p. 659 (March 2004)
16. Muthukrishnan, K., Koprinkov, G.T., Meratnia, N., Lijding, M.E.M.: Using time-of-flight for WLAN localization: feasibility study. Technical Report TR-CTI, Enschede (June 2006)
17. Chiang, J.T., Haas, J.J., Hu, Y.-C.: Secure and Precise Location Verification Using Distance Bounding and Simultaneous Multilateration. In: Proceedings of the Second ACM Conference on Wireless Network Security, WiSec 2009, pp. 181–192. ACM, New York (2009)
18. Thrun, S., Burgard, W., Fox, D.: Probabilistic robotics. MIT Press (2005)
19. Marchetti, L., Grisetti, G., Iocchi, L.: A comparative analysis of particle filter based localization methods. In: Lakemeyer, G., Sklar, E., Sorrenti, D.G., Takahashi, T. (eds.) RoboCup 2006: Robot Soccer World Cup X. LNCS (LNAI), vol. 4434, pp. 442–449. Springer, Heidelberg (2007)
20. González, J., Blanco, J.L., Galindo, C., Ortiz-de Galisteo, A., Fernández-Madrigal, J.A., Moreno, F.A., Martínez, J.L.: Mobile robot localization based on Ultra-Wide-Band ranging: A particle filter approach. Robotics and Autonomous Systems 57(5), 496–507 (2009)
21. Dahlgren, E., Mahmood, H.: Evaluation of indoor positioning based on BluetoothR Smart technology. Master's thesis (2014)
22. Android Beacon Library, http://altbeacon.github.io/android-beacon-library/
23. Doucet, A., de Freitas, N., Murphy, K.P., Russell, S.J.: Rao-Blackwellised Particle Filtering for Dynamic Bayesian Networks. In: Proceedings of the 16th Conference on Uncertainty in Artificial Intelligence, UAI 2000, pp. 176–183. Morgan Kaufmann Publishers Inc., San Francisco (2000)
24. MRPT Application: ro-localization, http://www.mrpt.org/list-of-mrpt-apps/application-ro-localization/

Development of a Web Platform for On-line Robotics Learning Workshops Using Real Robots

R. García-Misis[1], J.M. Cuadra-Troncoso[1(✉)], F. de la Paz López[1],
and J.R. Álvarez-Sánchez[1]

Departmento de Inteligencia Artificial, UNED, Madrid, Spain
jmcuadra@dia.uned.es

Abstract. In this work a tool for on-line robotics learning workshops is introduced, with this tool students can test from home their robotics subjects exercises on real robots situated in a remote laboratory. It is composed by a set of e-puck robots, a server application to manage user requests, a server application for video recording and transmission and a web application for user interaction with robot and video. Students can see in real time robot movements in their own computer and send commands to the robot.

Keywords: Distance learning · Virtual learning environments · E-learning · Robotics

1 Introduction

Distance learning [1,2,3] concept and its implementation have evolved widely. Before Internet arriving the students made homework and send it to the teachers via physical mail. In last years we have seen a genuine revolution in information technologies, fortunately distance education has not been indifferent it. At the beginning with the use of e-mail as a faster communication mean, saving time and trips. And recently web applications have enabled the use of cooperative environments with shared documents, now students deliver homework to the platform and have a fluent communication with their teachers and classmates. In addition virtual libraries and many other extra resources have proliferate in order to facilitate students work.

However in the current distance learning model, and even in the pure on-line or virtual one [4,5,6], there exists a set of activities being impossible to carry out at home, as for example laboratory tasks, and thus, forcing students to travel to the faculty. This fact complicates distance learning and it distorts the concept of on-line learning.

A step further is needed: to transform, as much as possible, distance learning in virtual learning. This is not a new idea and it has been discussed in many publications [7,8,9]. In this paper we propose the evolution towards a model of virtual robotics workshops, so that students are able to manage a robot situated in a properly equipped room from anywhere in the world, using only a computer with Internet connection having a common web navigator. Our wish is to facilitate students skills acquisition and, thus, to improve their scores[10].

© Springer International Publishing Switzerland 2015
J.M. Ferrández Vicente et al. (Eds.): IWINAC 2015, Part II, LNCS 9108, pp. 232–239, 2015.
DOI: 10.1007/978-3-319-18833-1_25

In our web platform development we set following goals:

- Build a complete and functional platform. This is the main goal: students should carry out their activities as they really were in the same room as robot is and they could control every activity aspect.
- Avoid specific support software installation in students computers. Installation procedures could take long and be complicate for robotics novices, they only should to be worried about to start working with robots.
- Minimize platform installation costs: using open license software and low cost devices.
- Platform robustness: changes on every component, robot, camera, etc., at any time must not affect the whole system structure.
- Platform reusability: it has to be capable to allow the realization of different types of activities, other than those for which it was initially designed.

We have split our solution to the posed problem in five software components:

- Robot controller: software for interaction with robot, at the moment of writing these lines only for e-puck robot.
- Video server: an application being able to work with every generic camera and it has to be capable of video transmission and recording in real time.
- Web server: this component has to allow students to manage robot while they visualize robot movements.
- Database: in order to maintain web server integrity and coherence a database scheme has been created, it uses several parametrization tables.
- Integration with Cybersim simulator: Cybersim is an application in development by our team. This work is focused in e-puck integration in Cybersim. Students are able to control robot and see its movements in real time, as they do using the web server.

Each solution component has been designed to be independent from the others. They can run in the same server or in different computers. They has been designed for being dispensable and they can be replaced by new implementations without to change anything of the others components.

The paper is structured as follows: section 2 describes features of physical devices used, section 3 describes the structure of the software components and in section 4 conclusions and possible extensions of this work are commented.

The relation between software components, physical devices and students is shown in figure 1.

2   Devices

Other than normal computers, we use two specific physical devices: the e-puck robot and the video camera.

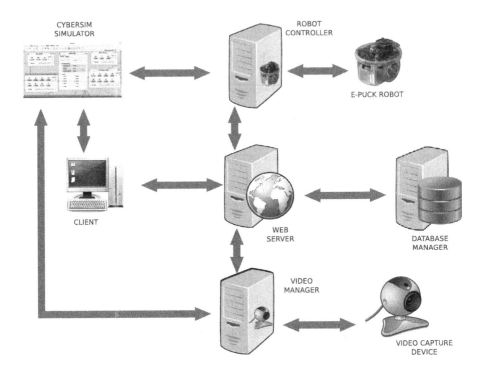

Fig. 1. Solution components design

2.1 The E-puck Robot

E-puck robot is a small programmable mechanical device. It was originally designed by Michael Bonani and Franchesco Mondada to be used at professor Roland Siegwart's laboratory at École Polytechnique Fédérale de Laussane. It is figured out for educational environments [11,12,13]. E-puck hardware and software are under open license, it was designed in base to tow important criteria: minimum size and flexibility. In order to reduce production costs it is shipped with cheap components and it is been manufactured using mass production technologies.

This robot interacts with the environment in several ways. On the one hand, it perceives what is surrounding it through eight infrared sensors, three ground sensors and three omni-directional microphones. On the other hand, it is capable of moving using two directional wheels, it can reproduce sounds, and it emits light from leds situated at robot ring and robot body.

It is a simple machine being able to execute instructions stored in its memory or commands sent via bluetooth. Everything depends on the program loaded in robot memory: it can work only instructions loaded in memory or it can accept commands from outside, given multiple work possibilities.

It is really configurable and It can be equipped with other sensors configurations, even we can design ours own. These sensors are placed on the top part of the robot and connected to robot via interface RS-232. In robot's web page [14] more information about robot features and last novelties can be founded.

2.2 Video Capture Device

In order to see on real time what robot is doing any conventional camera for PC with USB connection can be used. The only one restriction is a minimum resolution of 640x480 pixels and a angle of view wide enough to include the whole work area. We can obtain images from three different sources. The most common source is a web camera. A Kinect device [15] can be used also, it has a wide focus and covers as much area as conventional cameras, its movement and range sensors make it really interesting for future development. The third source can be an video file, it could be useful for testing or for new devices using a temporal file as a intermediate storage for communication.

3 Software Components

Five software components integrate our solution: the e-puck robot controller, the video manager, a database, the web server and its GUI.

3.1 E-puck Robot Controller

The controller has been designed to allow users connection and driving in a transparent way for the users, preventing them to have to deal with communication complexities. It is written in Python language and its operation follows the client-server model. Normally only one user would working with a robot, however multiple connections per robot are allowed, although this is not recommended given that it affects robot performance.

The controller can manage several robots via bluetooth connections, storing a list of robot-client pairs. Once the client is connected four request types can be done: sensors readings, mechanical operations such as running or turning leds on, file reception for memory uploading and connection termination with robot resources release.

3.2 Video Manager

A generic capture application has been developed. Images source is not fetter neither to robot nor to client. That is, video clients, as the web server, Cybersim or others, always get the same video format independently of video source.

Video manager has been implemented using C++ and it uses OpenCV [16] and freenect [17,15] libraries. It follows the client-server model as the robot controller does and only one video source can be manage per server instance. Ir works in a loop executing always the same three steps: image capture in an unified image format [18], scanned image transformation in an appropriate way for sending to the client and image delivery, the server asynchronously sends a package with an identification header and later the image divided in fix sized packages.

3.3   Database Design

Three groups of tables configure the database. Users, this group stores users personal data, video recording files attached to users, programs loaded by users and planned sessions. Robots, this group keep information about tables storing sensors configurations, how sensors are arranged in the screen and connections to video manage and robot controller. And the third group is named system settings, stores information about images sizes, GUI colors, etc.

3.4   Web Server

The web server is the communication mean between users and the workshops system. This application connects with database managers, video managers and robots. It runs on every of the most known web navigators: Chrome, Firefox, Explorer and Safari. It allows users to manage the robot from everywhere without the need of installing specific software.

Three types of server users are defined. The first group is formed by administrators, they can access to every server aspect and can manage tables in database, their main work is to configure workshop sessions. The second group is the students group, they carry out activities with robots, interacting with robots in a remote way. And finally observers group, they can only visualize users robot sessions, they cannot interact given that they see data an images in an indirect way, through system shared memory.

In order to get an efficient sessions assignment system we have supply web server with a calendar tool. Administrators create sessions, other users can access to the calendar and sing up for sessions.

Robot Session GUI. Special interest in web server implementation has been given to the graphical user interface (GUI) in order to users can interact with robots. Efforts has been done in order to be able to show every interaction aspect at a time in one screen. The GUI has five main views, see figure 2, but only four are displayed at the same time, they can be changed by clicking them. Each view has several functioning modes or/and information display possibilities. The views are:

- Zenithal camera: this view show in real time images captured by video manager. Video is sent as a image sequence, rather than a video stream, in order to enable Internet connection speed adjust: the higher is the speed fluent is the video. Images quality is also configurable from 80x60 to 640x480 pixels.
- Robot interaction: robot can be moved by hand or control programs can be uploaded to robot. A default program is running from robot boot enabling control by hand, other programs should add default program capabilities if they need control by hand.
- Numeric information: this view shows numeric information about sensor readings, velocities, etc. Administrators can configure what kind of information is shown in this view.

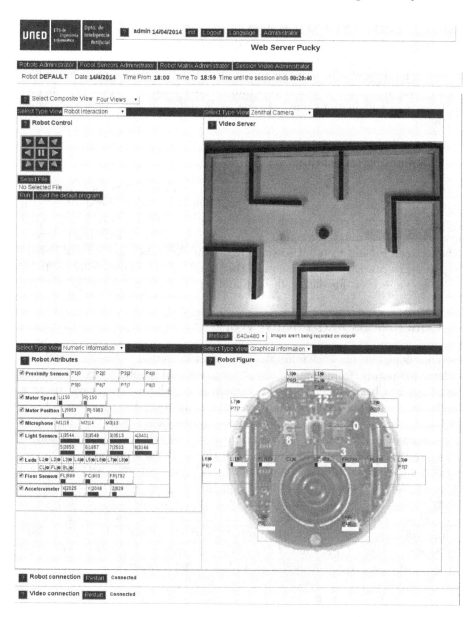

Fig. 2. Robot session GUI

- Graphical information: this view shows the same information as the previous view but in a graphical way, showing sensors positions and numeric readings over a zenithal robot image. This view is also configurable by administrators.
- Log: a view with textual information about action and responses, mainly connection status and uploading actions. Usually this view is hidden in normal operation and is normally used for debugging purposes.

In order to assure web server operation independence from robot controller and video server, the integration is done via plugins located at web server side. If controller or video server are changed only is necessary to change plugins, the server implementation remains unchanged. For this work e-puck controller, generic computer cameras and Kinect plugins have been developed. Servers and controller can run in the same machine or in different ones, allowing system scalability.

3.5 Cybersim Integration

Cybersim simulator is multiplatform. It implements controllers for several well known robots and it also can simulate them. For this work Cybersim has been expanded to interact with e-puck robot in a similar way as robot session GUI does: video, robot information, etc. Video is displayed by an external program. Cybersim can replace to robot session GUI. It implements a lot of robot behaviors, students can use this behaviors if they need, saving implementation time. It also provides tools for data collection and representation.

4   Conclusions and Future Work

Tools developed in this work fulfill the targets set in section 1. A system enabling students to operate, from home or any other place, real robots in virtual robotics workshops has been done. Users interaction is possible in two ways: using a standalone web application or using Cybersim simulator. Web application has a quick and simply installation. System has low establishment costs as it uses open license software and not really expensive hardware: e-puck robot, zenithal camera and computer. The system developed is not closed, it has a modular design and can grow by means of adding support for new robots or cameras, without system structure changes or with minor ones. The system also is scalable in order to satisfy variations in connections demand by means of adding or removing servers.

We expect our system can increase students opportunities for carrying out on-line workshops working with real robots, and also it can help teachers with workshops creation. These kind of workshops can increase students curiosity and to cause students knowledge and practical skills improvement.

As future work lines we can mention the following. Integrate new robots models, or camera models with special features. Integrate in the platform structure the virtual laser system for e-puck based on Kinect developed in our laboratory [19]. Other types of workshops and experiments can follow the paradigm described in this paper: to control remote objects via web and at the same time to see what is happen. So experience and development obtained in this work can be exportable to other educational fields in addition to Robotics.

References

1. Webster, J., Hackley, P.: Teaching effectiveness in technology-mediated distance learning. Academy of Management Journal 40(6), 1282–1309 (1997)
2. Phipps, R., Merisotis, J.: What's the difference? A review of contemporary research on the effectiveness of distance learning in higher education (1999)
3. Bernard, R.M., Abrami, P.C., Lou, Y., Borokhovski, E., Wade, A., Wozney, L., Wallet, P.A., Fiset, M., Huang, B.: How does distance education compare with classroom instruction? A meta-analysis of the empirical literature. Review of Educational Research 74(3), 379–439 (2004)
4. Chou, S.-W., Liu, C.-H.: Learning effectiveness in a Web-based virtual learning environment: a learner control perspective. Journal of Computer Assisted Learning 21(1), 65–76 (2005)
5. Warburton, S.: Second Life in higher education: Assessing the potential for and the barriers to deploying virtual worlds in learning and teaching. British Journal of Educational Technology 40(3), 414–426 (2009)
6. Mikropoulos, T.A., Natsis, A.: Educational virtual environments: A ten-year review of empirical research (1999–2009). Computers & Education 56(3), 769–780 (2011)
7. Dede, C.: The evolution of distance education: Emerging technologies and distributed learning. American Journal of Distance Education 10(2), 4–36 (1996)
8. Beldarrain, Y.: Distance education trends: Integrating new technologies to foster student interaction and collaboration. Distance Education 27(2), 139–153 (2006)
9. García Aretio, L.: De la educación a distancia a la educación virtual. Ariel (2007)
10. Goldberg, H.R., McKhann, G.M.: Student test scores are improved in a virtual learning environment. Advances in Physiology Education 23(1), 59–66 (2000)
11. Mondada, F., Bonani, M., Raemy, X., Pugh, J., Cianci, C., Klaptocz, A., Magnenat, S., Zufferey, J.-C., Floreano, D., Martinoli, A.: The e-puck, a robot designed for education in engineering. In: Proceedings of the 9th Conference on Autonomous Robot Systems and Competitions, vol. 1, pp. 59–65. IPCB: Instituto Politécnico de Castelo Branco (2009)
12. Micael, S., Couceiro, C.M., Figueiredo, J., Miguel, A., Luz, N.M.: Ferreira, and Rui P Rocha. A low-cost educational platform for swarm robotics. International Journal of Robots, Education and Art (2011)
13. Petrovic, P.: Having Fun with Learning Robots. In: Proceedings of the 3rd International Conference on Robotics in Education, RiE 2012, pp. 105–112 (2012)
14. Robot Educacional e-puck (2013), http://www.e-puck.org/
15. Kramer, J., Burrus, N., Echtler, F., Daniel Herrera, C., Parker, M.: Hacking the Kinect. Apress (2012)
16. Brahmbhatt, S.: Practical OpenCV. Apress (2013)
17. Demaagd, K., Oliver, A., Oostendorp, N., Scott, K.: Practical Computer Vision with SimpleCV. O'Reilly Media, Inc. (2012)
18. Bradski, G., Kaehler, A.: Learning OpenCV. O'Reilly Media, Inc. (2008)
19. Martín-Ortiz, M., de Lope, J., de la Paz, F.: Hardware And Software Infrastructure To Provide To Small-Medium Sized Robots With Advanced Sensors. In: Proceedings of the Workshop of Physical Agents (WAF), XIII World Congress (2012)

A Growing Functional Modules
Learning Based Controller Designed
to Balance of a Humanoid Robot on One Foot

Jérôme Leboeuf-Pasquier[✉]

Department of Computer Science
Guadalajara University, Guadalajara, Mexico
jerome.lepg@mail.com

Abstract. The purpose of this paper is to exhibit the process of a Growing Functional Modules (GFM) controller designed for a humanoid robot balance learning. This learning based controller is graphically generated by interconnecting and configuring four kinds of components: Global Goals, Acting Modules, Sensing Modules and Sensations. Global Goals specify the intrinsic motivations of the controller. Acting and Sensing Modules develop their acting and respectively, sensing functionalities while interacting with the environment. Sensations provide the controlled system's feedback that renders the effects produced by the previous command. These characteristics together with an endless learning process allow the controller to perform as an 'artificial brain'. The present paper describes the design, functioning and performance of a humanoid equilibrium subsystem that learns balancing the robot on one foot meanwhile a disequilibrium is artificially produced by moving the opposite leg.

1 Controllers Endowed with Learning Abilities

In order to deal with the huge complexity of the real world, programmers drastically reduce the intricacy of the robot's environment otherwise, the number of lines of code required to deal with countless situations, would increase drastically. Classical artificial intelligence does not offer much help because replacing sequential-imperative programming by another paradigm - for example, functional or logical - implies a different but equally vast coding effort. On the opposite side, biological entities successfully deal with the complexity of the real world thanks to their learning abilities.

These considerations have led to develop cognitive architectures that try to reproduce the learning abilities of the human brain. Sometimes labeled as Biologically Inspired Cognitive Architectures (BICAs) - see [1] for a review of some leading proposals - these BICAs share the purpose of achieving brain-like functionalities. The Growing Functional Modules (GFM) should be classified in a subcategory of the BICAs that is focused on Developmental Robotics, a concept introduced by several authors during the 1990's including Brooks [2]. The IM-CLeVeR European project [3] illustrates the research efforts following

© Springer International Publishing Switzerland 2015
J.M. Ferrández Vicente et al. (Eds.): IWINAC 2015, Part II, LNCS 9108, pp. 240–250, 2015.
DOI: 10.1007/978-3-319-18833-1_26

such approach. Some examples of corresponding architectures could include Self-Adaptive Goal Generation Robust Intelligent Adaptive Curiosity (SAGG-RIAC) Baranes and Oudeyer [4] and Qualitative Learner of Action and Perception (QLAP) [5]. Another recent review of adaptive control architectures [6] makes a special emphasis on the hierarchical organization of behavior. This characteristic together with modularity are identified as organizational principles of the brain; in fact, both characteristics are incorporated to the GFM proposal. Moreover, other characteristics that GFM controllers provide as a developmental robotics architectures, include:

- Self exploration of the robot sensing and acting abilities.
- Introduction of motivations through the predefined internal global goals.
- Integration of neural networks structures to adjust the modules internal representations.
- Building a representation of the environment based on a bottom-up approach.

Developmental robotics architectures like GFM should be focused on robots control because only real world interaction can offer a consistent feedback to the learning process. Concerning biped robots balancing or walking, many machine learning techniques have been applied since the middle of the 1990's. A review of state-of-the-art learning algorithms applied to bipedal robots control is given in [7]. These techniques commonly try to avoid the use of kinematics or of dynamic models in the same way that the GFM approach does - but they rarely offer to replace the programming task by a manual design of the controller. Further on, the GFM application fields are not restricted to the control of mechanical systems but embrace unusual fields from a control point of view such as artificial vision or audition [8].

The present paper introduces a humanoid robot and its corresponding GFM controller designed to perform as an equilibrium subsystem balancing the robot on one leg meanwhile a disequilibrium is artificially produced by learning to move the other leg. The content is organized as follows: section 2 explains the GFM controllers principles, section 3 describes the humanoid robot, finally section 4 details the controller in charge of learning to balance the humanoid on one leg.

2   GFM Controllers

Biological brains may be associated to learning-based controllers: To perform in the real world, both must provide real time response to events sensed by their control system. Thus, the GFM paradigm has been implemented as a feedback control system composed by an Acting Area and a Sensing Area. Both Areas are provided with learning capacities. The Acting Area and the Sensing Area host interconnected Acting and Sensing modules, respectively. Initially, these modules do not integrate any knowledge but gradually elaborate it as the controller interacts with the environment. Consequently, such controllers are expected to perform as artificial brains.

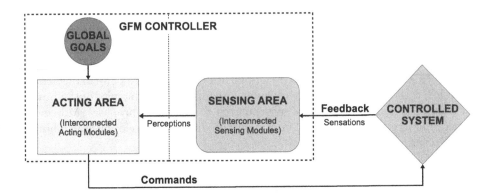

Fig. 1. The GFM controller's structure (delimited with dotted lines) and its control loop

As shown in figure 1, the GFM control loop shares many similarities with a standard one. At each cycle, the controller - delimited by a dotted line - sends an output command to trigger a specific actuator and then receives a feedback composed of a sequence of sensors values. Despite these similarities, two main differences stand out:

- Firstly, the concept of 'Reference Value' used in a classical controller, is replaced by the one of 'Global Goals' which is integrated into the controller. More specifically, Global Goals refer to a set of motivations that induces the controller's activities and, consequently, its behavior.
- Secondly, the GFM controller's structure is divided in two areas: the 'Sensing' and the 'Acting'. The Sensing Area is in charge of interpreting the feedback sensed from the controlled system in its environment. This feedback is specified by the 'Sensation' component (see figure 1). Then, Sensations are processed by a set of interconnected Sensing Modules that produces 'Perceptions'. Next, Perceptions, and eventually some Sensations, are submitted to a set of Acting Modules located in the Acting Area. Depending these Perceptions values and the input requests specified by the Global Goals, one of these Acting Modules will trigger a command to the controlled system. Triggering this command initiates the next cycle of the control loop.

For the equilibrium subsystem, subject of this paper, no sensing module is required meaning that Acting Modules exclusively process Sensations. A detailed description of the components and processes is given in [9] together with an illustration of these learning control abilities. Moreover, an editor described in [10], has been programmed to visualize the aforementioned components, to display their configurations and interconnections and to run the controller conjointly with its associated application. The configuration and functionality of each component are described in the next subsections.

2.1   Sensations

Sensations (see figure 2.a) are represented by a green rectangle and situated in the Sensing Area of the editor panel. A field named 'size' allows to specify the number of integer values that compose a Sensation. The declared sequence of Sensations must correspond to the system's feedback; their values are to be fed to the Sensing Modules input, and will optionally comprise the Acting Modules feedback.

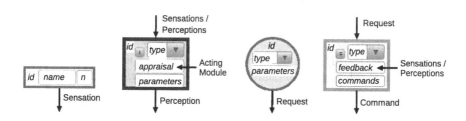

Fig. 2. GFM controller's components: a) Sensation; b) Sensing Module; c) Global Goal; d) Acting Module

2.2   Sensing Modules

Sensing Modules (see figure 2.b) are symbolized by a blue rectangle and situated in the Sensing Area of the editor panel. They are in charge of processing a subset of sensations/perceptions in order to produce a single perception. Sensing module's structure will not be further described because the two controllers designed to balance the humanoid do not require them. Sensations alone are sufficient to provide feedback to the Acting Modules.

2.3   Global Goals

Global Goals (see figure 2.c) are symbolized by a red circle and situated in the Acting Area of the editor panel. The upper field, initially filled with question marks, allows the definition of the Global Goal's type. This type determines the value of the request, for example a constant value or an integer randomly extracted from a specified interval. This request is sent to a unique Acting Module indicating the value that the first feedback element must reach. The lowest field allows capturing some extra parameters.

2.4   Acting Modules

Acting Modules (see figure 2.d) are symbolized by an orange rectangle and situated in the Acting Area of the editor panel. The upper field, initially filled with question marks, indicates the type (or functionality) of the component. For example, the Real Time Regulation acting module (Rtr) has been designed to learn how to maintain the controlled system variables close to a requested value. At each cycle, the requested value is specified by a single Global Goal or eventually, another Acting Module connected to input. The middle field specifies the feedback which is a subset of Sensations and/or Perceptions that reflect some effects of the prior command. The lowest field indicates the set of commands that the module may trigger. Finally, a hidden field contains some extra parameters whose number and meaning are specific to each type of module.

2.5 Learning and Adapting

Sensing and Acting Modules are the two components that integrate a dynamic internal structure designed to gradually learn a specific functionality which is specified by the type of the module (indicated in the upper field). For any module, learning is an endless process characterized by the creation of a new internal structure which induces memory allocation. Meanwhile, the adaptation processes of any module does not require extra memory because they just modify the values of their attributes.

To design the current controller, two acting modules are required; their functionalities are described in the next subsections.

2.6 The Rtr Module

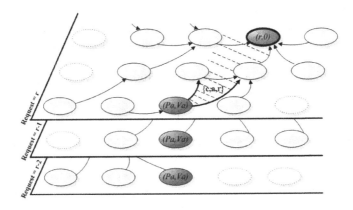

Fig. 3. Representation of the internal structure of the Rtr Acting Module where (Pa, Va) stands respectively for position and velocity

The functionality of the Rtr acting module is to perform real time control for example, maintaining an inverted pendulum in equilibrium. The internal dynamic structure of a Rtr Module is represented by a kind of state graph (see figure 2.6) where each state corresponds to a pair of variables (*position*, *velocity*); states are created when the feedback contains them for the first time. The input request indicates the internal state that should be reached, eventually a non-zero position. The (c, a, r), associated to internal command, amplitude and reliability, allows shifting from one state to another. These values are gradually discovered thanks to the adaptation process which is guided by the sensors feedback. In practice, as the module performs within non-deterministic or unstable environments, the (c, a, r) is not expected to reach a precise state but any of the states closer to the requested one. A complete description of the Rtr Module is given in [11], since then no significant improvement has been performed to the module's internal structure.

2.7   The Nac Module

The Nac module has been designed to learn a relation between a variable z and two others (x, y), that is: $z = f(x, y)$; the values (x, y, z) are given as feedback. More specifically, this module must be able to trigger some commands (c_x, c_y) in order to reach the value z_0 specified by an input request. Similarly to the previous Rtr Acting Module, the Nac Module internal structure may be represented by a state graph that is dynamically built. To reach a requested state, the internal propagation searches for the most secure and economic path through the graph; 'secure' according to the memorized frequencies of success and 'economic' considering the previous costs of state shifts. The only publication describing the Nac Module dates from 2004 [12], no improvement has been performed on this module since then. For the first experiment, the Rtr Module is in charge of learning how to move the knee and hip actuators in order to reach a random position of the foot inside the range $[-70, 140]$.

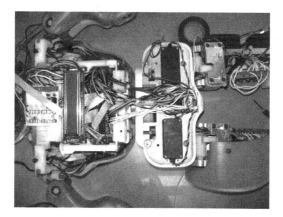

Fig. 4. Closeup of the humanoid robot showing the embedded electronics

Fig. 5. Four force sensors are located under each foot

3   The Humanoid Robot

Figure 4 shows the humanoid robot conceived and built in our laboratory. The robot is 80 centimeters high and all its body's parts are proportional to those of a four year old child including the size of its feet which strongly contributes to harden its equilibrium. The robot possesses 27 servomotors, 8 force sensors and a 2-D accelerometer. To perform equilibrium experiments, the robot stands on one foot or two feet and then moves a part of its body, for example its hips, in order to produce some random perturbation. For the mentioned experiments, only ten servomotors are required: two for each hip, one for each knee and two for each ankle. To achieve these experiments, the trunk of the humanoid has been removed to reduce the total weight and thus to allow more extreme positions, producing a higher torque, on its ankles [13]. Concerning feedback, only the values produced by the eight force sensors situated under its feet (see figure 5) are required; they measure differential pressure. Some feedback values delivered to the controller are expressed as 'virtual' sensors that are computed as a formula of other sensors values; for example, subtracting the pressure values measured on the left side of the foot from those measured on the right side.

The robot embedded system is situated in the abdomen of the robot just below the LCD display as shown in figure 4. This system has been implemented on a TS-7800 single board computer operated by a Debian Linux operating system. This implementation allows the communication between the embedded system of the robot and the controller running on an external PC through a TCP/IP over a Wifi connection. The sensors are connected to a TS-ADC24 extension board plugged in the TS-7800. The servomotors are controlled by a SSC-32 board, itself connected to the TS-7800 through a RS-232 serial communication. The robot hardware architecture including boards, connections, actuators, physical and virtual sensors, is declared in a XML file that is loaded by the embedded system when the board is powered on. A complete description of this embedded system and its functionalities is given in [14].

4   Learning to Balance on One Foot

The design of the GFM subsystem in charge of learning to balance on the left foot did not offer a mayor difficulty. The resulting controller is given figure 6 and may be interpreted as follows:

Firstly, in green color on the right panel of the editor canvas, a sequence of 48 feedback values. In this experiment, only the feedback values no.21, 35, 37, 42, 43 and 45 are taken into account by the the Acting Module and are consequently declared in their middle fields.

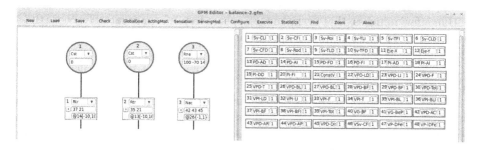

Fig. 6. The GFM controller design to learn balance on one foot

Secondly, the three Global Goals declared in the left panel of the editor canvas specify that Acting Modules no.1 and 2 must both try to reach a constant value 0 and that Acting Module no.3 must reach a random value between −70 and 140 [1], this value is recomputed at least once every 100 cycles.

Thirdly, the three Acting Modules express the following: the type of modules no.1 and no.2 is Real Time Regulation (Rtr) which has been designed to learn how to maintain a controlled system close to a requested value. The feedback values of the corresponding controlled system, typically an inverted pendulum, are classically defined by their error and the difference in this error. Presently, the error is given by the virtual sensor no.37 that at each cycle, computes the difference between the two front force sensors and the two rear ones, all belonging to the left foot. Then, in order to nullify this difference the acting module no.1 learns to trigger the 'pitch' servomotor @14 with an amplitude belonging to [-10; 10]. Similarly, the acting module no.2 learns to nullify the virtual sensor no.35 that computes the difference between left and right pressure sensors, triggering the 'roll control' servomotor @13 with an amplitude belonging to [−10, 10]. The control loop period has been set to a tenth of a second avoiding to deal with the 'change in time' error. Consequently, the difference in error is set to a constant value 0 through virtual sensor no.21.

---

[1] The last zero is not displayed in figure 6 due to the caption size.

Finally, to produce some disequilibrium, the robot is motivated to learn moving the right leg which is not in contact with the floor. In order to achieve this, the module no.3 with the Nac type shown in figure 6, is in charge of learning how to move this leg activating the servomotors @26 and @27 [2] that correspond respectively to the hip and the knee. The motivation is given by the global goal no.3 and consists in requesting a random position of the foot in the range $[-70, 140]$. The feedback includes three values: The angle of the hip (sensor no.42), the angle of the knee (sensor no.43) and the position of the foot (sensor no.45). Learning to move this leg alone takes a few minutes depending on the required precision and the control loop period.

Experimental proofs show that the equilibrium subsystem is able to learn within a few minutes, that is the robot appears to perform accurately. Nevertheless, as stated previously when experimenting control with an inverted pendulum [11], the subsystem is never fully reliable: An inappropriate adaptation may sporadically lead to a consequent failure; a phenomenon that must be accepted as a drawback of endless learning.

Figure 7 displays the curves corresponding to the evolution over time of lateral and frontal balance that is, the difference between left-right and front-rear sensors, respectively. Each peak of these curves corresponds to a disequilibrium induced by the movement of the opposite leg. One may appreciate that after the occurrence of a disturbance, the controller succeeds in quickly reaching a new equilibrium state.

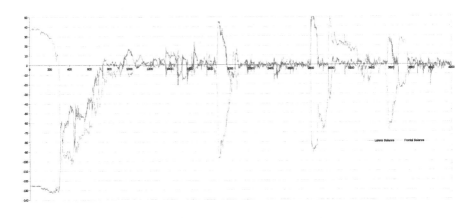

Fig. 7. Time monitoring of the two feedback values associated with balance

---

[2] Note that only the first command @26 is displayed in this figure due to the caption size.

5 Conclusions

The aim of the present experiment is to illustrate the ability of GFM controllers to deal with simple motor subsystems. Though the resulting control performances are probably not superior to those of more classical approaches, some valuable achievements should be emphasized: First, the approach allows to design a controller only by interconnecting and configuring a few components. Secondly, this controller exempts implementing kinematics or dynamic models; the behavior is fully induced by feedback. Thirdly, endless learning is a key factor that should allow the control system to adapt to dynamic or noisy environments.

Moreover, the control paradigm offers higher flexibility because it may actually be applied to such unusual fields concerning control, as artificial vision or audition [8]. Nevertheless, there is presently no estimation of the learning time a controller requires to be effective and its resulting behavior may not be predicted. Therefore, the present experiment in a real environment significantly contributes to the evaluation and development of GFM controllers able to learn how to accurately perform within a unknown environment.

Acknowledgment. This paper is based upon work supported by the Jalisco State Council of Science and Technology, Mexico (project Coecytjal-UdG 5-2010-1-821).

References

1. Goertzel, B., Lian, R., Arel, I., de Garis, H., Chen, S.: World survey of artificial brains, part ii: Biologically inspired cognitive architectures. Neurocomputing 74, 30–49 (2010)
2. Brooks, R., Breazeal, C., Irie, R., Kemp, C., Marjanovic, M., Scassellati, B., Williamson, M.M.: Alternative essences of intelligence. In: Proceedings - Electrochemical Society of 15th Nat. Conf. on Artificial Intelligence, pp. 961–978 (1998)
3. Baldassarre, G., et al.: The im-clever project - progress of last reporting year, Tech. rep., Lugano (Switzerland), Second Review Meeting (2011)
4. Baranes, A., Oudeyer, P.-Y.: Active learning of inverse models with intrinsically motivated goal exploration in robots. Robotics and Autonomous Systems 61(1), 49–73 (2013)
5. Mugan, J., Kuipers, B.: Autonomous learning of high-level states and actions in continuous environments. IEEE Transactions on Autonomous Mental Development (TAMD) 4, 70–86 (2012)
6. Marcos, E., Ringwald, M., Duff, A., Snchez Fibla, M., Verschure, P.: The hierarchical accumulation of knowledge in the distributed adaptive control architecture. In: Baldassarre, G., Mirolli, M. (eds.) Computational and Robotic Models of the Hierarchical Organization of Behavior, pp. 213–234. Springer, Heidelberg (2013)
7. Wang, S., Chaovalitwongse, W., Babuska, R.: Machine learning algorithms in bipedal robot control. IEEE, Transactions on Systems, Man, and Cybernetics, Part C: Applications and Reviews 42(5) (2012)

8. Leboeuf-Pasquier, J., Gómez Ávila, G.F., González Pacheco Oceguera, J.E.: A preliminary auditory subsystem based on a growing functional modules controller. In: Ferrández Vicente, J.M., Álvarez Sánchez, J.R., de la Paz López, F., Toledo Moreo, F. J. (eds.) IWINAC 2013, Part II. LNCS, vol. 7931, pp. 81–91. Springer, Heidelberg (2013)

9. Leboeuf-Pasquier, J.: A basic growing functional modules "artificial brain", Neurocomputing 151(pt.1(0)), 55–61 (2015), doi:http://dx.doi.org/10.1016/j.neucom.05.080

10. Leboeuf-Pasquier, J., González Pacheco, J.E.: Designing a growing functional modules artificial brain. Broad Research in Artificial Intelligence and Neuroscience 3(2), 5–16 (2012)

11. Leboeuf-Pasquier, J.: Improving the rtr growing functional module. In: Proceedings of the 32nd Annual Conference of the IEEE Industrial Electronics Society (2006)

12. Leboeuf-Pasquier, J.: Facing combinatory explosion in nac networks. In: Ramos, F.F., Unger, H., Larios, V. (eds.) ISSADS 2004. LNCS, vol. 3061, pp. 252–260. Springer, Heidelberg (2004)

13. Galván Zárate, R.: Sistema de equilibrio de un robot humanoide basado en el paradigma growing functional modules, Master's thesis, in Information Technology, CUCEA, Guadalajara University, México (2012)

14. Leboeuf-Pasquier, J., Gónzalez Villa, A., Herrera Burgos, K., Carr, D.: Implementation of an embedded system on a ts7800 board for robot control. In: IEEE International Conference on Electronics, Communications and Computers, pp. 135–141 (2014)

Optimized Representation of 3D Sequences Using Neural Networks

Sergio Orts-Escolano[1](✉), José García Rodríguez[1], Vicente Morell[2],
Miguel Cazorla[2], Alberto Garcia-Garcia[1], and Sergiu Ovidiu-Oprea[1]

[1] Department of Computing Technology, University of Alicante, Alicante, Spain
{jgarcia,sorts,agarcia,sovidiu}@dtic.ua.es
[2] Instituto de Investigación en Informática, University of Alicante, Alicante, Spain
{miguel,vmorell}@dccia.ua.es

Abstract. We consider the problem of processing point cloud sequences. In particular, we represent and track objects in dynamic scenes acquired using low-cost 3D sensors such as the Kinect. A neural network based approach is proposed to represent and estimate 3D objects motion. This system addresses multiple computer vision tasks such as object segmentation, representation, motion analysis and tracking. The use of a neural network allows the unsupervised estimation of motion and the representation of objects in the scene. This proposal avoids the problem of finding corresponding features while tracking moving objects. A set of experiments are presented that demonstrate the validity of our method to track 3D objects. Favorable results are presented demonstrating the capabilities of the GNG algorithm for this task.

Keywords: GNG · Point cloud sequence · Object tracking · Non-stationary distributions · Trajectory estimation

1 Introduction

Visual tracking is a very challenging problem due to some constraints of current acquisition systems like: the loss of information caused by the projection of the 3D world on a 2D image, noise in images, cluttered-background, complex object motion, partial or full occlusions, illumination changes as well as real-time processing requirements, etc. [1]. The availability of both depth and visual information provided by low-cost 3D sensors such as the Kinect sensor [1] opens up new opportunities to solve some of the inherent problems in 2D based vision systems. Objects tracking and motion estimation are some of the topics that can take advantage using these devices. A large number of researchers are adapting previous 2D tracking algorithms exploiting the combination of RGB and depth information to overcome some of the previously mentioned inherent problems in 2D vision systems.

A group of methods of particular interest to track objects and estimate its motion are the ones based on input data neural representation. Neural networks

---

[1] Kinect v2 for Windows: http://www.microsoft.com/en-us/kinectforwindows/

© Springer International Publishing Switzerland 2015
J.M. Ferrández Vicente et al. (Eds.): IWINAC 2015, Part II, LNCS 9108, pp. 251–260, 2015.
DOI: 10.1007/978-3-319-18833-1_27

have been extensively used to represent objects in scenes and estimate their motion. In particular, there are several works that use the self-organizing models for the representation and tracking of objects. Fritzke [2] proposed a variation of the original Growing Neural Gas (GNG) model [3] to map non-stationary distributions that in [4] were applied to people representation and tracking. In [5], self-organizing models are extended for the characterization of the movement. In a recent work, Frezza-Buet [6] modifies the original GNG algorithm to compute non-stationary distributions velocity fields but it is only applied to 2D data. [7] modifies the original SOM algorithm to adapt and track 3D hand and body skeleton structures. However, none of them exploit previous obtained maps to solve the correspondence problem or to improve segmentation. In addition, the structure of the neural network is not used to solve the feature correspondence problem along a whole sequence of frames.

Considering our previous works in the area [8,9] that demonstrated object representation and tracking capabilities of self-growing neural models for 2D data. We propose a neural method capable of segmenting and representing objects of interest in the scene, as well as analyzing the evolution of these objects over time to estimate their motion. The proposed method creates a flexible model able to characterize morphological and positional changes over the sequence. The representation identifies objects over time and establish feature correspondence through the different observations. This allows the estimation of motion based on the interpretation of the dynamics of the representation model. It has been demonstrated that architectures based on GNG [3] can be applied on problems with time restrictions such as object tracking. These systems have the ability to process sequences of images and thus offer a good quality of representation that can be refined very quickly depending on the time available.

The rest of the paper is organized as follows: first, Section 2 briefly introduces the GNG algorithm and presents our novel extension for managing point cloud sequences. Next, Section 3 presents some experiments and discuss results obtained using the proposed technique. Finally, in Section 4, we give our conclusions and directions for future work.

2   GNG Algorithm

The Growing Neural Gas (GNG) [3] is an incremental neural model able to learn the topological relations of a given set of input patterns by means of Competitive Hebbian Learning (CHL). From a minimal network size, a growth process takes place, where new neurons are inserted successively using a particular type of vector quantization [10].

In the remaining of this section we describe the growing neural gas algorithm. The network is specified as:

- A set A of nodes (neurons). Each neuron $c \in A$ has its associated reference vector $w_c \in \mathbb{R}^d$. The reference vectors are regarded as positions in the input space of their corresponding neurons.

– A set of edges (connections) between pairs of neurons. These connections are not weighted and its purpose is to define the topological structure. An edge aging scheme is used to remove connections that are invalid due to the motion of the neuron during the adaptation process.

The GNG learning algorithm is as follows:

1. Start with two neurons a and b at random positions w_a and w_b in R^d.
2. Generate at random an input pattern ξ according to the data distribution $P(\xi)$ of each input pattern.
3. Find the nearest neuron (winner neuron) s_1 and the second nearest s_2.
4. Increase the age of all the edges emanating from s_1.
5. Add the squared distance between the input signal and the winner neuron to a counter error of s_1 such as:

$$\triangle error(s_1) = \|w_{s_1} - \xi\|^2 \tag{1}$$

6. Move the winner neuron s_1 and its topological neighbors (neurons connected to s_1) towards ξ by a learning step ϵ_w and ϵ_n, respectively, of the total distance:

$$\triangle w_{s_1} = \epsilon_w(\xi - w_{s_1}) \tag{2}$$

$$\triangle w_{s_n} = \epsilon_n(\xi - w_{s_n}) \tag{3}$$

For all direct neighbors n of s_1.
7. If s_1 and s_2 are connected by an edge, set the age of this edge to 0. If it does not exist, create it.
8. Remove the edges larger than a_{max} . If this results in isolated neurons (without emanating edges), remove them as well.
9. Every certain number λ of input patterns generated, insert a new neuron as follows:
   – Determine the neuron q with the maximum accumulated error.
   – Insert a new neuron r between q and its further neighbor f:

$$w_r = 0.5(w_q + w_f) \tag{4}$$

   – Insert new edges connecting the neuron r with neurons q and f, removing the old edge between q and f.
10. Decrease the error variables of neurons q and f multiplying them with a consistent α. Initialize the error variable of r with the new value of the error variable of q and f.
11. Decrease all error variables by multiplying them with a constant γ.
12. If the stopping criterion is not yet achieved (in our case the stopping criterion is the number of neurons), go to step 2.

This method offers further benefits due to the incremental adaptation of the GNG. Input space denoising and filtering is performed in such a way that only concise properties of the point cloud are reflected in the output representation [11].

2.1   Point Cloud Sequences Processing

GNG capabilities for representing static 3D observations have been already shown in previous works [12], but we propose the extension of this method to allow representing continuous observations. In this way, GNG representation can be used in complex computer vision problems such as scene understanding by long-term observation, Simultaneous Localization and Mapping (SLAM) or 3D object tracking.

To obtain the representation of the first point cloud, the entire algorithm presented in Section 2 is computed, performing the complete learning process of the network. However, for the following point clouds, only reconfiguration of the network is processed (stages 2nd to 7th). In this way, neurons that are already placed in the map are adjusted and moved towards changes in the observation (partial learning). Only weight modification and spatial relationship steps are processed keeping stable the number of neurons of the GNG. Figure 1 shows the proposed work-flow for processing point cloud sequences using the GNG algorithm.

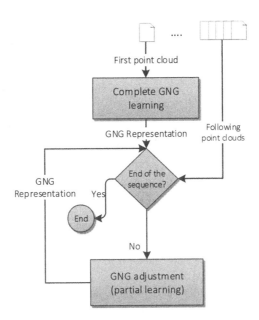

Fig. 1. An improved work-flow to manage point cloud sequences using the GNG algorithm

This extension of GNG algorithm has some advantageous features for processing sequences of point clouds. The most interesting is that restarting the learning of the network for each new point cloud that is captured from the scene is not required. Once the initial map that represent objects of interest in the scene is generated, this map can be used as the initial topology to represent and track previously specified objects in the next point cloud. In order to take advantage of this feature a good acquisition rate is required, 20-30 frames per second, so that the changes between different captured point clouds should be small. Moreover, this adaptive method is able to face real-time constraints, because the number of λ times that the reconfiguration module is performed can be chosen according to the available time between two successive frames that depends on the acquisition rate (see example of GNG adaptation process in Figure 2).

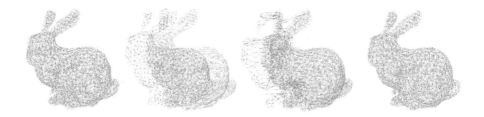

Fig. 2. From left to right: GNG representation of the initial input space. Input space translated over the x-axis. Intermediate state of the GNG structure during the adaptation process to fit the new input data (GNG extension for managing point clouds). Final GNG representation after activating 200000 λ patterns.

2.2 Exponential Time-decaying Adaptation

In the original GNG algorithm it is considered a stationary input space distribution where neurons are placed directly over the input data. However, considering sequences of non-stationary data, neurons need to move towards (step 6) a changing input space caused by the motion of the scene (camera motion) or objects motion. Therefore, the use of the original constant values for ϵ_w and ϵ_n parameters takes long time (λ iterations) to move the whole map to the new input space (next frame). In this work, we propose the modification of these constant ϵ_w and ϵ_n parameters so neurons move fast towards the selected input patterns using an exponential decaying function based on a time function (λ iterations). A similar approach was previously used in the Neural Gas (NG) algorithm proposed in [10]. Thanks to this modification, the neural map considerably accelerates the learning step fitting new input data. The learning step is defined as a single iteration over the steps detailed above. The ϵ_w and ϵ_n parameters are computed every iteration as follows:

$$\epsilon_w(t) = \epsilon_{w_i}(\epsilon_{w_f}/\epsilon_{w_i})^{t/t_{max}} \tag{5}$$

$$\epsilon_n(t) = \epsilon_{n_i}(\epsilon_{n_f}/\epsilon_{n_i})^{t/t_{max}} \tag{6}$$

where ϵ_{w_i}, ϵ_{w_f}, ϵ_{n_i} and ϵ_{n_f} are the initial and final epsilon values for adapting winning neurons during the adaptation step.

3 Experiments

In this section, different experiments are shown validating the capabilities of our extended GNG-based method for managing point cloud sequences. First, we studied the input space representation error of the proposed method for tracking a 3D object along a sequence. Next, we show how the proposed method is able to estimate the trajectory of a 3D non-stationary distribution along the time. Moreover, we tested the proposed method using synthetic (noise-free) and real point clouds. 3D non-stationary distributions were simulated using the Blensor software [13]. Blensor allowed us to generate synthetic scenes and to obtain partial views of the generated scene simulating real 3D sensors, such as the Kinect or a Time-of-Flight camera. The main advantage of this software is that it provides ground truth (absence of noise) information. Moreover, it allows to capture point cloud sequences simulating the movement of an object in the scene or the camera movement.

3.1 Representation Error

In this section, we show the input adaptation error of the generated map along some point cloud sequences. The adaptation of the self-organizing neural networks is often measured in terms of preservation of the topology of the input space, so it is needed to know the real distance from the generated structure to the input data. This measure specifies how close our generated model is from the original input data. In order to obtain a quantitative measure of the input space adaptation of the generated map, we computed the Mean Square Error (MSE) of the map against sampled points (input space).

Figure 3 (left) shows the input space adaptation error computed for an increasing number of iterations (λ patterns). The representation error is computed for the network adjustment between two different frames. In addition, Figure 3 (left) shows the obtained representation error using three different strategies for selecting the ϵ_w and ϵ_n parameters of the GNG algorithm. We can observe how the time-decaying strategy introduced in our proposal is able to obtain a low representation error using the lowest number of iterations. It also shows how the original GNG scheme, which uses low ϵ values is not suitable for managing point clouds sequences, since it takes a large number of iterations for getting a good representation error. Using large ϵ values a fast adaptation is achieved but the final error is worse compared to the time-decaying epsilon.

Figure 3 (right) shows the representation error for a whole sequence of 30 frames. MSE was computed for the three different strategies, which were previously commented, selecting ϵ values. 100000λ patterns were activated at each

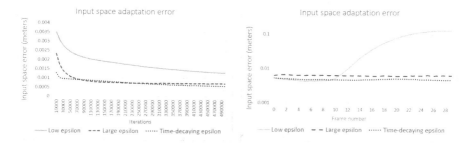

Fig. 3. Left: GNG representation error (MSE) between two different frames of a sequence. Right: Representation error along a whole sequence of 30 frames.

frame. We used a fixed network size of 2000. Large ϵ strategy uses $\epsilon_w = 0.15$, $\epsilon_n = 0.005$; low epsilon uses $\epsilon_w = 0.05$, $\epsilon_n = 0.0005$; time-decaying epsilon uses $\epsilon_{w_i} = 0.15$, $\epsilon_{w_f} = 0.05$, $\epsilon_{n_i} = 0.005$ and $\epsilon_{n_f} = 0.0005$. We animated the 3D Stanford bunny, translating the model over the x axis a known distance. This ground truth information about the object movement is later used to study the capabilities to track the trajectory of an object along the sequence. In addition, Figure 3 (right) shows how the 'low epsilon' strategy is not able to track the input distribution and soon starts producing a non accurate representation of the input data. In particular, in the middle of the sequence, when the motion velocity increases, the representation error dramatically raises. As it was previously demonstrated, the time-decaying strategy obtains the best adaptation error for a fixed number of iterations (100000λ per frame).

Figure 4 shows some frames of the generated point cloud sequence using the Blensor software for simulating a Kinect device. On the left side it is shown the initial acquired point cloud (red), whereas on the right side it is shown (blue) the last acquired point cloud.

3.2 Tracking Error

In this section, we study the capability of the proposed method to compute the translational movement of a non-stationary distribution along a sequence. For that purpose, we compute the centroid of the generated map using the GNG algorithm at each new frame. To evaluate the neural network capacity to follow the input space motion, we compare the movement performed by the whole object using the centroid trajectory. Thanks to the Blensor software, we are able to obtain ground truth information about the object position in the scene. This information is used for obtaining the translational error between the real movement and the computed one using the GNG.

For evaluating the tracking capabilities of the presented GNG extension, we used the sequence presented in Figure 4. Figure 5 shows the translational error obtained using the proposed GNG extension for point clouds sequence

Fig. 4. Point cloud sequence of a moving object (3D Stanford bunny) generated using the Blensor software for simulating a Kinect device

processing. Best results were obtained by using the exponential time-decaying strategy, which obtained for most frames of the sequence a translational error below 0.005 meters. Moreover, the proposed GNG modification was able to follow the trajectory of the non-stationary distribution obtaining a global translation error for the whole trajectory below 0.04 meters in a complete movement of 1.1 meters along the x axis.

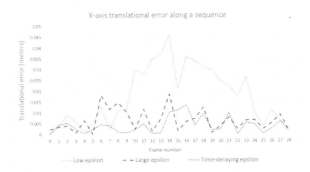

Fig. 5. x-axis translational error for the simulated sequence presented in Figure 4. Translational error was evaluated using the three different strategies for choosing the ϵ values. A constant number of 200000λ input patterns were activated between each frame of the sequence for adjusting the topology of the previous existing GNG structure.

Finally, some experiments were carried out using real information from a Microsoft Kinect v2 sensor. Figure 6 shows the result of applying the proposed method to a sequence of real 3D data. The GNG extension is able to represent and track the two moving objects in the scene. The proposed method takes advantage of motion difference from previous frames to focus on moving objects and depth information for separating object representations within the same scene. No qualitative results are presented since real sensors do not provide us with ground truth information for error validation. That is the mean reason why first experiments were performed using a simulated 3D sensor.

Fig. 6. Multiple object representation and tracking from real data obtained using the Kinect v2. The proposed method was used to represent and track objects in the scene using as a priori knowledge motion from previous frames. GNG representation (green wire-frame structure) is shown over the tracked objects: Creeper figure and the air freshener.

4 Conclusions and Future Work

In this paper we have presented a novel computing method to manage complete sequences of noisy 3D data. Previous knowledge about the sensor is not necessary. It has been demonstrated how our GNG-based technique is capable to deal with point cloud sequences by adjusting its topology and therefore reducing the computational time taken by creating a new structure for each new frame. Moreover, the proposed method is able to estimate the trajectory of the moving objects along the sequence with a low translational error. Furthermore, The method was validated with various 3D models ranging from synthetic objects (simulated sensor) to real ones acquired using the new Kinect v2 sensor. Obtained models were able to accurately represent objects with an error below 5 millimeters.

Future work includes the evaluation of the proposed method using existing data sets and performance comparison with other state-of-the-art 3D object tracking techniques.

Acknowledgments. This work was partially funded by the Spanish Government DPI2013-40534-R grant.

References

1. Yang, H., Shao, L., Zheng, F., Wang, L., Song, Z.: Recent advances and trends in visual tracking: A review. Neurocomput. 74(18), 3823–3831 (2011)
2. Fritzke, B.: A self-organizing network that can follow non-stationary distributions. In: Gerstner, W., Hasler, M., Germond, A., Nicoud, J.-D. (eds.) ICANN 1997. LNCS, vol. 1327, pp. 613–618. Springer, Heidelberg (1997)
3. Fritzke, B.: A Growing Neural Gas Network Learns Topologies, vol. 7, pp. 625–632. MIT Press (1995)
4. Frezza-Buet, H.: Following non-stationary distributions by controlling the vector quantization accuracy of a growing neural gas network. Neurocomput. 71, 1191–1202 (2008)
5. Cao, X., Suganthan, P.N.: Hierarchical overlapped growing neural gas networks with applications to video shot detection and motion characterization. In: Proc. Int. Joint Conf. Neural Networks IJCNN 2002, vol. 2, pp. 1069–1074 (2002)
6. Frezza-Buet, H.: Online computing of non-stationary distributions velocity fields by an accuracy controlled growing neural gas. Neural Networks 60, 203–221 (2014)
7. Coleca, F., State, A., Klement, S., Barth, E., Martinetz, T.: Self-organizing maps for hand and full body tracking. Neurocomputing 147, 174–184 (2015)
8. Garcia-Rodriguez, J., Garcia-Chamizo, J.M.: Surveillance and human-computer interaction applications of self-growing models. Appl. Soft Comput. 11(7), 4413–4431 (2011)
9. Garcia-Rodriguez, J., Orts-Escolano, S., Angelopoulou, A., Psarrou, A., Azorin-Lopez, J., Garcia-Chamizo, J.: Real time motion estimation using a neural architecture implemented on gpus. Journal of Real-Time Image Processing, 1–19 (2014)
10. Martinetz, T.M., Berkovich, S.G., Schulten, K.J.: 'Neural-gas' network for vector quantization and its application to time-series prediction 4(4), 558–569 (1993)
11. Orts-Escolano, S., Morell, V., Garcia-Rodriguez, J., Cazorla, M.: Point cloud data filtering and downsampling using growing neural gas. In: The 2013 International Joint Conference on Neural Networks, IJCNN 2013, Dallas, TX, USA, August 4-9, pp. 1–8 (2013)
12. Orts-Escolano, S., Garcia-Rodriguez, J., Moreli, V., Cazorla, M., Garcia-Chamizo, J.M.: 3d colour object reconstruction based on growing neural gas. In: 2014 International Joint Conference on Neural Networks (IJCNN), pp. 1474–1481 (July 2014)
13. Gschwandtner, M., Kwitt, R., Uhl, A., Pree, W.: BlenSor: Blender Sensor Simulation Toolbox. In: Bebis, G., Boyle, R., Parvin, B., Koracin, D., Wang, S., Kyungnam, K., Benes, B., Moreland, K., Borst, C., DiVerdi, S., Yi-Jen, C., Ming, J. (eds.) ISVC 2011, Part II. LNCS, vol. 6939, pp. 199–208. Springer, Heidelberg (2011)

Object Recognition in Noisy RGB-D Data

José Carlos Rangel[1(✉)], Vicente Morell, Miguel Cazorla, Sergio Orts-Escolano, and José García Rodríguez

Institute for Computer Research, University of Alicante, Alicante, Spain
jcrangel@dccia.ua.es

Abstract. The object recognition task on 3D scenes is a growing research field that faces some problems relative to the use of 3D point clouds. In this work, we focus on dealing with noisy clouds through the use of the Growing Neural Gas (GNG) network filtering algorithm. Another challenge is the selection of the right keypoints detection method, that allows to identify a model into a scene cloud. The GNG method is able to represent the input data with a desired resolution while preserving the topology of the input space. Experiments show how the introduction of the GNG method yields better recognitions results than others filtering algorithms when noise is present.

Keywords: Growing neural gas · 3D object recognition · Keypoints detection

1 Introduction

3D object recognition is a growing research line which has been impulsed by the notorious advantages that offers the use of 3D sensors against the 2D based recognition methods. However, it is a topic that presents difficulties to achieve an effective recognition. Some of these difficulties are: noise, occlusions, rotations, translations, scaling and holes that are present in the 3D raw point clouds provided by the nowadays RGB-D sensors like Microsoft Kinect. Therefore, new algorithms are required to handle these problems and perform a correct object recognition process.

There exist several works in the field of 3D object recognition, some of these make a survey, review or evaluation of the existing 3D object recognition methods while other works focus on the proposal of new methods and approaches for the recognition process. In [5] a survey of 3D object recognition methods based on local surface features is presented. They divided the recognition process in three basic phases: 3D keypoint detection, feature description, and surface matching. It also describes existing datasets and the algorithms used by every phase of the whole process. Other studies like [13] focus on the evaluation of stereo algorithms. It makes an evaluation in terms of the recognition ability of this kind of algorithms. Using a different approach, [2] evaluates the different 3D shape descriptors for object recognition to study the feasibility of such descriptors in 3D object recognition.

J.M. Ferrández Vicente et al. (Eds.): IWINAC 2015, Part II, LNCS 9108, pp. 261–270, 2015.
DOI: 10.1007/978-3-319-18833-1_28

There are some works that propose novel object recognition pipelines, like [6], where using depth maps and images it achieves good recognition results on heavy cluttered scenes. Using a different approach, [12] proposes a novel Hough voting algorithm to detect free-form shape in a 3D space, which produces good recognition rates. [8] describes a general purpose 3D object recognition framework that combines machine learning procedures with 3D local features, without a requirement for a priori object segmentation. This method detects 3D objects in several 3D point cloud scenes, including street and engineering scenes. [1] proposes a new method called Global Hypothesis Verification (Global HV). This work adds the Global HV algorithm to the final phase of the recognition process to discard false positives. Our approach is based in the pipeline presented in this work, introducing noise into the original point cloud to test the effect of that noise in the recognition process.

In this paper we propose the use of a Growing Neural Gas (GNG) to represent and reduce the raw point clouds. These self-organizing maps learn the distribution of the input points and adapting their to topology. This feature allows to get a compact and reduced representation of the input space in a set of 3D neurons and their connections. In addition, we test different keypoints detectors to determine which obtain betters recognition results. Other papers like[17] use a GNG algorithm to filter and reduce single frontal point clouds. This GNG reduction improves the recognition process and reduces noisy 3D values. We will also compare our proposal against other reduction/filtering methods like Voxel Grid. Hence we present experiments that test a 3D object recognition pipeline with both the raw point cloud, the GNG and Voxel Grid filtered point clouds. Besides, we describe the selected dataset for the experiments and show the results.

The rest of this work is organized as follows. First, we introduce and describe in Section 2 the GNG and Voxel Grid methods that we will use in the experimentation. Then, in Section 3 the pipeline is explained. Next, Section 4 describes the dataset and how the recognition experiments are carried out. After that, in Section 5 we present the results and discussion of our experiments and, finally, conclusions are drawn.

2   3D Filtering Methods

One way of selecting points of interest in 3D point clouds is to use a topographic mapping where a low dimensional map is fitted to the high dimensional manifold of the model, whilst preserving the topographic structure of the data. In this section, we review some typical methods to represent and reduce 3D data. First, we describe the Growing Neural Gas algorithm and how it works. Then, we briefly describe the Voxel Grid method, which is other commonly used data structure, in order to compare our proposal method.

2.1 GNG Method

In GNG, nodes in the network compete for determining the set of nodes with the highest similarity to the input distribution. In our case, the input distribution is a finite set of 3D points that can be extracted from different types of sensors. The highest similarity reflects which node together with its topological neighbors is the closest to the input sample point which is the signal generated by the network. The n-dimensional input signals are randomly generated from a finite input distribution.

The nodes move towards the input distribution by adapting their position to the input data geometry. During the learning process local error measures are gathered to determine where to insert new nodes. New nodes are inserted near the node with the highest accumulated error. At each adaptation step a connection between the winner and its topological neighbors is created as dictated by the competitive Hebbian learning method. This is continued until an ending condition is fulfilled, as for example evaluation of the optimal network topology, a predefined networks size or a deadline.

Using a Growing Neural Gas model to represent 3D data has some advantages over the traditionally used methods like Voxel Grid. For example, we specify the number of neurons (representative points of the map), while other methods like the Voxel Grid get different number of occupied cells depending on the distribution and resolution of the cells (voxels).

Figure 1 shows an example of a GNG representation of one of the objects we use in the experimentation. The GNG forms a map and we use only the neurons as a new filtered and reduced representation of the object.

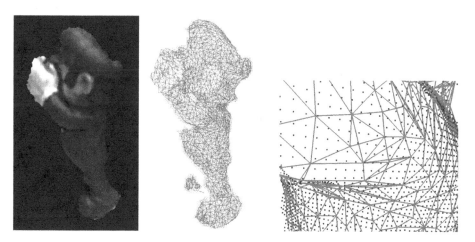

Fig. 1. Left: Original object. Center: GNG representation of one of the objects we use in the experimentation. Right: Zoom of the arm of the mario.

2.2   Voxel Grid Method

The Voxel Grid (VG) down-sampling technique is based on the input space sampling using a grid of 3D voxels[18]. The VG algorithm defines a voxel grid in the 3D space and for each voxel, a point is chosen as the representative of all points that lie on that voxel. It is necessary to define the size of the voxels as this size establishes the resolution of the filtered point cloud and therefore the number of points that form the new point cloud. The representative of each cell is usually the centroid of the voxel's inner points or the center of the voxel grid volume. Thus, a subset of the input space is obtained that roughly represents the underlying surface. The VG method presents the same problems than other sub-sampling techniques: it is not possible to define the final number of points which represents the surface; geometric information loss due to the reduction of the points inside a voxel and sensitivity to noisy input spaces.

3   3D Object Recognition

Our object recognition pipeline (Figure 2) is based on the proposed in [1]. However, we have introduced some changes in the original method, like adding more keypoints detectors. The recognition pipeline is as follow:

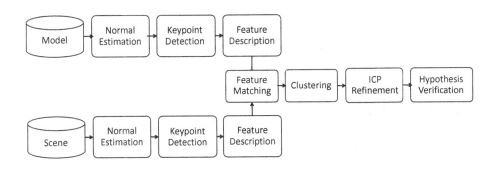

Fig. 2. Recognition Pipeline scheme

Keypoints Detection. After the normals are estimated, the next step is to extract the keypoints for the model and scene point cloud using the Uniform Sampling, Harris3D or Intrinsic Shape Signature(ISS) method.

Uniform Sampling: This method builds a 3D grid over the input point cloud. This grid of 3D cuboids, which are called voxels, are located upon the point cloud and only one point is used to represent all the points inside each voxel [9]. Usually this representative point is the centroid of the voxel inner points.

Harris 3D: [5][11] Harris 3D is an extension of the 2D Harris detector method which uses the normals instead of the image gradients. It uses a Gaussian function to smooth the derivative surfaces to mitigate the effect of local deformations introduced by noise, holes, etc. It also proposes an adaptive neighborhood selection which improves the feature detection.

Intrinsic Shape Signatures(ISS):[19][5][16] It is based on the Eigenvalue Decomposition of the scatter matrix of the points belonging to the support of a point. This method employs the ratio of two successive eigenvalues (λ_1,λ_2,λ_3) to prune the points. Only the points which ratio between two successive eigenvalues is below a threshold (τ) are retained. Among the remaining points, the salience is determined by the magnitude of the smallest eigenvalue λ_1, $\lambda_2/\lambda_1 < \tau_{21}$ and $\lambda_3/\lambda_2 < \tau_{32}$.

Feature Description using SHOT Descriptor. A descriptor codifies the underlining information in a certain neighborhood around a keypoint. Once the keypoints are computed, we need to extract the descriptors. There are several 3D descriptors, but the original work uses the Unique Signatures of Histograms for Local Surface Description (SHOT) [14] descriptor. We have tested different descriptors and, in fact, this is the one providing better results for this task. This descriptor is an intersection between signatures and histograms. It takes each detected keypoint and builds a local reference frame and then it divides the neighborhood space into 3D spherical volumes to compute the signature or histogram of that point. The SHOT descriptor is highly descriptive, computationally efficient and robust to noise.

Feature Matching and Clustering. To determine the correspondences between the model and the scene descriptors, we used the KDTreeFLANN method. This method of the Fast Library for Approximate Nearest Neighbors (FLANN) library, uses a kd-tree and an approximate nearest neighbor scheme to find a closer feature (the closest is not guaranteed) in a quick way [7]. After that, we group the correspondences found into smaller clusters using the Geometric Consistency (GC)[5] grouping method.

ICP Refinement and Hypothesis Verification. Using the Iterative Closest Point (ICP) method [3], we refine the 6 DoF (degrees of freedom) pose given by the absolute orientation of the subsets provided by the clustering stage. The Hypothesis Verification algorithm was proposed in [1]. This method uses a Simulated Annealing Meta-heuristic algorithm to solve the cost function it uses to determine if the hypothesis is valid or not.

4   Experimentation

This section briefly describes the features of the selected dataset, the library used for the 3D object recognition comparison and how the experiments were performed.

4.1   Dataset

To test the different approaches, we used the SHOT Dataset[1] (University of Bologna)[15][14][4]. This dataset has been acquired by means of the Spacetime Stereo (STS) technique and consists of 7 models, with different views of each model, and 17 scenes for a total of 49 object recognition instances. Using these point clouds we will test the noise influence on the recognition rate. Figure 3 shows two models of the dataset and two scenes where they appear.

Fig. 3. On the left, two models of the dataset, the Mario and the Peter Rabbit. On the right, two scenes from the dataset.

4.2   Experimentation Setup

For the experiments we used PCL[2]. The experiment consists of searching a selected model in a scene with the recognition pipeline above described. We take a model and seek it in one of the noised/filtered scenes. First, using the raw point clouds, and the filtered point clouds with GNG and Voxel Grid.

We filter the scene dataset cloud with the GNG and the Voxel Grid method with 10000, 15000, 17500 and 20000 representatives points and the models to 2000 representative points. Besides, we applied five different levels of Gaussian noise with 0.001, 0.0025, 0.005, 0.0075 and 0.01 meters of standard deviation only to the scene, because the stored models are supposed to be in a higher quality level.

---

[1] http://www.vision.deis.unibo.it/research/80-shot
[2] The Point Cloud Library (or PCL) is a large scale, open project[10] for 2D/3D image and 3D point processing.

At last, we have a dataset with six different set of point clouds and five levels of noise over the scene datasets. We decided to test every combination of model and scene onto the different datasets to get the most representative values. Table 1 shows all the possible combinations of our datasets, the first word of each pair is the method applied to the model and the second one the scene method.

Table 1. List of the experiments combinations

GNG Models	RAW Models	Voxel Grid Models
GNG_GNG	RAW_GNG	VoxelGrid_GNG
GNG_GNGNoise	RAW_GNGNoise	VoxelGrid_GNGNoise
GNG_RAW	RAW_RAW	VoxelGrid_RAW
GNG_RAWNoise	RAW_RAWNoise	VoxelGrid_RAWNoise
GNG_VoxelGrid	RAW_VoxelGrid	VoxelGrid_VoxelGrid
GNG_VoxelGridNoise	RAW_VoxelGridNoise	VoxelGrid_VoxelGridNoise

The experiment consists of comparing the results obtained by the object recognition pipeline using three different keypoints detectors: Uniform Sampling, Harris 3D and ISS. For every experiment combination in the Table 1, we test the system with the 49 recognition instances available in the dataset.

To measure the performance of the recognition pipeline we use the Hypothesis Verification algorithm which analyzes the results and provides us the true positives of the recognition method over the different datasets. When the system finds a true positive, it only takes the instance with more matched points between the model and the scene that have been located and shows a screen with the model superimposed in the scene, in the place where the instance has been located Figure 4. When the system produces a false positive it is because the instance found could not be right located in the scene. Next section will show the results of the experiments described above.

5   Results

This section presents the results obtained after the execution of the different set of experiments. The percentage of true positives obtained of the whole experiments set is 72,8%. After analysis of the data we compare the result for each keypoint detector, getting Uniform Sampling the highest true positive recognition percentage with 45,8% followed by Harris 3D with 44,9% and finally ISS with 9,3%. Another tested keypoint detectors were SUSAN, SIFT and Harris 6D, but without positives results since our GNG clouds were filtered without the RGB data. Figure 5 shows the percentage of true positives obtained from Uniform Sampling experiments. There, we can see that independently of the combination model/scene the better results was observed when we used the noisy scenes filtered by the GNG method. The Figure 6 shows the mean of the recognition results obtained for the evaluated keypoints detectors. Here we can

Fig. 4. Recognition Result obtained by the pipeline

see that the higher values were obtained using the scenes filtered by the GNG method. In Figure 5 and Figure 6 the highest values in the set are shaded.

In the presence of noise, the higher recognition rate obtained was 83%, using the scenes with 0.005 noise level and filtered using the GNG method with 10000 representatives points. By using a Raw cloud with the same noise level do not produce recognition results. This result supports our proposal that the use of the GNG improves the results of recognition on noisy clouds. Another remarkable result is that the GNG outperformed the rate obtained for the clouds filter with the Voxel Grid Method.

						Scenes					
		RAW	**GNG**				**VoxelGrid**				
Models	**Scene's Noise**	**All Points**	**10000**	**15000**	**17500**	**20000**	**10000**	**15000**	**17500**	**20000**	
RAW	0	86	50	47	53	65	9	31	23	31	
	0.001	82	45	49	50	57	17	23	27	28	
	0.0025	20	58	67	74	73	9	16	22	36	
	0.005	0	75	77	69	73	0	0	0	0	
	0.0075	0	37	43	31	22	0	0	0	0	
	0.01	0	3	0	0	0	0	0	0	0	
GNG	0	84	52	58	57	70	23	29	26	42	
	0.001	69	59	64	67	72	18	38	40	49	
	0.0025	9	59	67	79	73	32	24	22	40	
	0.005	0	83	71	73	76	0	0	3	0	
	0.0075	0	49	31	14	10	0	0	0	0	
	0.01	0	0	0	0	0	0	0	0	0	
VoxelGrid	0	78	51	66	64	70	27	38	42	47	
	0.001	65	48	64	62	70	27	38	37	45	
	0.0025	0	63	74	76	66	24	35	35	41	
	0.005	0	60	74	63	69	0	0	3	8	
	0.0075	0	36	24	26	13	0	0	0	0	
	0.01	0	0	0	0	0	0	0	0	0	

Fig. 5. Results for the Experiments using Uniform Sampling

Models	Scene's Noise	RAW	Scenes GNG				VoxelGrid			
		All Points	10000	15000	17500	20000	10000	15000	17500	20000
RAW	0	85	45	48	48	53	4	21	16	27
	0.001	83	46	52	46	51	7	20	22	26
	0.0025	11	52	57	63	70	6	16	26	38
	0.005	0	57	70	63	66	0	0	1	2
	0.0075	0	24	26	17	12	0	0	0	0
	0.01	0	1	0	0	0	0	0	0	0
GNG	0	80	52	60	56	67	15	29	25	38
	0.001	63	53	60	63	67	14	33	37	47
	0.0025	3	56	64	68	70	18	24	34	42
	0.005	0	60	67	63	60	0	0	3	0
	0.0075	0	32	19	10	6	0	1	1	0
	0.01	0	0	0	0	0	0	0	0	0
VoxelGrid	0	72	49	59	56	67	23	42	40	48
	0.001	59	49	57	64	72	22	45	44	47
	0.0025	0	53	60	70	68	19	37	46	52
	0.005	0	50	63	56	55	0	0	1	4
	0.0075	0	19	16	10	5	0	0	0	0
	0.01	0	0	0	0	2	0	0	0	0

Fig. 6. The mean of the results for the Experiments using the evaluated keypoint detectors

6   Conclusions

The presence of noise in a 3D point cloud could made impossible a recognition process. Therefore applying a filtering method that reduce the noise will improve the recognition on noisy clouds. This paper presented an approach that reduces the noise level of the clouds and improves the recognition rate on noisy clouds, using the GNG or Voxel Grid algorithm. Besides, we identify that the Uniform Sampling is the keypoint detector that achieve better rates of recognition for the used pipeline. Our results show that the use of the GNG method improve the recognition rates outperforming the result for Voxel Grid and for the clouds without a filtering process. The GNG reduces the noise without losing significant information and making possible a good recognition results.

As future work, we propose to perform a deep evaluation of the existing 3D detection and description methods in order to find the best combination of performance and noise robustness.

Acknowledgments. This work was funded by the Spanish Government DPI2013-40534-R grant.

References

1. Aldoma, A., Tombari, F., Di Stefano, L., Vincze, M.: A global hypotheses verification method for 3D object recognition. In: Fitzgibbon, A., Lazebnik, S., Perona, P., Sato, Y., Schmid, C. (eds.) ECCV 2012, Part III. LNCS, vol. 7574, pp. 511–524. Springer, Heidelberg (2012)
2. As'ari, M.A., Sheikh, U.U., Supriyanto, E.: 3D shape descriptor for object recognition based on Kinect-like depth image. Image and Vision Computing 32(4), 260–269 (2014)

3. Besl, P.J., McKay, N.D.: A method for registration of 3-d shapes. IEEE Trans. on Pattern Analysis and Machine Intelligence 14(2), 239–256 (1992)
4. Computer Vision LAB: SHOT: Unique signatures of histograms for local surface description - computer vision LAB, http://www.vision.deis.unibo.it/research/80-shot
5. Guo, Y., Bennamoun, M., Sohel, F., Wan, J., Lu, M.: 3d object recognition in cluttered scenes with local surface features: A survey. IEEE Transactions on Pattern Analysis and Machine Intelligence PP(99), 1 (2014)
6. Hinterstoisser, S., Holzer, S., Cagniart, C., Ilic, S., Konolige, K., Navab, N., Lepetit, V.: Multimodal templates for real-time detection of texture-less objects in heavily cluttered scenes. In: 2011 IEEE International Conference on Computer Vision (ICCV), pp. 858–865 (November 2011)
7. Muja, M.: FLANN - fast library for approximate nearest neighbors: FLANN - FLANN browse, http://www.cs.ubc.ca/research/flann/
8. Pang, G., Neumann, U.: Training-based object recognition in cluttered 3d point clouds. In: 2013 International Conference on 3D Vision - 3DV 2013, pp. 87–94 (June 2013)
9. Radu Bogdan Rusu: Point cloud library (PCL): pcl::UniformSampling< PointInT> class template reference
10. Rusu, R.B., Cousins, S.: 3D is here: Point Cloud Library (PCL). In: Proceedings of the IEEE International Conference on Robotics and Automation (ICRA), Shanghai, China, May 9-13 (2011)
11. Sipiran, I., Bustos, B.: Harris 3d: a robust extension of the harris operator for interest point detection on 3d meshes. The Visual Computer 27(11), 963–976 (2011)
12. Tombari, F., Di Stefano, L.: Object recognition in 3d scenes with occlusions and clutter by hough voting. In: 2010 Fourth Pacific-Rim Symposium on Image and Video Technology (PSIVT), pp. 349–355 (November 2010)
13. Tombari, F., Gori, F., Di Stefano, L.: Evaluation of stereo algorithms for 3d object recognition. In: 2011 IEEE International Conference on Computer Vision Workshops (ICCV Workshops), pp. 990–997 (November 2011)
14. Tombari, F., Salti, S.: A combined texture-shape descriptor for enhanced 3d feature matching. In: 2011 18th IEEE International Conference on Image Processing (ICIP), pp. 809–812 (September 2011)
15. Tombari, F., Salti, S., Di Stefano, L.: Unique signatures of histograms for local surface description. In: Daniilidis, K., Maragos, P., Paragios, N. (eds.) ECCV 2010, Part III. LNCS, vol. 6313, pp. 356–369. Springer, Heidelberg (2010)
16. Tombari, F., Salti, S., DiStefano, L.: Performance evaluation of 3d keypoint detectors. International Journal of Computer Vision 102(1-3), 198–220 (2013)
17. Viejo, D., Garcia, J., Cazorla, M., Gil, D., Johnsson, M.: Using GNG to improve 3d feature extraction-application to 6dof egomotion. Neural Networks (2012)
18. Xu, G., Mourrain, B., Duvigneau, R., Galligo, A.: Analysis-suitable volume parameterization of multi-block computational domain in isogeometric applications. Computer-Aided Design 45(2), 395–404 (2013), solid and Physical Modeling 2012
19. Zhong, Y.: Intrinsic shape signatures: A shape descriptor for 3d object recognition. In: 2009 IEEE 12th International Conference on Computer Vision Workshops (ICCV Workshops), pp. 689–696 (September 2009)

Topology Preserving Self-Organizing Map of Features in Image Space for Trajectory Classification

Jorge Azorin-Lopez[1]([✉]), Marcelo Saval-Calvo[1], Andres Fuster-Guillo[1], Higinio Mora-Mora[1], and Victor Villena-Martinez[1]

Department of Computer Technology, University of Alicante, 03080 Alicante, Spain
{jazorin,msaval,fuster,hmora,vvillena}@dtic.ua.es

Abstract. Self-Organizing maps (SOM) are able to preserve topological information in the projecting space. Structure and learning algorithm of SOMs restrict the topological preservation in the map. Adjacent neurons share similar vector features. However, topological preservation from the input space is not always accomplished. In this paper, we propose a novel self-organizing feature map that is able to preserve the topological information about the scene in the image space. Extracted features in adjacent areas of an image are explicitly in adjacent areas of the self-organizing map preserving input topology (SOM-PINT). The SOM-PINT has been applied to represent and classify trajectories into high level of semantic understanding from video sequences. Experiments have been carried out using the Shopping Centre dataset of the CAVIAR database taken into account the global behaviour of an individual. Results confirm the input preservation topology in image space to obtain high performance classification for trajectory classification in contrast of traditional SOM.

Keywords: Self-organizing maps · Image topology preservation · Trajectory classification

1 Introduction

A Self-Organizing Map (SOM) is one of the most popular and used artificial neural network model. It was introduced by Kohonen in 1982 [8] and nowadays it remains been used and applied in many areas. It mainly converts a high dimensional input into a low dimensional map of features, being the two-dimensional map the most used representation. The map structure varies in some characteristics (lattice shape, neighbourhood connections, etc.) but generally it conforms a grid of interconnected neurons with a specific neighbourhood. This interconnection assures a topological preservation in the map space. The SOM uses a unsupervised learning algorithm based on a competition where nodes of the grid compete to become the winning neuron (the most similar neuron to the input vector) in order to adjust its vector components and those of the nearest neighbour neurons. This competition allows to project the input space into the map.

© Springer International Publishing Switzerland 2015
J.M. Ferrández Vicente et al. (Eds.): IWINAC 2015, Part II, LNCS 9108, pp. 271–280, 2015.
DOI: 10.1007/978-3-319-18833-1_29

Different variants have been proposed to the classical SOM for many years [7]. For example, variants proposed by Fritzke as the Growing Cell Structure, GC [4] and Growing Grid, GG [5] are able to automatically find a suitable network structure and size by a controlled growth process, while maintain the network topology. Other networks as Neural Gas, NG [11] and Growing Neural Gas, GNG [6] are able to reconfigure the neighbourhood relationships to avoid a predefined network topology. They are able to make explicit the important topological relations in a given distribution of input data. Moreover, the Growing Neural Gas eliminates the need to predefine the network size in the similar way as the Growing Cell Structure. In consequence, although SOM, GG and GC are able to preserve the topological constraints in the projected space (in the map) [18], they are not always able to preserve the input space as GNG and NG.

In this paper, we are interested in representing features extracted from an image for a posterior classification but preserving the topological relations of an image in which the features have been extracted. In consequence, the SOM preserving topology for the projected space is necessary. Moreover, the two-dimensional grid is the topology that fits an image. However, the classical SOM does not take into account the location of the extracted features. Hence, the learning process does not assure the topological input preservation in the image.

We propose a variant of the self-organizing feature map that is able to preserve the topological information about the scene in the image space. Extracted features in adjacent areas of an image are explicitly in adjacent areas of the self-organizing map preserving the input topology (SOM-PINT). The SOM-PINT has been applied to represent and classify trajectories into high level of semantic understanding from video sequences. The neural network is able to deal with the big gap between human trajectories in a scene and the global behaviour associated to them preserving the spatial information about trajectories. As example of features, we use the Activity Descriptor Vector [1] as inputs of the network in order to incorporate the trajectory information of people in a scene. The trajectory provides very interesting data for a wide rage of activities, and it is also easy to obtain due to the large number of cameras available nowadays, mainly CCTV.

Self-Organizing networks have been widely used in Computer Vision. Concretely, in trajectory analysis, various studies are present in the state-of-the-art. Morris and Trivedi [13] present a framework in which the behaviours are described using motion patterns, for real-time characterization and prediction of future activities, as well as the detection of abnormalities. Trajectories are utilized to automatically build activity models in a 3-stage hierarchical learning process. Martinez-Contreras et al. [12] use SOMs only for motion (trajectory) sampling. A SOM is trained with different motions, and then a new motion is classified and the template is used in a Hidden Markov Model to determinate the action. Schreck et al. [16] developed a framework to classify trajectories using SOMs, scaling the paths into unit square values and sampling them in a predefined number of parts. Madokoro et al. [10] extracts typical behaviour patterns and specific behaviour patterns from human trajectories quantized using

One-Dimensional Self-Organizing Maps (1D-SOMs). Subsequently, they apply Two-Dimensional SOMs (2D-SOMs) for unsupervised classification of behaviour patterns. A hierarchical SOM is presented in [15] to detect abnormal human activities based on trajectories, body features and directions, showing accurate results in different scenarios and sensors.

Normally, raw trajectories are not studied directly using pattern analysis due to the varying in length of data (same trajectory pattern can be done slower or making small variations of the path). Therefore, a normalization of data has to be done. In our previous work [1,2], we proposed a descriptor of the behaviour using the trajectory information. The Activity Description Vector (ADV) was proposed to equally divide the scene in cells, and estimates the up, down, left, right and frequency of the person in each cell. We evaluated this using different pattern recognition techniques obtaining high accuracy in classification. Moreover, this descriptor has been proved to obtain prediction capabilities for early recognition [2].

The remainder of the paper is organized as follows. Section 2 presents the novel Self Organizing Map Preserving the Input Topology of the image proposed in this research. Experimental results obtained by applying the network for classify trajectories into human behaviours are presented in Section 3. They are discussed and compared to other approaches in the same Section. Finally, conclusions about the research are presented in Section 4.

2 Self Organizing Map Preserving Input Topology

Self Organizing Map Preserving Input Topology (SOM-PINT) of images is a novel neural network able to represent features and classify them preserving the image topology. It is based on the classical SOM. Briefly, a SOM neural network involves two phases: a training/learning and a classification process. For the training phase using a sequential learning, in each step, one sample (a vector) is selected from the input data set to calculate a distance measure between the sample and each neuron (the corresponding reference vector for the neuron). The neuron whose reference vector is closer to the input sample is selected as the winning neuron. The neighbourhood of the winning neuron is adapted to the input sample. In this paper, self organizing basis are considered to represent features preserving the image topology in a map. The whole map is used in classification tasks. Some variants have been introduced in the learning and classification process to preserve the scene topology in the network.

Specifically, the SOM-PINT uses as input samples a collection of descriptors D extracted from an image I. As we are interested in preserving the topological information, a matrix is the specific collection of descriptors. We assume that the image I of size uxv could be discretized into cells C of size mxn that represent regions in the image (being $mxn \leq uxv$). Each component of the matrix, Descriptor Matrix (DM) represents a region of the scene, a specific cell, by means of the descriptor D calculated in that area of the image.

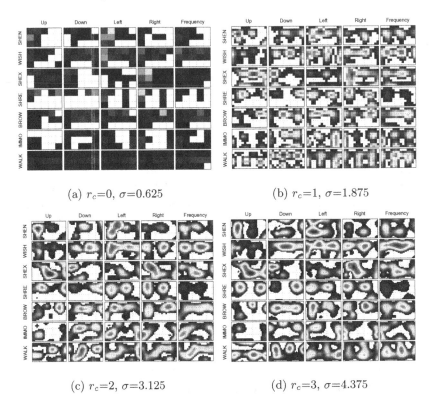

(a) $r_c=0$, $\sigma=0.625$ (b) $r_c=1$, $\sigma=1.875$

(c) $r_c=2$, $\sigma=3.125$ (d) $r_c=3$, $\sigma=4.375$

Fig. 1. SOM-PINTs calculated from a 3x5 discretized grid for different behaviours, neighbour radius r_c and σ for Gaussian funtion

The network structure is completely related to the Descriptor Matrix and is specified as:

- A set N of neurons that represent cells of the grid C in which the image has been discretized. Each neuron $\nu \in N$ stores an associated reference vector $w_\nu \in R^n$. The reference vectors are related to the descriptors D of the region assigned to the neurons.
- A structure about connections between adjacent neurons. It defines the topological structure of the map that represents the geometry of the image. Usually, a 2-dimensional regular grid of neurons.

The size of the map is determined by the number of cells in C. Specifically, a SOM-PINT establishes a subset of neurons $N_{i,j}$ to represent a specific cell $C_{i,j}$. Each $N_{i,j}$ contains a neuron and a neighbourhood of neurons within a radius r_c.

2.1 Learning Process

Although we are interested in an unsupervised learning, SOM-PINT considers specific labels for classification purposes. The label is the higher level of semantic that a DM represents. Hence, a different SOM-PINT$_b$ is trained for a specific label b. In consequence, the input dataset $X = \{DM_1, DM_2, \ldots, DM_n,\}$ is divided into different groups according to samples pertaining to a single label $X_b \in X$. The training process for each SOM-PINT$_b$ considers a set of winning neurons as opposite of a winner for an input DM_i (as in the original SOM). Each descriptor D in DM_i activate a neuron. In consequence, the whole map is adapted considering all winning neurons to represent the DM for a specific label. For classification purposes, each SOM-PINT$_b$ is compared to a new input DM to establish the minimum distance for the whole map. Next, learning and classification steps are explained in detail.

The learning algorithm for a single SOM-PINT$_b$ having an input dataset $X_b = \{DM_1, DM_2, \ldots, DM_n,\}$ of a specific label is as follows:

1. Initialize each neuron $\nu \in N$ with random values w_ν in R^n.
2. Choose randomly a sample pattern DM from the input data set X.
3. For each element $D_{i,j}$ in the DM matrix, find the nearest neuron, winning neuron, $s_{i,j}$ in the corresponding $N_{i,j}$ set of neurons associated to the cell $C_{i,j}$. In consequence, a set of mxn winning neurons $S \in N$ will be associated to the sample pattern DM.

$$\|w_{s_{i,j}} - D_{i,j}\|^2 = \min_{\nu \in N_{i,j}} \{\|w_\nu - D_{i,j}\|^2\} \tag{1}$$

4. Update the map
   – Determine the adapted neighbours and the strength of the adaptation by the neighbourhood function $h(d,t)$ for each winning neuron $s_{i,j}$. It depends on the distance d of each neuron in N to the winning neuron and on the training time t.

$$a_{s_{i,j}} = h_{s_{i,j}}(d,t) * \|w_{s_{i,j}} - D_{i,j}\| \tag{2}$$

   – Update the reference vector w_ν for each $\nu \in N$ according to all adapted neighbours $a_{s_{i,j}}$. A neuron ν could be affected by the adaption for different neighbourhoods. Finally, the strength of the update for each reference vector w is established by the learning rate α. It depends also on the training time t.

$$w_\nu(t+1) = w_\nu(t) - \alpha(t) * \sum_{\forall s_{i,j} \in S} a_{s_{i,j}} \tag{3}$$

5. If the training time t is not yet achieved, go to step 2.

2.2 Classification Process

Finally, the classification process of a new input data DM, for a set of k labels represented by the set $B = \{\text{SOM-PINT}_1, \text{SOM-PINT}_2, \ldots, \text{SOM-PINT}_k\}$ is as follows:

1. Determine the distance η from the DM to each $SODM_i \in B$

- For each element $D_{i,j}$ in the DM matrix, find the nearest neuron, winning neuron, $s_{i,j}$ in the corresponding $N_{i,j}$ set of neurons associated to the cell $C_{i,j}$ as in the step 3 of the training process.

$$\|w_{s_{i,j}} - D_{i,j}\|^2 = \min_{\nu \in N_{i,j}} \left\{ \|w_\nu - D_{i,j}\|^2 \right\} \tag{4}$$

- Determine the sum of distances of the winning neurons S

$$\eta_{\text{SOM-PINT}_i} = \sum |w_{s_{i,j}} - D_{i,j}\|^2 \tag{5}$$

2. Select the SOM-PINT$_i$ with minimum distance $\eta_{\text{SOM-PINT}_i}$ as the label associated to the DM input

$$label = \min_{\forall \text{SOM-PINT}_i \in B} (\eta_{\text{SOM-PINT}_i}) \tag{6}$$

3   Experiments

3.1   Trajectory Representation and Classification

In this section, SOM-PINT has been applied to represent trajectories in order to classify them into global human behaviour. Experiments have been carried out using the CAVIAR database [3]. Specifically, validation of the representation of trajectories to recognize human behaviours makes use of the 26 clips from the *Corridor* dataset of Shopping Centre in Portugal. The dataset contains 1500 frames on average tracking 235 individuals performing 255 different trajectories (some of them have different contexts during a sequence): shop enter (55 samples), windowshop (18 samples), shop exit (63 samples), shop reenter (5 samples), browsing (10 samples), immobile (22 samples) and walking (82 samples). As the samples are imbalanced, the Synthetic Minority Over-Sampling Technique (SMOTE) [14] has been applied to obtain the same number of samples for each context. Also, for the Walking context, a subset of samples is randomly taken with 60 samples. All these samples have been used for training and classification steps.

For each frame, the Activity Descriptor Vector (ADV) [1] has been calculated to extract the collection of descriptors DM. This method divides the scenario in cells, C, to discretize it. Each cell of the grid has information about the movements performed it. Specifically, 1x1, 3x5, 5x7, 7x11 and 9x13 are used as grid cells. Each cell and its corresponding ADV represent a region of the scene. Additionally, the radius r_c of neighbourhood in N_i, j has been selected from 0 to 4, conforming from 1 to 81 neurons to represent each region of the scene. According to the descriptor D used, the reference vector for each neuron $w_\nu \in R^5 = \{U, D, L, R, F\}$.

For the learning process, all samples in X have been normalized to the range (0,1) dividing each component of the DM matrix by the maximum value for each component. Two learning rates are established according two tuning steps: first stage as rough training phase for α in range (0.5, 0.1) for 20 epochs and α in range (0.05, 0.01) for 50 epochs. The neighbourhood function to determine the strength of the adaptation radius is a Gaussian function, being $\sigma = r_c = 1$, and d_{ci} the distance from the neuron w_c to w_i on the map grid.

3.2 Results and Discussion

For each grid selection of the discetized scene and for each radius r_c, a 10-fold cross validation has been performed to analyse the representation capabilities of the SOM-PINT proposal for trajectories and its classification performance for human behaviour (see Table 1). The performance of the SOM-PINT for classifying human behaviour increases with the number of neurons and grid cells. The radius of neighbourhood r_c to represent a cell in the scene affects the performance of the classifier. In general, the larger the radius r_c, the better performance according to sensitivity, specificity and accuracy. This occurs because more neurons in the cell can represent more samples in the input for a specific cell. For example, in Table 1 we can see larger the radius, a more detailed map representing the trajectory. Moreover, a SOM-PINT with $r_c = 0$ (Fig. 1a it is not able to represent the variability for a specific descriptor representing a behaviour using only one neuron. The worst performance is obtained with a radious $r_c = 0$ for the different grid sizes. However, using a grid cell of 7x11 and 9x13, increasing the radius does not assure a better performance. The discretized scene is representative enough to get a very good performance with a $rc = 1$ (i.e. 9 neurons for cell). The SOM-PINT gets the best performance using a grid cell of 9x13 and $r_c = 3$, in consequence, 63x91 neurons. These results show that, for human behaviour analysis, even a 1x1 (no sampling) grid size, the SOM-PINT provide good results. Hence, our SOM-PINT is able to represent and recognize behaviour of persons in the Shopping Centre with great accuracy.

It is important to compare the SOM-PINT proposal with our previous results using a SOM for analysing human behaviour (that had the worst results classifying with ADV compared to other classifiers). In Fig. 2 mean and standard deviation of sensitivity and specificity for SOM-PINT with r_c from 0 to 4 and the corresponding for classical SOM is presented. It is important to highlight that the SOM-PINT is able to detect about 70% (in average) of behaviour using a 1x1 grid (this means the whole ground plane was sampled in 1 ADV). Comparatively, SOM-PINT and SOM obtain similar results. The SOM classifier is able to achieve the best results for a 5x7 grid size (71.43% sensitivity and 95.71% specificity). However, it is the worst result for this grid size for the SOM-PINT. For the best radius, SOM-PINT is almost 20% and 2% better than SOM for probability of detection and probability of false alarm. SOM performance increases up to a 5x7 grid size. However, the performance decreases as grid size increases. Comparatively, the performance of SOM-PINT increases as grid size does. The problem is that the SOM is not able to preserve the topological information about cell grids as they are converted to a vector for SOM training and classification. The neighbourhood for a cell is not preserved due to each

Table 1. Classification performance for different radius r_c for a $r_n=1$

Cells	r_c	Neur.	Sensit.	Specif.	Accur.	Cells	r_c	Neur.	Sensit.	Specif.	Accur.
1x1	0	1x1	0.5262	0.9210	0.8646	7x11	0	7x11	0.6619	0.9437	0.9034
	1	3x3	0.6905	0.9484	0.9116		1	21x33	**0.8667**	**0.9778**	**0.9619**
	2	5x5	0.7238	0.9540	0.9211		2	35x55	0.8500	0.9750	0.9571
	3	7x7	0.7357	0.9560	0.9245		3	49x77	0.8548	0.9758	0.9585
	4	9x9	**0.7571**	**0.9595**	**0.9306**		4	63x99	0.8333	0.9722	0.9524
3x5	0	3x5	0.6500	0.9417	0.9000	9x13	0	9x13	0.6619	0.9437	0.9034
	1	9x15	0.7976	0.9663	0.9422		1	27x39	0.8667	0.9778	0.9619
	2	15x25	0.8190	0.9698	0.9483		2	45x65	0.8548	0.9758	0.9585
	3	21x35	0.8333	0.9722	0.9524		3	63x91	**0.8738**	**0.9790**	**0.9639**
	4	27x45	**0.8476**	**0.9746**	**0.9565**		4	81x117	0.8571	0.9762	0.9592
5x7	0	5x7	0.6500	0.9417	0.9000						
	1	15x21	0.8262	0.9710	0.9503						
	2	25x35	0.8429	0.9738	0.9551						
	3	35x49	**0.8452**	**0.9742**	**0.9558**						
	4	45x63	**0.8452**	**0.9742**	**0.9558**						

component of the vector reference associated to each neuron is compared to corresponding component of the input vector. Learning process for the SOM takes into account the neighbourhood close in the space $R^{n \times m \times 5}$ but does not take into account the neighbourhood in the ground space. However, the structure of the SOM-PINT assure the neighbours for a specific neuron are close to adjacent cells in the ground plane.

The SOM-PINT proposal has been compared (see Table 2) to other contemporary methods in order to show the high accuracy of the proposed representation and classification method to include behaviour information. Sensitivity and specificity results of context classification have been calculated from reported success rates in [3], [17] and [9] of comparable experiments on the same dataset. Additionally, SOM-PINT has been compared to our previous work [1] in order to show the advantages of the topology preservation using the self-organizing proposal.

Table 2. Classification performance comparison

Method	Sensitivity	Specificity	Dif. Sens	Dif. Spe
Rule-based [3]	0,570	N/A	31%	N/A
HSMM [17]	0,651	0,987	23%	-1%
PN [9]	0,809	0,968	7%	1%
MC 5x7 [1]	0,814	**0,988**	6%	-1%
SOM-PINT 5x7	0,845	0,974	2%	0%
SOM-PINT 9x13	**0,873**	0,979		

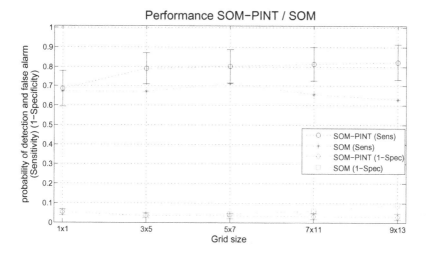

Fig. 2. Comparative performance for average and standard deviation of SOM-PINTs results for different radius r_c and classical SOM

4   Conclusion

In this paper a Self-organizing neural network has been proposed to preserve the topological information about the scene in the image space: SOM-PINT. Extracted features in adjacent areas of the image are explicitly in adjacent areas of the proposed map. The SOM-PINT contains specific neurons for each region of the image preserving the topology of the scene. Each input is able to activate a neuron of each area associated to the cell at the same time. It allow us to adapt the whole map according to an input sample. After training, a specific SOM-PINT represents a higher level of understanding about the scene. In classification, a sample is compared in parallel to each SOM-PINT to establish the winner SOM-PINT and, consequently, the associated high-level of understanding. The neural network has been applied to represent and classify human behaviour from video sequences. The network is able to learn the behaviour by means of learning activities happening in each specific area of the scene. Experimental results show how SOM-PINT proposal is able to classify input samples into human behaviour in complex situations with great accuracy. Moreover, it is able to outperform previous methods using the same dataset. We are currently exploring the feasibility of SOM-PINT to represent and recognize 3D trajectories to analyse the generality of the neural network.

References

1. Azorin-Lopez, J., Saval-Calvo, M., Fuster-Guillo, A., Garcia-Rodriguez, J.: Human Behaviour Recognition based on Trajectory Analysis using Neural Networks. In: International Joint Conference in Neural Networks (2013)

2. Azorin-Lopez, J., Saval-Calvo, M., Fuster-Guillo, A., Oliver-Albert, A.: A predictive model for recognizing human behaviour based on trajectory representation. In: 2014 International Joint Conference on Neural Networks, IJCNN 2014, Beijing, China, July 6-11, pp. 1494–1501 (2014)

3. Fisher, R.B.: The PETS04 Surveillance Ground-Truth Data Sets. In: Sixth IEEE Int. Work. on Performance Evaluation of Tracking and Surveillance (PETS 2004), pp. 1–5 (2004)

4. Fritzke, B.: Growing cell structures a self-organizing network for unsupervised and supervised learning. Neural Networks 7(9), 1441–1460 (1994)

5. Fritzke, B.: Growing grid a self-organizing network with constant neighborhood range and adaptation strength. Neural Processing Letters 2(5), 9–13 (1995)

6. Fritzke, B., et al.: A growing neural gas network learns topologies. Advances in Neural Information Processing Systems 7, 625–632 (1995)

7. Kangas, J.A., Kohonen, T.K., Laaksonen, J.T.: Variants of self-organizing maps. IEEE Transactions on Neural Networks 1(1), 93–99 (1990)

8. Kohonen, T.: Clustering, taxonomy, and topological maps of patterns. In: Proceedings of the 6th International Conference on Pattern Recognition, pp. 114–128. IEEE (1982)

9. Lavee, G., Rivlin, E., Rudzsky, M.: Understanding video events: a survey of methods for automatic interpretation of semantic occurrences in video. IEEE Transactions on Systems, Man, and Cybernetics, Part C: Applications and Reviews 39(5), 489–504 (2009)

10. Madokoro, H., Honma, K., Sato, K.: Classification of behavior patterns with trajectory analysis used for event site. In: The 2012 International Joint Conference on Neural Networks (IJCNN), pp. 1–8 (June 2012)

11. Martinetz, T., Schulten, K.: A "Neural-Gas" Network Learns Topologies. Artificial Neural Networks I, 397–402 (1991)

12. Martinez-Contreras, F., Orrite-Urunuela, C., Herrero-Jaraba, E., Ragheb, H., Velastin, S.A.: Recognizing Human Actions Using Silhouette-based HMM. In: 2009 Sixth IEEE International Conference on Advanced Video and Signal Based Surveillance, pp. 43–48 (September 2009)

13. Morris, B., Trivedi, M.: Trajectory learning for activity understanding: Unsupervised, multilevel, and long-term adaptive approach. IEEE Transactions on Pattern Analysis and Machine Intelligence 33(11), 2287–2301 (2011)

14. Chawla, N.V., Bowyer, K.W., Hall, L.O., Kegelmeyer, W.P.: Smote: Synthetic minority over-sampling technique. Journal of Artificial Intelligence Research 16, 321–357 (2002)

15. Parisi, G., Wermter, S.: Hierarchical som-based detection of novel behavior for 3d human tracking. In: The 2013 International Joint Conference on Neural Networks (IJCNN), pp. 1–8 (August 2013)

16. Schreck, T., Bernard, J., von Landesberger, T., Kohlhammer, J.: Visual cluster analysis of trajectory data with interactive Kohonen maps. Information Visualization 8(1), 14–29 (2009)

17. Tweed, D., Fisher, R., Bins, J., List, T.: Efficient hidden semi-markov model inference for structured video sequences. In: 2nd Joint IEEE International Workshop on Visual Surveillance and Performance Evaluation of Tracking and Surveillance, pp. 247–254 (2005)

18. Uriarte, E.A., Martín, F.D.: Topology preservation in som. International Journal of Applied Mathematics and Computer Sciences 1(1), 19–22 (2005)

A Comparative Study of Downsampling Techniques for Non-rigid Point Set Registration Using Color

Marcelo Saval-Calvo[1], Sergio Orts-Escolano[1], Jorge Azorin-Lopez[1(✉)],
José García Rodríguez[1], Andres Fuster-Guillo[1],
Vicente Morell-Gimenez[1], and Miguel Cazorla[1]

University of Alicante, 03690, Alicante, Spain
{msaval,sorts,jazorin,jgarcia,fuster}@dtic.ua.es,
{vmorell,miguel}@dccia.ua.es

Abstract. Registration of multiple sets of data into a common coordinate system is an important problem in many areas of computer vision and robotics. Usually a large set of data is involved in the process. Moreover, the sets are in general composed by a large number of 3D points. The input for registration techniques based on point set as inputs make sometimes intractable the process due to time needed to provide a feasible solution to the transformation between data. This problem is harder when the transformation is non-rigid. Correspondence estimation and transformation is usually done for each point in the data set. The size of the input is critical for the processing time and, in consequence, a sampling technique is previously required. In this paper, a comparative study of five sampling techniques is carried out. Specifically, is considered a bilinear sampling, a normal-based, a color-based, a combination of the normal and color-based samplings, and a Growing Neural Gas (GNG) based approach. They have been evaluated to reduce the number of points in the input of two non-rigid registration techniques: the Coherent Point Drift (CPD) and our proposal of a non-rigid registration technique based on CPD that includes color information.

Keywords: Downsampling · Non-rigid registration · CPD

1 Introduction

Nowadays, many problems in computer vision use three dimensional data. Moreover, those data come from 3D sensors which provide points in the space, commonly coined as point cloud. Therefore, a wide variety of methods make use of this type of data. Furthermore, with the advent of new low-cost RGB-D sensors as Microsoft Kinect, color information is also available. Registration techniques, which treat the problem of aligning two or more input data, use point clouds to estimate the transformation that maps one into another. The registration could be applied rigidly or non-rigidly, according to the transformation nature. Centering the problem in non-rigid, or deformable registration, traditional approaches

© Springer International Publishing Switzerland 2015
J.M. Ferrández Vicente et al. (Eds.): IWINAC 2015, Part II, LNCS 9108, pp. 281–290, 2015.
DOI: 10.1007/978-3-319-18833-1_30

tend to be time consuming, as an individual transformation should be found for each point in the point set. Hence, sampling techniques must be applied to reduce the input data (downsampling).

Regarding the non-rigid registration, Chui and Rangarajan [1,2] proposed the TPS-RPM non-rigid registration method for 3D point cloud based on Thin Plate Splines to stabilize the displacement of the points. One of the most popular algorithms currently used for non-rigid registration is the Coherent Point Drift (CPD) proposed by Myronenko et al. in [10]. This method is based on Gaussian Mixture Models (GMM) to model the target point set and Expectation Maximization (EM) to compute the registration transform. In order to constrain the movement they make use of the Coherent Motion Theory that helps the translation of points to be regular. Yawen et al. [15] and Gao et al. [6], proposed variants of CPD by means of a method to evaluate the outliers. Other methods have been proposed. Jian et al. in [9] presented GMMreg where they align two GMMs to each other using the L2 distance. Zhou et al. [16] uses the Student's probability in the mixture model. Yang et al. proposed [14] GLMD, a two step non-rigid registration method for point sets. They used combined local and global distances. In [8] Ge et al. presented a similar approach, called Global-Local Topology Preservation (GLTP).

The non-rigid registration methods usually iteratively find the best transformation for each point, therefore, for large data sets the time exponentially increases. Thus, sampling methods are usually applied. The result of the registration is highly dependent on the sampling quality. For example, is the registration uses color as a feature, then a uniform sampling will not work as good as a color-based sampling. There exists several different approaches for performing data sampling reducing the amount of information that has to be processed. In this work, we applied different approaches to perform data downsampling before applying the proposed non-rigid registration technique.

Apart from evaluating traditional downsampling techniques like uniform sampling, bilinear interpolation, color-based or normal-based sampling, etc., in this work we evaluate a novel technique based on the Growing Neural Gas (GNG) [5] algorithm. From a machine learning point of view, this neural network approach based on Self-Organizing Maps (SOM) is evaluated for point cloud downsampling. The GNG algorithm has been successfully applied in different applications like hand and full body tracking [3], 3D surface reconstruction from unorganized data [4], feature descriptor extraction [13] and 3D colour object reconstruction [11]. Recently, in [12] and [7] it has also been demonstrated that the GNG algorithm presents some beneficial attributes for performing noise filtering while keeping the topology of the input space.

2   Deformable Registration

Many studies in computer vision are focused nowadays in non-rigid registration methods. One of the most common is the Coherent Point Drift (CPD) [10]. This is based on Gaussian Mixture Model to describe the target point set and

Expectation-Maximization technique to iteratively evaluate the parameters of the GMM to eventually align the point sets. As we have mentioned before, when the input data is large, the method needs long time to execute all iterations. Then a sampling previous steps are necessary.

In this section we introduce a variant of CPD which includes color in the probability evaluation of the GMM parameters. Therefore, a sampling method which stores the color information is necessary. Growing Neural Gas artificial neural network has been implemented for 3D data sampling. Moreover, a variant which estimates the color of the neurons has been proposed and used in [11]. In the subsection 2.1 the GNG is briefly described. Next, in subsection 2.2 the variant of CPD with color information is presented.

2.1 Sampling GNG-based

In this section it is presented an explanation of the GNG for sampling purposes. The network is specified as:

- A set N of nodes (neurons). Each neuron $c \in N$ has its associated reference vector $w_c \in R^d$. The reference vectors can be regarded as positions in the input space of their corresponding neurons.
- A set of edges (connections) between pairs of neurons. These connections are not weighted and its purpose is to define the topological structure. An edge aging scheme is used to remove connections that are invalid due to the motion of the neuron during the adaptation process.

The GNG learning algorithm is as follows:

1. Start with two neurons a and b at random positions w_a and w_b in R^d.
2. Generate at random an input pattern ξ according to the data distribution $P(\xi)$ of each input pattern.
3. Find the nearest neuron (winner neuron) s_1 and the second nearest s_2.
4. Increase the age of all the edges emanating from s_1.
5. Add the squared distance between the input signal and the winner neuron to a counter error of s_1 such as:

$$\triangle error(s_1) = \|w_{s_1} - \xi\|^2 \tag{1}$$

6. Move the winner neuron s_1 and its topological neighbors (neurons connected to s_1) towards ξ by a learning step ϵ_w and ϵ_n, respectively, of the total distance:

$$\triangle w_{s_1} = \epsilon_w(\xi - w_{s_1})$$
$$\triangle w_{s_n} = \epsilon_n(\xi - w_{s_n}) \tag{2}$$

For all direct neighbors n of s_1.
7. If s_1 and s_2 are connected by an edge, set the age of this edge to 0. If it does not exist, create it.

8. Remove the edges larger than a_{max}. If this results in isolated neurons (without emanating edges), remove them as well.
9. Every certain number λ of input patterns generated, insert a new neuron as follows:
   - Determine the neuron q with the maximum accumulated error.
   - Insert a new neuron r between q and its further neighbor f:

$$w_r = 0.5(w_q + w_f) \tag{3}$$

   - Insert new edges connecting the neuron r with neurons q and f, removing the old edge between q and f.
10. Decrease the error variables of neurons q and f multiplying them with a consistent α. Initialize the error variable of r with the new value of the error variable of q and f.
11. Decrease all error variables by multiplying them with a constant γ.
12. If the stopping criterion is not yet achieved (in our case the stopping criterion is the number of neurons), go to step 2.

The GNG method has been modified regard original version, considering also original point cloud color information. Once GNG network has been adapted to the input space and it has finished learning step, each neuron of the net takes color information from nearest neighbors in the original input space. Color information of each neuron is learned during the adaptation process obtaining a interpolated value of the surrounding points. Color downsampling is performed for allowing later post-processing steps that deal with color information.

2.2 Non-rigid Registration Using Color

In this paper we present a variant of CPD introducing color information in the calculus of the probability. Original CPD uses point set space position to register both point clouds. Using Gaussian Mixture Models with Expectation-Maximization, the algorithm iteratively aligns them.

Given two point sets where $Y_{MxD} = (y1, \cdots, y_M)$ is to be aligned with the reference point set $X_{NxD} = (x1, \cdots, x_N)$, where D is the dimensionality. Each point in Y is the centroid of a distribution in a Gaussian Mixture Model (GMM), all with equal isotropic covariance σ^2. The objective is to find the parameters of the GMM that make Y best fit X by maximizing the likelihood, or equivalently, minimizing the negative log-likelihood. To calculate the parameters, Expectation-Maximization is used. Using Bayes' theorem the posterior probability is used to find the current, or old, parameters.

CPD uses for the posterior probability (P) the space position of the points. Here, the color is introduced along with the space position to calculate this probability. In order to do this, we use another GMM which brings independence of the two spaces, instead of concatenating color and position data. CY and CX represent color information associated to Y and X respectively. The union of both probabilities are then performed and weighted to let the algorithm adapt different situations. Hence, the final posterior probability is:

$$P_{(m|x_n)} = \frac{(PL_{m|x_n})^{wl} * (PC_{m|x_n})^{wc}}{(\sum_{k=1}^{M} PL_{k|x_n})^{wl} * (\sum_{j=1}^{M} PC_{j|x_n})^{wc} + o_L + o_C}$$

PL represents the posterior probability of the original CPD. PC is the new posterior probability of the color space. wl and wc, $wl + wc = 1$ are the weights to balance the effect of each probability. o_L is another gaussian probability to take into account outliers in 3D position, being $o_L = (2\pi\sigma_j^2)^{D/2}\frac{wM}{(1-w)N}$

$$PL_{m|x_n} = exp^{-\frac{(\|x_n - y_m\|^2)}{(2\sigma^2)}} \qquad PC_{m|x_n} = exp^{-\frac{(\|cx_n - cy_m\|^2)}{(2\sigma_c^2)}}$$

σ_c^2 represents the covariance of the color space, different from σ^2.

Originally, the outliers were modeled with a different gaussian distribution with a $\frac{1}{N}$ probability. Here, the outliers are modeled with a gaussian using the color distribution.

$$o_C = \frac{M}{\sigma_{ot}\sqrt{2\pi}} \cdot exp^{-\frac{\left\|\frac{\sum_m^M PC_{m|x_n}}{M}\right\|^2}{2\sigma_{ot}^2}}$$

Lastly, it is very important to highlight that the method uses Coherent Motion Theory to regularize the movement of the points, avoiding non-sense displacements.

3   Experiments

In this section we present the experimentation to evaluate the method for non-rigid registration as well as the sampling technique. The dataset includes two different objects, a flower and a face. The synthetic models have been acquired using the tool Blensor, a Blender plugin which simulates a Microsoft Kinect RGB-D sensor. The models have been deformed with Blender to have both origin a target point sets.

For the non-rigid registration experimentation, the original CPD implementation in Matlab has been used. For the color variant of CPD, we have modified the toolbox provided by the authors of the original CPD. For the GNG implementation, it has been used C++ implementation.

3.1   Downsampling

Five different techniques of sampling have been evaluated, for each sampling rate (250, 500 and 1000 points). Depending on the feature and process of the sampling method, a different result is provided. Bilinear sampling uses bilinear interpolation to remove data, and the value of the new points is the average color of original points used in the interpolation. Normal-based downsampling calculates the angle between each normal and a reference vector in order to bring a 3D data (normals) in 1D data (angles). All angles are place in a common space

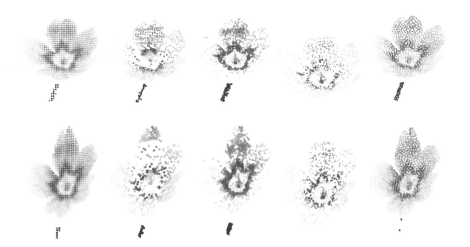

Fig. 1. Sampling examples for 1000 data points. First row shows the original and the second the deformed flower. From left to fight the sampling techniques are bilinear, normals, color, normals+color, and GNG.

which is divided in isometric parts, and then sampled uniformly in a fix number of data, i.e. independently the number of angles in a division, a fix number is selected. Therefore, smooth regions will have less samples and irregular parts will keep more data. Color-based is similar to the previous in the color space. The combination of normal and color-based techniques results in an intermediate solutions, that have color and irregular parts well detailed, and those smooth and homogeneous parts with less data.

From the data presented in Figure 1, bilinear and GNG downsampling techniques provide homogeneous data spread all over the shape for both flowers. This effect is interesting when the purpose of the method that uses these data needs to have information about the whole space, such as rendering or surface extraction. Moreover, GNG has demonstrated noise reduction capabilities in some works as [12]. The normal-based sampling returns higher density in those parts which orientation angle is different from the rest. In the second flower, the top part has higher density because it look frontally and only a yellow part of the rest looks to the same orientation. Similar situation occurs with the right part. The color-based sampling has a higher density in the center region due to it is the most different in the flower, similar that the stem. The combination provides a hybrid solution of the two previous methods. The last three provide very detailed data in those parts with a specific characteristics, however they penalize the general sampling of the object.

3.2 Non-rigid Registration Evaluation

A comparative of the original CPD and a variant with color information is here presented. The main difference is the use of color in the core of the process when the posterior probability is calculated. The color information gives to the registration method allows to achieve good results when the surface is not very detailed so the drift of the points is not constrained by the irregularities of the shape.

For this experimentation the ROI (Region of Interest) is the red color of the center, because in this part the registration of the color is critical. To evaluate how the sampling affects the registration, a general set of parameters is used in the registration method same for all.

Using the data sampled in the subsection 3.1, here the non-rigid registration methods are evaluated qualitatively by visual inspection. Figure 2, 2 and 4 show the flower shape for the color CPD and the original CPD. For each figure, the first row presents the color variant of CPD and the second the original version. From left to right, the sampling techniques bilinear, normal-based, color-based, combination of normal and color-based, and GNG.

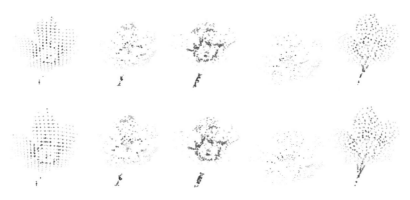

Fig. 2. Non-rigid registration result for the color CPD in the top row, and the original CPD in the second, for a 250 points sampling. Columns are from left to right, bilinear, normal-based, color-based, normal and color-based, GNG.

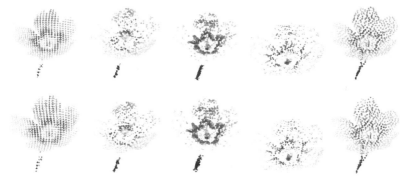

Fig. 3. Non-rigid registration result for the color CPD in the top row, and the original CPD in the second, for a 500 points sampling. Columns are from left to right, bilinear, normal-based, color-based, normal and color-based, GNG.

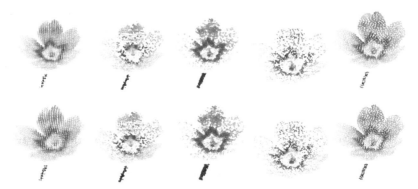

Fig. 4. Non-rigid registration result for the color CPD in the top row, and the original CPD in the second, for a 1000 points sampling. Columns are from left to right, bilinear, normal-based, color-based, normal and color-based, GNG.

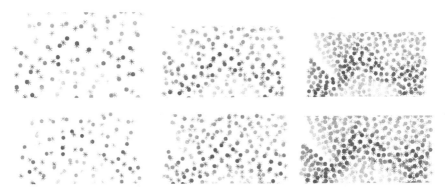

Fig. 5. Enlarged example of the ROI for the flower sampled with GNG. The first row shows the color variant CPD and the second the original algorithm. The data is from left to right, 250 points, 500, and 1000 for the GNG.

Studying in depth the results of the experimentation, different conclusions can be extracted. The most important one, which is the relation between the sampling methods and the non-rigid registration technique, is that the data provided to the registration methods has a direct effect in the final alignment. The registration result for both color and original CPD, is very similar for the color-baser, normal-based and combination of both. The color CPD is slightly better because it uses color information and then some points move to a better location. This situation is presented in Figure 5 and Figure 6. For the GNG and the bilinear results, the color CPD achieves better alignment. Due to the leaves are larger in the deformed shape but the ROI is not deformed in the scale, the CPD moves coherently the points and the the color is not properly aligned. However, the presented color variant of CPD result in an accurate registration.

Figures 5 and 6, it is easy to appreciate how the presented color CPD achieves better results than the original version in the alignment of the color.

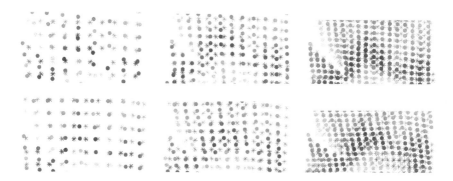

Fig. 6. Enlarged example of the ROI for the flower sampled with bilinear. The first row shows the color variant CPD and the second the original algorithm. The data 250 points, 500, and 1000 for the bilinear.

4   Conclusion

Different problems in computer vision and robotics have to deal with point clouds. Usually the solutions are time consuming. Specifically, the problem of registration point sets into a common coordinate system requires high processing time solutions. The input for registration techniques make sometimes intractable the process due to time needed to provide a feasible solution to the transformation between data. This problem is harder when the transformation is non-rigid due to the process has to be performed for each point in the data set. In this paper, we provide a comparative study of downsampling techniques to reduce the set of points for non-rigid registration techniques. Five methods have been evaluated, including bilinear filtering, normal-based, color-based, a combination of normal and color-based, and lastly, the artificial neural network GNG. In order to make the comparative analysis, the Coherent Point Drift (CPD) and our proposal of a non-rigid registration technique based on CPD that includes color information (CCPD) have been take into account. Bilinear and the GNG provide a homogeneous sampling, which is interesting in problems where the whole space is taken into account, as in the flower experimentation where the deformation is global. Color-based method and its combination with normal-based techniques provide interesting results for the non-rigid registration using color. Experiments showed that the best results are achieved using data provided by the color-based sampling technique because this method fits perfectly the problem of color registration. Since the regions of interest are highly detailed, being point density higher in those parts.

We are working in modifying the GNG algorithm for considering color distribution during the neuron insertion process. Thanks to this modification we expect to achieve similar results that the ones obtained using the color-based sampling method. Moreover, in the registration method, we are working in generalizing the method for including other visual and geometrical descriptors.

References

1. Chui, H., Rangarajan, A.: A New Algorithm for Non-Rigid Point Matching. CVPR 2, 44–51 (2000)
2. Chui, H., Rangarajan, A.: A new point matching algorithm for non-rigid registration. Computer Vision and Image Understanding 89(2-3), 114–141 (2003)
3. Coleca, F., State, A., Klement, S., Barth, E., Martinetz, T.: Self-organizing maps for hand and full body tracking. Neurocomputing 147, 174–184 (2015); advances in Self-Organizing Maps Subtitle of the special issue: Selected Papers from the Workshop on Self-Organizing Maps 2012 (WSOM 2012)
4. Do Rêgo, R.L.M.E., Araújo, A.F.R., De Lima Neto, F.B.: Growing self-reconstruction maps. Neural Networks 21(2), 211–223 (2010)
5. Fritzke, B.: A Growing Neural Gas Network Learns Topologies, vol. 7, pp. 625–632. MIT Press (1995)
6. Gao, Y., Ma, J., Zhao, J., Tian, J., Zhang, D.: A robust and outlier-adaptive method for non-rigid point registration. Pattern Analysis and Applications 17(2), 379–388 (2013)
7. Garcia-Rodriguez, J., Cazorla, M., Orts-Escolano, S., Morell, V.: Improving 3d keypoint detection from noisy data using growing neural gas. In: Rojas, I., Joya, G., Cabestany, J. (eds.) IWANN 2013, Part II. LNCS, vol. 7903, pp. 480–487. Springer, Heidelberg (2013)
8. Ge, S., Fan, G., Ding, M.: Non-rigid Point Set Registration with Global-Local Topology Preservation. In: The IEEE Conference on Computer Vision and Pattern Recognition (CVPR) Workshops (Ml), pp. 245–251 (2014)
9. Jian, B., Vemuri, B.C.: Robust Point Set Registration Using Gaussian Mixture Models. IEEE Transactions on Pattern Analysis and Machine Intelligence 33(8), 1633–1645 (2010)
10. Myronenko, A., Song, X.: Point set registration: coherent point drift. IEEE Transactions on Pattern Analysis and Machine Intelligence 32(12), 2262–2275 (2010)
11. Orts-Escolano, S., Garcia-Rodriguez, J., Morell, V., Cazorla, M., Garcia-Chamizo, J.: 3d colour object reconstruction based on growing neural gas. In: 2014 International Joint Conference on Neural Networks (IJCNN), pp. 1474–1481 (July 2014)
12. Orts-Escolano, S., Morell, V., Garcia-Rodriguez, J., Cazorla, M.: Point cloud data filtering and downsampling using growing neural gas. In: The 2013 International Joint Conference on Neural Networks, IJCNN 2013, Dallas, TX, USA, August 4-9, pp. 1–8 (2013)
13. Viejo, D., Garcia-Rodriguez, J., Cazorla, M.: Combining visual features and growing neural gas networks for robotic 3d {SLAM}. Information Sciences 276, 174–185 (2014)
14. Yang, Y., Ong, S.H., Foong, K.W.C.: A robust global and local mixture distance based non-rigid point set registration. Pattern Recognition (June 2014)
15. Yawen, Y., Peng, Z.P., Yu, Q., Jie, Y., Zheng, W.S.: A Robust CPD Approach Based on Shape Context. In: 33rd Chinese Control Conference, Nanjing, China, pp. 4930–4935 (2014)
16. Zhou, Z., Zheng, J., Dai, Y., Zhou, Z., Chen, S.: Robust non-rigid point set registration using student's-t mixture model. PloS One 9(3), e91381 (2014)

Robust Control Tuning by PSO of Aerial Robots Hose Transportation

Julian Estevez[1(✉)] and Manuel Graña[1,2]

[1] Computational Intelligence Group, University of the Basque Country,
UPV/EHU, San Sebastian, Spain,
`julian.estevez@ehu.es`
[2] ENGINE project, Wroclaw University of Technology (WrUT), Wroclaw, Poland

Abstract. This work presents a method to build a robust controller for a hose transportation system performed by aerial robots. We provide the system dynamic model, equations and desired equilibrium criteria. Control is obtained through PID controllers tuned by particle swarm optimization (PSO). The control strategy is illustrated for three quadrotors carrying two sections of a hose, but the model can be easily expanded to a bigger number of quadrotors system, due to the approach modularity. Experiments demonstrate the PSO tuning method convergence, which is fast. More than one solution is possible, and control is very robust.

1 Introduction

Unmanned Aerial Vehicles (UAV) and, more specifically, aerial robotics, is a growing industry field, with many civil and military applications. The most popular models are variations of multicopters, such as quadcopters or hexacopters. A recently proposed challenging task is the transportation of wires or hoses by a team of quadrotors. A hose can be considered as a deformable linear object (DLO) which acts as a passive object with strong dynamics linking the quadrotors, and influencing their dynamics. There are several examples of DLO models [14],[25] such as wires, hoses and ropes, which have different applications in industry [15], i.e. medical robots [20]. DLO geometrical and kinematics modeling is a complex task and requires a compromise between feasibility of the model and computational cost. Models of DLO have been developed specifically to model rope manipulation to produce knots [18], [22]. Vibration damping of DLO has been dealt with fuzzy control and sliding mode control [9],[8]. Reproducing the motion of the DLO in response to a force involves careful modeling and optimization techniques [19],[6]. Linked Multicomponent Robotic Systems (L-MCRS) [10] are a special case of DLO manipulated by robots, whose dynamics have studied in the case of ground mobile robots [11], achieving control by reinforcement learning [13], [17].

Recently, a model for the transportation of a hose with a system of quadrotors ha been proposed [12]. A physical realization of this task is shown in figure 1. Criteria for the transportation of this DLO were specified, proposing the DLO modeling by catenary sections, which lead to study the dynamics of the system

J.M. Ferrández Vicente et al. (Eds.): IWINAC 2015, Part II, LNCS 9108, pp. 291–300, 2015.
DOI: 10.1007/978-3-319-18833-1_31

in a quasi-stationary state. Though the work dealt with 2D catenary curves, the extension to 3D is immediate [5]. Collision with ground surfaces render the catenary model inaccurate. These situations happen in the phases of take off and landing. Quadrotors control is achieved by heuristically tuned Proportional Integral Derivative (PID) controllers [1]. In the present article we demonstrate the effectiveness and robustness of particle swarm optimization (PSO) for the selection of the controller parameters in PID tuning.

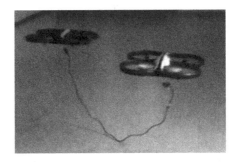

Fig. 1. Physical representation of drones carrying a DLO

The structure of the paper is as follows: in Section 2 presents the catenary model of the hose in quasi-stationary state, and the characterization of the desired state with equal distribution of the weight forces on all quadrotors. Section 3 reviews the dynamics of quadrotor UAVs and presents the quadrotor control. Section 4 defines the computational experiment carried out. Section 5 shows the robustness results. Finally, Section 6 provides conclusions and directions for future work.

2   System Geometrical and Dynamical Model

2.1   DLO Modeling

For lack of space, we refer the reader to [12] for explanations of the hose catenary model. Hence, our problem is restricted to quasi-stationary state study of DLOs modeled as 2D catenary curves [3], or a collection of catenary curves sharing extreme points. Therefore, hose transportation can be modeled by transformations between catenary curves. We made the assumption that a long hose divided in two sections might be transported by three drones ($n = 3$), so that always a team n quadrotor UAVs carry a hose composed of $n - 1$ sections. For experimentation requiring visual feedback, we need to adapt the catenary model in terms of visually measurable parameters [24]. This model is illustrated in Figure 2, where F_1, F_2, F_3, and F_4 are the projected components of cable tension in the (X, Y) axes at the extreme points. Our model has the advantage of providing visually measurable parameters l_x and l_y allowing for visual servoing experiments.

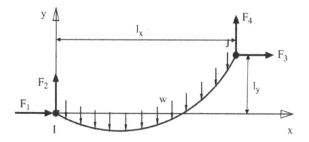

Fig. 2. Catenary parameters used in the model

2.2   Desired State

The energy preservation criteria for the transportation of the catenary is that all the quadrotors in the system support the same vertical load of the catenary, thus all drones consume the same energy. If we consider the case of 3 robots, the hose shape can be modeled by two catenaries sharing one of the extreme points, as shown in 3. To achieve equal distribution of loads in the system, we need to make node B hold one third of the total weight of both catenaries. The equations and resolution method to find the height difference between nodes $A - C$ and B are specified in [12], having as free parameters the catenary total length (L_0) and its weight per unit length (w).

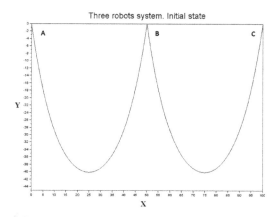

Fig. 3. Catenary and three robots system. Initial state

3   Quadrotor Control System

This section describes the equations and parameters needed to control the quadrotor and manage them to get to the final equilibrium position according to criteria resumed in this article. Though there are proposals for advanced adaptive control using artificial neural networks [7], or optimal robust control [23], in this section we will approach the problem from the point of view of classical PID controllers [1]. We follow coordinate system conventions and the modeling approach of quadrotor dynamics and kinematics proposed in [4] consisting of a collection of equations, involving spatial coordinates, local angles, torques, angular speed of rotors. Here, electric modeling will not be taken into consideration as our focus is on the mechanical aspects of the system.

Equation (1) defines how the basic movements are related to the propellers' squared speed, assuming a cross configuration of the rotors.

$$
\begin{aligned}
U_1 &= b\left(\Omega_1^2 + \Omega_2^2 + \Omega_3^2 + \Omega_4^2\right), \\
U_2 &= l \cdot b\left(-\Omega_2^2 + \Omega_4^2\right), \\
U_3 &= l \cdot b\left(-\Omega_1^2 - \Omega_3^2\right), \\
U_4 &= d\left(-\Omega_1^2 + \Omega_2^2 - \Omega_3^2 + \Omega_4^2\right).
\end{aligned}
\tag{1}
$$

where U_2, U_3 and U_4 are, respectively, roll, pitch and yaw torques, and Ω_i^2 is the squared speed of the i-th propeller. The parameter b is the propeller thrust coefficient, while parameter d is the its drag.

We formulate four controllers for the roll, pitch, yaw and height, respectively based on the equations in [4]. Angles ϕ, θ, and ψ represent the Euler angles referring the Earth inertial frame with respect to the body-fixed frame (ϕ, θ, and ψ represent the rotation along axis Y, X and Z respectively, as shown in figure 4). The following equations

$$
\begin{aligned}
\ddot{z} &= -g - \frac{(T_{V,A} + T_{V,B})}{m} + (\cos\theta\cos\phi)\frac{U_1}{m}, \\
\ddot{\phi} &= \frac{U_2}{I_{xx}}, \\
\ddot{\theta} &= \frac{U_3}{I_{yy}} + \frac{(T_{H,A} - T_{H,B}) \cdot d_{GC}}{I_{yy}}, \\
\ddot{\psi} &= \frac{U_4}{I_{zz}}.
\end{aligned}
\tag{2}
$$

are the linear and angular accelerations of the system including the tensions of catenaries A and B represented in figure 4 (V, H subindexes represent vertical and horizontal components respectively), which are obtained by the classical formulas for catenaries from applied mechanics. Note that the approximation in Equation (2) is valid only for small angles and small rotational velocities, and that equations are specified for the central drone, as the remaining drones are affected only by tension components on their own catenary section. Parameter d_{GC} is the vertical distance from the attachment point of the catenary to the gravitational center of the drone, which is assumed in this paper to be $10cm$ without loss of generality in the results. Taking into account that Z coordinate for the drone is Y coordinate for the catenary, while X remains the same for both systems.

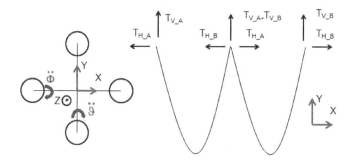

Fig. 4. Coordinate systems and catenary tensions

In order to obtain the torque values, different PID control blocks are implemented. The block diagram for roll control is shown in Figure 5. The variable $\phi^d[rad]$ represents the desired roll angle, $\phi[rad]$ is the measured roll angle, $e_\phi[rad]$ is the roll error and $U_2[N \cdot m]$ is the required roll torque. K_P, K_I, and K_D are the three control parameters. Finally, $I_{XX} [N \cdot m \cdot s^2]$ is the body rotational moment of inertia around the X axis. This contribution comes from equation (2) and is necessary to relate the roll control to U_2. The block diagram for pitch (θ), yaw (ψ) and height control are similar to this one.

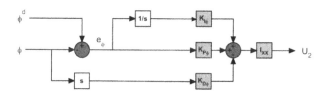

Fig. 5. PID diagram for roll control

3.1 Controller Tuning

Tto tune the PID parameters we use the particle swarm optimization (PSO) [21]. PSO is a relatively recent heuristic search method whose mechanics are inspired by the swarming or collaborative behavior of biological populations, which looks to search the objective value taking into account a cost function. In PSO, a set of particles are distributed randomly in a function and share information among each other, letting the rest of the particles direct to the particle that find the best solution in each iteration at a configurable speed, which makes reference to the mutation velocity of the algorithm. PSO system is appropriate in order to avoid local minimums in a function.

The main steps in the PSO and selection process are described as follows [21]:

1. Initialize a population of particles with random positions and velocities in d dimensions of the problem space.
2. Evaluate the fitness of each particle in the swarm.
3. For every iteration, compare each particle's fitness with its previous best fitness (P_{best}) obtained. If the current value is better than P_{best}, then set P_{best} equal to the current value and the P_{best} location equal to the current location in the d-dimensional space.
4. Compare P_{best} of particles with each other and update the swarm global best location with the greatest fitness (G_{best})
5. Change the velocity and position of the particle according to equations 3 and 4 respectively,

$$V_{id} = \omega \cdot V_{id} + C_1 \cdot rand_1 \left(P_{id} - X_{id} \right) + C_2 \cdot rand_2 \left(P_{gd} - X_{id} \right) \quad (3)$$

$$X_{id} = X_{id} + V_{id} \quad (4)$$

where:

(a) V_{id} and X_{id} represent the velocity and position of the ith particle with d dimensions, respectively. $rand_1$ and $rand_2$ are two uniform random values, and ω is the inertia weight.
(b) The constants C_1 and C_2 in equations 3 and 4 are acceleration constants which changes the velocity of a particle towards P_{best} and G_{best}. This is, the mutation velocity.

In this study we tuned the controller for P-D parameters, following the example in [2].

4   Experiment

We carried out a single type experiment with varying boundary conditions of the problem. The aim is to move a system of three drones charging two sections of catenary starting from a situation where three robots are at same height to a configuration of equal forces supported by all robots. Thus, the equilibrium is achieved descending the central drone. Specific values used in the experiment are these: L_0 of each catenary: $200cm$; horizontal distance between drones: $120cm$; $w = 0,005kg/cm$. We adopt standard[1] values for mass and inertia moments of the quadrotors: $m = 0.5kg$, $l = 25cm$, $I_{xx} = I_{yy} = 5 \cdot 10^{-3}[N \cdot m \cdot s^2]$, and $I_{zz} = 10^{-2}[N \cdot m \cdot s^2]$. Finally, $b = 3 \cdot 10^{-6}[N \cdot s^2]$, and $d = 1 \cdot 10^{-7}[N \cdot m \cdot s^2]$. Initial angles of three drones are 0.

In these conditions, the desired state of the system is that the drone in the middle of the team descends $64.74cm$, so that all the quadrotors charge the same

---

[1] https://github.com/gibiansky/experiments/blob/master/quadcopter/matlab/simulate.m

vertical load. In order to get the parameters of the controller, PSO methodology was used. The objective sought is the P-D combination that makes the quadrotor reach the equilibrium height, D_F. And the chosen cost function is the overshot of the system. Once P-D combination values are chosen, sensitivity experiments are performed with different height objectives and quadrotor dynamic parameters. Experiments were carried out using a discretization of the time variable. Time increment used to compute simulation steps is 0.1 seconds. Speed is specified by an array of values corresponding to each time instant. Scilab 5.4 version simulation software was used to perform the experiment.

5   Results

Number of particles in the PSO algorithm was 10, $\omega = 1$, $C_1 = C_2 = 2.5$. In the majority of tests, most particles reached the optimum before 10 iterations. However, these particles present different velocities of convergence. In figure 6, we plot the evolution of the central drone position for three different P-D tunings.

We choose the third combination of PD values ($K_p = 4.6521$ $K_d = 10.1373$) in figure 6 for ensuimg tests, because it has the fastest convergence among the three alternatives. In this situation, different values of final height for the central drone were chosen, regardless of obtaining a statically balanced system. Results are shown in figure 7:

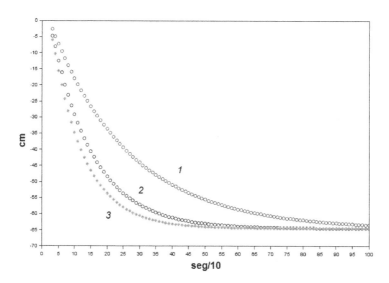

Fig. 6. Motion of the middle quadrotor to reach the equiload state controlled by PSO P-D tuning. 1) $K_p = 2.008$ $K_d = 4.9687$ 2) $K_p = 3.6796$ $K_d = 12.008$ 3) $K_p = 4.6521$ $K_d = 10.1373$

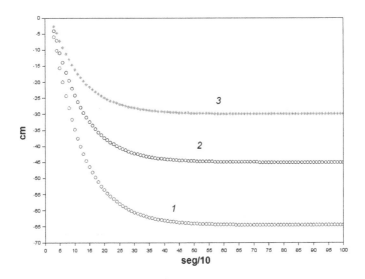

Fig. 7. Drone motion to reach different heights controlled by the PSO P-D : 1) $D_F = 64.74$ (balanced) 2) $D_F = 45cm$ 3)$D_F = 30cm$

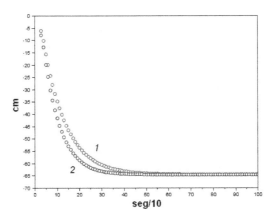

Fig. 8. Different drone performance under the PSO P-D controller: 1) Quadrotor from Github 2) AR Drone Parrot quadrotor

The next experiment tests the performance with different quadrotor dynamic parameters for a $D_F = 64.74cm$. The default parameter values in this article are standard[2]. The alternative parameters are commercial AR Drone Parrot's, reported in [16] ($m = 0.38kg$, $l = 17cm$, $I_{xx} = I_{yy} = 86 \cdot 10^-[kg \cdot m \cdot s^2]$,

---

[2] https://github.com/gibiansky/experiments/blob/master/quadcopter/matlab/simulate.m

and $I_{zz} = 172 \cdot 10^{-2}[kg \cdot m \cdot s^2]$. Finally, we set $b = 3.13 \cdot 10^{-5}[N \cdot s^2]$, and $d = 7.5 \cdot 10^{-7}[N \cdot m \cdot s^2]$). Convergence plot results are shown in figure 8.

6 Conclusions

In this work, a P-D tuning method with PSO was developed for a system of DLO transportation with aerial robots. Experiments proof that K_i is dispensable, which facilitates the implementation and computational time. Results show that the P-D tuning methodology converges fast enough, offering more than one possible values combinations. Besides, robustness of this P-D values election was demonstrated. Ongoing work consists on the development of a controller and a model for the transport of the DLO in an unspecified trajectory. Results reported here improve over [12] where the control model for the same problem follows Zieger-Nichols PID tuning.

References

1. Ang, K.H., Chong, G.C.Y., Li, Y.: Pid control system analysis, design, and technology. IEEE Trans. Control Systems Tech. 13(4), 559–576 (2005)
2. PSO based PID, Control Design, for the Stabilization, of a Quadrotor, Boubertakh, H., Bencharef, S., Labiod, S. Systems and control, algiers, algeria. In: Proceedings of the 3rd International Conference on WeAC.6. IEEE 4799 (October)
3. Bowden, G.: Stretched Wire Mechanics. eConf, C04100411:038 (2004)
4. Bresciani. Modelling, identification and control of a quadrotor helicopter. Master's thesis, Lund University, Department of Automatic Control (2008)
5. Chan, T.-O., Lichti, D.D.: 3d catenary curve fitting for geometric calibration. IS-PRS - International Archives of the Photogrammetry. Remote Sensing and Spatial Information Sciences, XXXVIII-5/W12, 259–264 (2011)
6. Denisov, G.G., Novilov, V.V., Smirnova, M.L.: The momentum of waves and their effect on the motion of lumped objects along one-dimensional elastic systems. Journal of Applied Mathematics and Mechanics 76(2), 225–234 (2012)
7. Dierks, T., Jagannathan, S.: Output feedback control of a quadrotor uav using neural networks. IEEE Transactions on Neural Networks 21(1), 50–66 (2010)
8. Ding, F., Huang, J., Wang, Y., Fukuda, T., Matsuno, T.: Adaptive sliding mode control for manipulating deformable linear object with input saturation. In: 2012 International Conference on Mechatronics and Automation (ICMA), pp. 1862–1867 (August 2012)
9. Ding, F., Huang, J., Wang, Y., Mao, L.: Vibration damping in manipulation of deformable linear objects using sliding mode control. In: 2012 31st Chinese Control Conference (CCC), pp. 4924–4929 (July 2012)
10. Duro, R.J., Graña, M., de Lope, J.: On the potential contributions of hybrid intelligent approaches to multicomponen robotic system development. Information Sciences 180(14), 2635–2648 (2010)
11. Echegoyen, Z., Villaverde, I., Moreno, R., Graña, M., d'Anjou, A.: Linked multicomponent mobile robots: Modeling, simulation and control. Robotics and Autonomous Systems 58(12), 1292–1305 (2010)

12. Estevez, J., Lopez-Guede, J.M., Grana, M.: Quasi-stationary state transportation of a hose with quadrotors. Robotics and Autonomous Systems 63, 187–194 (2015)
13. Fernandez-Gauna, B., Lopez-Guede, J.M., Zulueta, E.: Linked multicomponent robotic systems: Basic assessment of linking element dynamical effect. In: Graña Romay, M., Corchado, E., Garcia Sebastian, M.T. (eds.) HAIS 2010, Part I. LNCS(LNAI), vol. 6076, pp. 73–79. Springer, Heidelberg (2010)
14. Hirai, S.: Energy-based modeling of deformable linear objects. In: Henrich, D., Worn, H. (eds.) Robot Manipulation of Deformable Objects, pp. 11–27. Springer, London (2000)
15. Jian Huang, Pei Di, T. Fukuda, and T. Matsuno. Dynamic modeling and simulation of manipulating deformable linear objects. In *Mechatronics and Automation, 2008. ICMA 2008. IEEE International Conference on*, pages 858–863, 2008.
16. Koszewnik, A.: The parrot UAV controlled by PID controllers. Acta Mechanica et Automatica 8(2) (January 2014)
17. Lopez-Guede, J.M., Graña, M., Ramos-Hernanz, J.A., Oterino, F.: A neural network approximation of l-mcrs dynamics for reinforcement learning experiments. In: Ferrández Vicente, J.M., Álvarez Sánchez, J.R., de la Paz López, F., Toledo Moreo, F.J. (eds.) IWINAC 2013, Part II. LNCS(LNAI), vol. 7931, pp. 317–325. Springer, Heidelberg (2013)
18. Matsuno, T., Tamaki, D., Arai, F., Fukuda, T.: Manipulation of deformable linear objects using knot invariants to classify the object condition based on image sensor information. IEEE/ASME Transactions on Mechatronics 11(4), 401–408 (2006)
19. Menon, M.S., Ananthasuresh, G.K., Ghosal, A.: Natural motion of one-dimensional flexible objects using minimization approaches. Mechanism and Machine Theory 67, 64–76 (2013)
20. Moll, M., Kavraki, L.E.: Path planning for deformable linear objects. IEEE Trans. Robot. 22(4), 625–636
21. A Novel, Particle Swarm, Optimization PSO, Tuning Scheme, for, PMDC Motor, and Drives Controllers. Powereng 2009 lisbon, portugal. IEEE (March 2009)
22. Saha, M., Isto, P.: Manipulation planning for deformable linear objects. IEEE Transactions on Robotics 23(6), 1141–1150 (2007)
23. Satici, A.C., Poonawala, H., Spong, M.W.: Robust optimal control of quadrotor uavs. IEEE Access 1, 79–93 (2013)
24. Thai, H.-T., Kim, S.-E.: Nonlinear static and dynamic analysis of cable structures. Finite Elements in Analysis and Design 47(3), 237–246 (2011)
25. Wakamatsu, H., Hirai, S.: Static modeling of linear object deformation based on differential geometry. The International Journal of Robotics Research 23(3), 293–311 (2004)

Visual Bug Algorithm for Simultaneous Robot Homing and Obstacle Avoidance Using Visual Topological Maps in an Unmanned Ground Vehicle

Darío Maravall[1], Javier de Lope[1], and Juan Pablo Fuentes[2]([⊠])

[1] Department of Artificial Intelligence, Faculty of Computer Science,
Universidad Politécnica de Madrid, Madrid, Spain
dmaravall@fi.upm.es, javier.delope@upm.es
[2] Centro de Automática y Robótica (UPM-CSIC),
Universidad Politécnica de Madrid, Madrid, Spain
juanpablo.fuentes.brea@alumnos.upm.es

Abstract. We introduce a hybrid algorithm for the autonomous navigation of an Unmanned Ground Vehicle (UGV) using visual topological maps.
The main contribution of this paper is the combination of the classical bug
algorithm with the entropy of digital images captured for the robot. As the
entropy of an image is directly related to the presence of a unique object or
the presence of different objects inside the image (the lower the entropy of
an image, the higher its probability of containing a single object inside it;
and conversely, the higher the entropy, the higher its probability of containing several different objects inside it), we propose to implement landmark
search and detection using topological maps based on the bug algorithm,
where each landmark is considered as the leave point for guide to the robot
to reach the target point (robot homing). The robot has the capacity of
avoid obstacles in the enviroment using the entropy of images too. After
the presentation of the theoretical foundations of the entropy-based search
combined with the bug algorithm, the paper ends with the experimental
work performed for its validation.

Keywords: Bug algorithm · Unmanned ground vehicles · Entropy
search · Visual topological maps

1 Introduction

The bug algorithms are a simple and efficient family of techniques for obstacle
avoidance in robot navigation with metric maps [1,2]. The main drawback of the
bug algorithms is that they need the knowledge of the robot localization (the
hardest constraint) besides the coordinates of the goal (x, y).

In this paper we introduce a version of the bug algorithm for robot navigation
with visual topological maps[6], in which case the coordinates of both the robot
and the goal are not needed. More specifically, the proposed visual bug algorithm
is meant to drive the robot towards a target landmark while simultaneously
avoiding any existing obstacle using exclusively vision capacity.

© Springer International Publishing Switzerland 2015
J.M. Ferrández Vicente et al. (Eds.): IWINAC 2015, Part II, LNCS 9108, pp. 301–310, 2015.
DOI: 10.1007/978-3-319-18833-1_32

The pseudocode of the visual bug algorithm is as simple as this:

Algorithm 1. Visual bug algorithm

Given a visual topological map [nodes store the landmarks images and edges store the orientations linking the landmarks] or more specifically, given the image of the target landmark and given the orientation θ towards the next landmark provided by the topological map:

Begin

1. **While** { the robot does not detect the target landmark specified in the topological map } **do**
   (a) Activate the visual search procedure in a controlled way based on entropy maximization of the digital images captured (high entropy).
   (b) If { a landmark is visible} then Goto 3
   (c) If { an obstacle is encountered (low entropy) } then Go to 4
2. **End while**
3. Go forward to the landmark until either happens:
   (a) The target landmark is reached, then exit
   (b) Other landmark is reached, then the robot executes the orientation θ defined in its edge
4. Move following the obstacle contour until the robot meets a leaving point L (a landmark in the current obstacle which is linked to the goal without hitting the current obstacle itself) , then Go to 3

End Begin

Besides the basic procedure visual search, this apparently simple algorithm consists of three additional complex vision-based robot behaviors:

- B_1 : "go forward to the goal"
- B_2 : "circumnavigate an obstacle"
- B_3 : "execute the orientation θ"

We can also define three basic environment´s states or situations:

- S_1 : "the goal is visible"
- S_2 : "there is an obstacle in front of the robot"
- S_3 : "A landmark from the topological map is visible"

In words the visual bug algorithm can be summarized as follows:"search the target landmark and once it is visible then move always towards the target landmark and cicumnavigate any existing obstacle when necessary".

Besides the implementation of the visual search procedure (which is obviously less critical when the robot controller is provided with the orientation θ to the target landmark), the critical element is the robot odometry[3] and more specifically the robot´s dead reckoning or path integration, for this reason the robot planner is then based in the control of the orientation θ towards the target landmark using topological maps.

The visual bug algorithm, however simple may it appears, it sets out also hard computer vision problems for both the perception of the environment´s states S_1, S_2 and S_3 and for the implementation of the three basic robot´s behaviors B_1, B_2, and B_3, so that we devote the bulk of the remaining sections of the paper to describe our experiments aimed at solving these specific computer vision problems.

In the sequel, we describe the unmanned ground vehicle that we have built for experimenting with the proposed visual bug algorithm. Afterwards, we introduce a method for the search of the target visual landmark based on entropy maximization of digitial images captured by the robot.

The paper ends with the discussion of the experimental work carried out to test the visual bug algorithm.

2   Description of the Unmanned Ground Vehicle (UGV)

We have assembled an autonomous Unmanned Ground Vehicle (UGV) from scratch using onboard commercial microcontrollers, which has been used during ours experiments. The UGV (Figure 1) is structered in two different layers:

Fig. 1. Unmanned Ground Vehicle (UGV)

Main layer: This is the main controller of the robot which is based on a Raspberry Pi Model B+ as tiny computer, whose objective is generate all control signals of the robot and then send them to the motor layer. The robot has onboard a HD camera with 720p of resolution connected to the Raspberry Pi by USB port. This camera is used as vision sensor for its autonomous navigation. The communication between the main layer and motor layer is based in the I2C protocol, which uses 3 pins on the Pi (SDA, SCL and Ground 0V) connected to the similar pins on the microcontroller of the motor layer.

The main layer has its own independent power supply, which uses a battery (2500mAh) connected directly to the 5V microUSB of the Pi.

Motor layer: All control signals from main layer are executed through the 4 DC gear motors implemented in this layer. The motor layer is based on a microcontroller Arduino UNO with a Motor Shield for the coordination and regulation of signals through of each motors. These control signals are received from main layer through the I2C protocol, using the specific pins on the Arduino UNO like digital inputs. The 4 DC gear motors are a brushed DC motors with a built-in gear train, with deliver high torque at medium to low speeds. Each motor is fixed to one wheel with forward and backward direction. This layer needs two power supplies: one 6V battery for the Arduino UNO and other equal battery for the Motor Shield.

In addition, the robot has an ultrasonic sensor that could be use for calculate the distance between the robot and obstacles. (Note: this sensor is not used during the experiments in this paper).

This UGV has been mounted over a robust stackable chasis, and its autonomy time is approximately 2 hours and 30 minutes.

3 Visual Landmark Search and Detection: Entropy-Based Search and the Homing Modes

In previous works[4] we have introduced a novel method based on the use of image entropy as a means for the search and detection of visual objects and landmarks. The main idea behind the entropy-based search (shown in Figure 2) is the direct relation between the entropy of an image and its probability of containing a single object (in the case of low entropy), and conversely the probability of containing several different objects (in the case of high entropy).

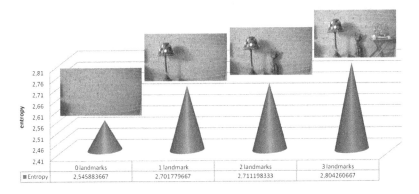

Fig. 2. This figure shows the empirical fact that the higher the number of objects (landmarks) inside an image, the higher its entropy

By using this simple idea, the critical procedure of visual object/landmark search and detection can be performed as a process of entropy maximization combined with a vision-based homing loop as explained above.

Given the image I_k in each k iteration, the following equation represents the standard definition [10] of the entropy $\mathcal{H}$ of the normalized histogram $Hist(I_k)$ (size $n = 256$) obtained by the UGV:

$$\mathcal{H}\left[Hist(I_k)\right] = -\sum_{i=1}^{n} Hist_i(I_k) \cdot log_2 Hist_i(I_k) \tag{1}$$

The entropy-based landmark search can be substituted by a random search, in which the robot also explores the environment to detect the target landmark although with a purely random strategy instead. The entropy strategy is more optimal than the random strategy [4].

After converging to an image with a high entropy that contains several objects candidates for the target landmark, and after recognizing the actual target landmark, the robot´s controller switches to a homing mode based on internal model [9] aimed at guiding the robot towards the target landmark by means of a vision-based control loop [8].

This process of switching between the robot´s landmark search mode [S] an the landmark´s homing mode [H] is modulated by the magnitude error [5] o difference between the current image I captured by the robot´s onboard camera[7] and the landmarks L provided by the robot´s navigation planner (based on a topological map) as shown in Figure 3.

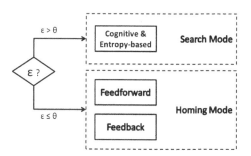

Fig. 3. Finite state automaton that models the robot´s controller between the search mode [S] and the homing mode [H] depending on the magnitude of the error ε or the difference images (respect to the landmarks defined in the topological map) and according to the simple heuristic rule If { error is big } Then { S } Else { H }

If the ε is low, the robot is near of the landmark L ([H] is activated), else the robot navigates using entropy-based procedure ([S]) described above. In the next paragraph we describe the experimental work tested with the visual bug algorithm detail in this paper.

4 Experimental Work: Robot Homing and Obstacle Circumnavigation

For experimenting in this paper we have used the UGV descripted in the paragraph 2. To test experimentally the proposed visual bug algorithm, we have defined a topological map (Figure 4), which defines tree spherical landmarks $(L_{1-yellow}, L_{2-blue}, L_{3-red})$ with colour yellow, blue and red respectively.

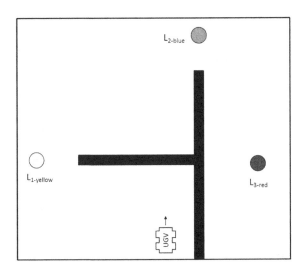

Fig. 4. This figure shows the enviroment used during experiments. The UGV starts from a initial point and must execute its behaviors B_x for reach the goal (L_{3-red}). The topological map defines a set of landmarks $(L_{1-yellow}, L_{2-blue}, L_{3-red})$, which correspond to the three spherical landmarks yellow, blue and red respectively.

The robot starts from a "start point" in the enviroment, and the L_{3-red} landmark is defined as the "target point". For guide the autonomous navigation of the UGV, this should visit sequentially all landmarks: first the $L_{1-yellow}$, then L_{2-blue} and finally L_{3-red} as goal in the navigation.

During the autonomous navigation of the UGV, the method of calculation for obtain its orientation θ along k iterations , is given by the following equation:

$$\theta_k = \theta_{k-1} + \frac{(v_r - v_l)}{I} \tag{2}$$

where v_r and v_l represents the control signals executed by right wheels and left wheels of the robot respectively. The I parameter is the distance between the wheels, in this case the UGV has an value of $I = 15$ centimeters. Each landmark represents a specific node of the topological map, and its edges store the orientation θ towards the next landmark provided by the topological map.

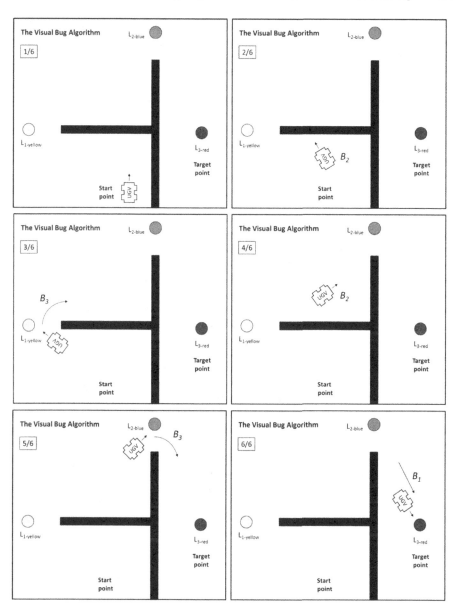

Fig. 5. The figure shows the sequential steps during the autonomous navigation of the UGV. From a "start point" the robot starts navigation through the execution of its behaviors B_x: B_1 : "go forward to the goal", B_2 : "circumnavigate an obstacle" and B_3 : "execute the orientation θ". Finally, the robot reaches the "target point" through the visual bug algorithm.

The B_3 behavior is executed when the landmark L_k is visible in the k iteration, and the robot must execute the orientation θ according to equation below:

$$\theta = \theta_k + \theta[L_k] \tag{3}$$

where the $\theta[L_k]$ represents the orientation θ stored in the Table 6. For this experiments, the three nodes store the following information:

Landmark	θ
$L_{1-yellow}$	$45°$
L_{2-blue}	$45°$
L_{3-red}	$0°$

Fig. 6. This table shows all orientations defined for each landmark in the topological map. Note the L_{3-red} is associated with $0°$ because is the "target point", and therefore the UGV is stopped when is reached it.

Based on the above topological map, we have tested all behaviors B_x described for the visual bug algorithm. The Figure 5 shows a sequence of the phases executed during the experiments with the UGV from "start point" (step 1).

(a)   (b)

Fig. 7. First Person View of the robot: (a) shows a snapshot of a landmark detected by the robot, and (b) a snapshot of the walls as obstacles perceived by the robot is shown

Depending on the situation S_x of the robot, it can activate a specific behavior B_x: when the UGV detects an obstacle in front it (through a low value of entropy in the image captured), the behavior B_2 (steps 2 and 4) activates a procedure based on entropy maximization for detect landmarks (high entropy ≈ 2.213943) and circumnavigate the obstacle (low entropy ≈ 2.005811) as is shown in Figure 7. Once the UGV detects a visible landmark, the behavior B_3 (steps 3 and 5) is activated, and then the robot executes the specific orientation θ (Figure. 6) through the motor layer for go straight on the next landmark defined in the topological map. We considered these landmarks ($L_{1-yellow}$ and L_{2-blue}) as leave point L based on the classical bug algorithm [1]. Finally, when the UGV see the target landmark, go forward to this "target point" through the behavior B_1 (step 6), and when the goal is reached, the visual bug algorithm stops the robot.

(a)   (b)

(c)   (d)

Fig. 8. The autonomous navigation of the UGV using the visual bug algorithm: (a) the robot executes a circumnavigate of the obstacle (low entropy), (b) approximation maneuver to the landmark (high entropy), (c) execution of the orientation θ in the edge and (d) the robot go forward to the goal following a straight line

From the experimental results obtained in our laboratory (Figure 8), we can conclude that the UGV is able to successfully perform all behaviors B_x defined in the visual bug algorithm using a specific topological map that has been defined previously. With this algorithm the iterations are reduced during the cirumnavigate of an obstacle, since if a landmark is visible, the procedure based on entropy maximization of the digital images is activated and the robot navigates toward to landmark directly.

5   Conclusions and Future Work

A hybrid algorithm for the autonomous navigation of a UGV based on the bug algorithm and the use of the entropy of digital images as method of search has been presented. The proposed algorihtm uses a visual topological map to autonomously navigate in the environment, where the nodes are the L leave-point or landmark in which the robot must re-oriented its navigation for reach the goal. This novel algorithm does not need know its coordinates in the enviroment unlike the classic algorithm bug, since in this case during the navigation, the robot uses the landmark detection for know its position. The paper finally presents the experimental work for the validation of the proposed visual bug algorithm, where it has defined a topological map for testing the homing robot and obstacle circumnavigation behaviors, obtaining good results for both maneuvers, and reducing the number of iterations during the navigation.

Future work is planned towards the UGV´s autonomous navigation in a more complex environment with no regular obstacles, also the implementation of a learning process as we have studied in previous works[5].

References

1. Lumelsky, V.J., Stepanov, A.A.: Path-planning strategies for a point mobile automaton moving amidst unknown obstacles of arbitrary shape. Algorithmica 2(1-4), 403–430 (1987)
2. Lumelsky, V.J.: Sensing, intelligence, motion: how robots and humans move in an unstructured world. John Wiley & Sons (2005)
3. Lumelsky, V.J., Stepanov, A.A.: Dynamic Path Planning for a Mobile Automaton with Limited Information on the Environment. UEEE Transactions on Automatic Control AC-31(11) (November 1986)
4. Brea, J.P.F., Maravall, D., de Lope, J.: Entropy-Based Search Combined with a Dual Feedforward-Feedback Controller for Landmark Search and Detection for the Navigation of a UAV Using Visual Topological Maps. ROBOT (2), 65–76 (2013)
5. Maravall, D., de Lope, J., Fuentes, J.P.: Brea: Vision-based anticipatory controller for the autonomous navigation of an UAV using artificial neural networks. Neurocomputing 151, 101–107 (2015)
6. Maravall, D., de Lope Asiaín, J., Fuentes, J.P.: Brea: A Vision-Based Dual Anticipatory/Reactive Control Architecture for Indoor Navigation of an Unmanned Aerial Vehicle Using Visual Topological Maps. IWINAC (2), 66–72 (2013)
7. Maravall, D., de Lope, J., Fuentes, J.P.: Brea: Fusion of probabilistic knowledge-based classification rules and learning automata for automatic recognition of digital images. Pattern Recognition Letters 34(14), 1719–1724 (2013)
8. Kawato, M.: Feedback-Error-Learning Neural Network for Supervised Motor Learning, Advanced Neural Computers (1990)
9. Kawato, M.: Internal models for motor control and trajectory planning. Neurobiology 9, 718–727 (1999)
10. Shannon, C.E.: A mathematical theory of communication. The Bell System Tech. J. 27, 379-423, 623–656 (1948)

Neural Modeling of Hose Dynamics to Speedup Reinforcement Learning Experiments

Jose Manuel Lopez-Guede[1](✉) and Manuel Graña[1,2]

[1] Computational Intelligence Group of the Basque Country University (UPV/EHU),
San Sebastian, Spain
jm.lopez@ehu.es

[2] ENGINE project, Wroclaw University of Technology (WrUT), Wroclaw, Poland

Abstract. Two main practical problems arise when dealing with autonomous learning of the control of Linked Multi-Component Robotic Systems (L-MCRS) with Reinforcement Learning (RL): time and space consumption, due to the convergence conditions of the RL algorithm applied, i.e. Q-Learning algorithm, and the complexity of the system model. Model approximate response allows to speedup the realization of RL experiments. We have used a multivariate regression approximation model based on Artificial Neural Networks (ANN), which has achieved a 90% and 27% of time and space savings compared to the conventional Geometrically Exact Dynamic Splines (GEDS) model.

1 Introduction

A Linked Multi-Component Robotic system (L-MCRS) is described as a set of autonomous robots which are linked using a flexible unidimensional element [1]. That link introduces additional non-linearities and uncertainty in the control of the robots that makes difficult to accomplish the tasks related to some function of the flexible link itself. L-MCRS are complex systems used to carry out tasks exemplified by the paradigmatic task of transporting a wire or hose-like object [2,3]. The case under consideration involves the transportation of the hose tip to a given position in the working space while the other extreme is attached to a source.

Achieving autonomous learning by the L-MCRS to solve this kind of tasks is still an open issue. Reinforcement Learning (RL) [4] is a set of machine learning algorithms that in some cases, such as Q-Learning, allow autonomous learning from real world experience, requiring the repetition of the experiments a large number of times. To avoid the waste of material resources and time, we have used accurate simulation of the hose dynamics based on a Geometrically Exact Dynamic Splines (GEDS) model [2]. However, even in this case the simulation of so large number of simulations and so complex model makes it a time consuming task [3,5,6,7,8,9,10].

Previous papers addressed the problem of the efficient simulation of the autonomous learning experiments of L-MCRS systems [11] from a time efficiency point of view [9] using Artificial Neural Networks (ANN) approximations to the

© Springer International Publishing Switzerland 2015
J.M. Ferrández Vicente et al. (Eds.): IWINAC 2015, Part II, LNCS 9108, pp. 311–319, 2015.
DOI: 10.1007/978-3-319-18833-1_33

GEDS model. In this paper we also pay attention to the space requirements in order to reduce the amount of space needed to store the learning information, so the obtained ANN model provides very fast responses with space saving with respect to previous approximations, allowing to perform exhaustive simulations for RL.

The content of the paper is as follows. Section 2 introduces the computational cost problem that we aim to solve from the time and the space point of view. Section 3 describes the proposed approach, oriented to speed up the computation of the system's next state and to reduce the space requierements, and the achieved results are also discussed. Finally, section 4 gives our conclusions.

2   The Problem

2.1   Hose-Robot System

The hose-robot system can be described as follows: an unidimensional object (the hose) that has one end attached to a fixed point (which is set as the origin of the ground working space), while the other end (the tip) is transported by a mobile robot which can perform a finite set actions. The task for the robot is to bring the tip of the unidimensional object to a designated destination point through discrete actions of a predefined duration (i.e., of a predefined length if they are movement actions with constant velocity). In this case, the working space is a perfect square of 2×2 m^2, and we have applied a spatial discretization of $0, 5\,m$, so each discrete action of the robot is the translation of the robot to a neighboring box. We have designed a procedure to generate a number of different initial configurations. Figure 1 illustrates a possible arbitrary initial configuration of the hore-robot system, where $P_{r\,Initial}$ and P_d are, respectivley, the initial and desired position of the robot. The main goal from a machine learning point of view is that the robot learns autonomously how to reach any arbitrary point carrying the hose attached. To achieve this goal, the RL algorithm has to perform multiple simulations of the system based on a GEDS model to meet the optimal convergence conditions of the RL algorithm. Figure 1 illustrates one of these simulations in which the robot carrying the tip of the unidimensional object goes from an initial position $P_{r\,initial}$ to a final position $P_{r\,Final}$, that corresponds with P_d, through five intermediate positions and corresponding hose configurations.

2.2   On the Computational Cost of Experimental Simulation

Due to the complexity of the hose-robot system GEDS model (implemented in Matlab), it's performance is acceptable only when few repetitions are needed, but not when much more simulations (several millions) must be done. Specifically, on a Dell Optiplex 760 personal computer, equipped with a processor Intel(R) Core(TM) 2 Duo CPU E8400 @ 3,00 Ghz with 3,00 GB of RAM memory and a Microsoft Windows XP Professional v. 2002 SP 3 operating system, a simulation of a movement of $0, 5\,m$ lasts 4 seconds. This response time is unacceptable under

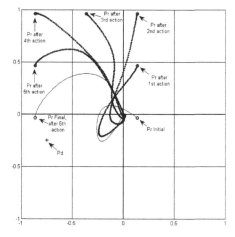

Fig. 1. Evolution of the hose starting from a different initial position. The tip of the unidimensional object robot reaches the goal. Arrows are used to indicate the motion of the hose.

the convergence conditions of the RL algorithm, because it requires to repeat all possible movements in each reachable situation of the system many times. This implies that a simulation of one million of movements may last over 45 days running 24 hours a day. Finally, both the GEDS model and the RL algorithm are not suitable for parallelization that could reduce the computational burden. On the other hand, the space requiremetns to save the learning process is relevant, and some data on this topic will be given on next sections.

3   Hose Model Learning

In this section we explain the procedure to learn the efficient hose model and we discuss the results obtained paying attention to several versants.

3.1   Learning Experimental Design

A general algorithmic specification of the learning procedure of an accurate approximation of the GEDS model by means of ANN is provided in Algorithm 1. The main idea is to obtain all feasible paths between any two points to obtain both the input and the output patterns of the train and test samples. In these experiments the inputs are the initial configuration of the hose and the action to be performed by the robot placed at the tip of the hose. On the other hand, the output is the hose configuration after the execution of the action by the robot. We have chosen a spatial discretization step of $0, 2\,m$ for both X and Y dimensions of the working plane. In all cases, the hose configuration is given by 8 control points in the 2D workspace. So, the input patterns are 17 dimensional

---

Algorithm 1. Hose model learning general procedure using control points

Given algorithm finding all feasible paths between any two arbitrarily designed points, without crossings, whose inputs are: exact path length, a set of actions, and the distance that the robot moves with each movement.

1. Use the path finding algorithm to produce all the paths from the fixed point of the hose at $(0, 0)$ to any point reachable by the tip of a flexible hose of $1\,m$.
2. for each path,
   (a) Generate 8 bidimensional control points to model the path
   (b) for each available robot action,
      i. Simulate the effect of the action on the system robot-hose by the GEDS initialized with selected path
      ii. Save as input patterns the control points that denote the initial position of the system robot-hose and the robot action
      iii. Save as output patterns the control points that denote the resulting position of the system robot-hose after and the robot action
3. Train a set of rtificial neural networks with the input/output patterns obtained in the previous step, partitioning all examples in three data sets (60% train, 20% validation and 20%test).
4. Test the model learned by each artificial neural network on the test data set, to validate that that model can be used instead the analytical model.

---

real vectors (16 values due to the 8 control points and 1 value due to the action) and the output patterns are 16 dimensional vectors (due to the 8 control points). The approximation of the GEDS model behavior is a multivariate regression problem with strong no-linearities. The ANNs that we have used are feedforward single hidden layer models with the size of the hidden layer varying between 1 and 300 with a step of 5. Three activation functions of the neural units have been tested, i.e., linear, tan-sigmoid and log-sigmoid. Regarding the training algorithm, it was used the Levenberg-Marquardt algorithm for speed reasons, presenting all the ANN input vectors once per iteration in a batch without normalization. Such training and test procedure was performed five times for each structure (combination of number of hidden units and activation function) to assess its generalization. Finally, the input vectors and target vectors have been divided into three sets using random indices as follows: 60% are used for training, 20% are used for validation, and the last 20% are used for testing.

3.2   Results

In this subsection we present results regarding the accuracy of the ANN approximation, the space saved with respect to other approximations, and, finally, the time gain achieved by this approximation.

ANN Learning Performance Versant. The best result (mean squared error, MSE) among the five initializations with diferent number of hidden nodes for each type of activation function (linear, tan-sigmoid and log-sigmoid) is shown in Figure 2. Figure 2(b) and Figure 2(c) show quite similar results, but in this last case an ANN of 170 neurons reachs a MSE of $1,44.10^{-5}$ on the training data, $2,86.10^{-3}$ on validation data, and $9,98.10^{-5}$ on test data.

Once the best initialization with log-sigmoid activation function and 170 neurons in the hidden layer has been found, figure 3 shows the linear regression of target values relative to output values for the training, validation and test datasets, as well as the whole datasets. The fit of the ANN is very good, with a value $R = 0.9899$ with the test dataset and $R = 0.9969$ with all patterns.

Since the previous MSE values are relative to control points, we have carried out a practical illustration of the results in terms of errors in the final configuration of the hose. We have use a test case and using the 170 hidden neurons ANN, we show in figure 4 the following results: (a) the original path corresponding to the initial configuration of the hose, (b) the result of the exact GEDS simulation of the action selected (90° turn of the lead robot), taken as the output hose configuration, which are the target ground truth values, (c) the output provided by the ANN, which is indistinguisable from the ground truth, and, finally, (d) the euclidean distance between corresponding 101 points of the ground truth discretization and the ANN output. The maximun error reached between two corresponding points is only the $4,5\%$ of the space discretization resolution, so it can be assumed a perfect approximation.

Time Reduction Versant. One of the two main goals of the paper was to provide a time-efficient approximation of the GEDS model by an ANN. To assess the time gain, we run the GEDS model and its approximation on a Dell Optiplex 760 personal computer, equipped with a processor Intel(R) Core(TM) 2 Duo CPU E8400 @ 3,00Ghz with 3,00 GB of RAM memory and a Microsoft Windows XP Professional v. 2002 SP 3 operating system, using Matlab for the implementation of both approaches.

Table 1 shows the times spent by each model to provide a response on the simulation of an isolated action and a sequence of 10^6 actions, which is the order of the number of actions in a conventional RL simulation. It can be said that the time gain is at least a 90% speedup of the ANN process against the GEDS model.

Space Reduction Versant. Some time ago, our group was dealing with the accurate and efficient learning of the robot-hose system dynamics, but only from a time point of view [9]. However, space efficiency regarding the data necessary

Table 1. Simulation time of both models

	GEDS model	ANN model
Time 1 action	> 2,5 s	< 0,25 s
Time 10^6 actions	> 28 days	< 70 hours

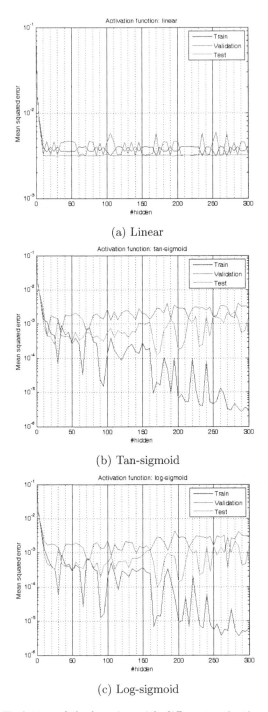

(a) Linear

(b) Tan-sigmoid

(c) Log-sigmoid

Fig. 2. Evolution of the learning with different activation functions

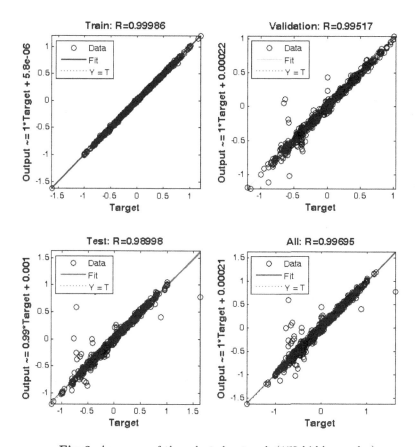

Fig. 3. Accuracy of the selected network (170 hidden nodes)

to store all episodes in a long simulation was neglected. So, the other main goal of this paper was to provide a space-efficient approximation of the GEDS model by an ANN. Table 2 shows the space needed by a model based on 11 discretization points [9] and by the model introduced in this paper based on 8 control points, in both cases to save a simulation episode of an isolated action and a sequence of 10^6 actions, which is the order of the number of actions in a conventional RL simulation. It can be seen that the space gain is more than 27%.

4   Conclusions and Future Work

In this paper we have described the Linked Multi-Component Robotic Systems (L-MCRS) and we have referenced some previous works on the autonomous task learning. We have addressed two of the main practical problems when dealing with this kind of systems and we have argued that an Artificial Neural Network (ANN) model is desirable to learning the Geometrically Exact Dynamic Splines

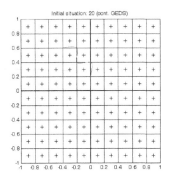

(a) Initial situation (continuous GEDS)

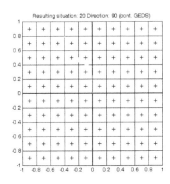

(b) Resulting situation (continuous GEDS)

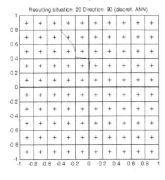

(c) Resulting situation (ANN)

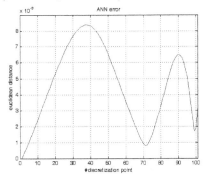

(d) Error of 101 discretization points

Fig. 4. Experiment with the hose situation #20 and the *North* action (test set)

(GEDS) model and overcome these problems (time and space consumption). We have described the training procedure by means a multivariate regression approximation combining several parameters. The trained model has reached a MSE below 10^{-4} on an independent test dataset of control points, which means an error of than 4,5% of the space discretization resolution (as explained in the discussion), so it can be assumed a perfect approximation. Regarding the time saving, we have achieved at least a 90% speedup of the ANN process against the GEDS model. Finally, the space saving is more than 27% against previous models. So, this approach will allow much more extensive simulations to obtain control processes with a reasonable cost (time and space) where the overhead of

Table 2. Space requirements of both models

	ANN model (discretized points)	ANN model (control points)
Space 1 action	352 bytes	256 bytes
Space 10^6 actions	335,69 MB	244,14 MB

the initial ANN training is compensated by the time reduction during the RL experiments execution.

Acknowledgements. The research was supported by Grant UFI11-07 of the Research Vicerectorship, Basque Country University (UPV/EHU). The Computational Intelligence Group is funded by the Basque Government with grant IT874-13. ENGINE project is funded by the European Commission grant 316097.

References

1. Duro, R., Graña, M., de Lope, J.: On the potential contributions of hybrid intelligent approaches to multicomponen robotic system development. Information Sciences 180(14), 2635–2648 (2010)
2. Echegoyen, Z., Villaverde, I., Moreno, R., Graña, M., d'Anjou, A.: Linked multicomponent mobile robots: modeling, simulation and control. Robotics and Autonomous Systems 58(12, SI), 1292–1305 (2010)
3. Fernandez-Gauna, B., Lopez-Guede, J.M., Zulueta, E., Graña, M.: Learning hose transport control with q-learning. Neural Network World 20(7), 913–923 (2010)
4. Sutton, R.S., Barto, A.G.: Reinforcement Learning: An Introduction. MIT Press (1998)
5. Fernandez-Gauna, B., Lopez-Guede, J.M., Graña, M.: Towards concurrent Q-learning on linked multi-component robotic systems. In: Corchado, E., Kurzyński, M., Woźniak, M. (eds.) HAIS 2011, Part II. LNCS, vol. 6679, pp. 463–470. Springer, Heidelberg (2011)
6. Lopez-Guede, J.M., Fernandez-Gauna, B., Graña, M., Zulueta, E.: Empirical study of Q-learning based elemental hose transport control. In: Corchado, E., Kurzyński, M., Woźniak, M. (eds.) HAIS 2011, Part II. LNCS, vol. 6679, pp. 455–462. Springer, Heidelberg (2011)
7. Lopez-Guede, J., Fernandez-Gauna, B., Graa, M., Zulueta, E.: Further results learning hose transport control with Q-learning. Journal of Physical Agents (2012) (in press)
8. Lopez-Guede, J.M., Fernandez-Gauna, B., Graa, M., Zulueta, E.: Improving the control of single robot hose transport. Cybernetics and Systems 43(4), 261–275 (2012)
9. Lopez-Guede, J.M., Graña, M., Ramos-Hernanz, J.A., Oterino, F.: A neural network approximation of L-MCRS dynamics for reinforcement learning experiments. In: Ferrández Vicente, J.M., Álvarez Sánchez, J.R., de la Paz López, F., Toledo Moreo, F. J. (eds.) IWINAC 2013, Part II. LNCS, vol. 7931, pp. 317–325. Springer, Heidelberg (2013)
10. Lopez-Guede, J.M., Fernandez-Gauna, B., Graña, M.: State-action value modeled by elm in reinforcement learning for hose control problems. International Journal of Uncertainty Fuzziness and Knowledge-Based Systems (2013) (submitted)
11. Lopez-Guede, J.M., Fernandez-Gauna, B., Zulueta, E.: Towards a real time simulation of linked multi-component robotic systems. In: KES, pp. 2019–2027 (2012)

Autonomous Robot Navigation Based on Pattern Recognition Techniques and Artificial Neural Networks

Yadira Quiñonez[1(✉)], Mario Ramirez[2], Carmen Lizarraga[1], Iván Tostado[1], and Juan Bekios[2]

[1] Facultad de Informática Mazatlán,
Universidad Autónoma de Sinaloa, Culiacán, Mexico
{yadiraqui,carmen.lizarraga,itostado}@uas.edu.mx
[2] Ingeniería Civil en Computación e Informática
Universidad Católica del Norte de Chile, Antofagasta, Chile
lmrs006@estudiantes.ucn.cl, juan.bekios@ucn.cl

Abstract. The autonomous navigation of robots is one of the main problems among the robots due to its complexity and dynamism as it depends on environmental conditions as the interaction between themselves, persons or any unannounced change in the environment. Pattern recognition has become an interesting research line in the area of robotics and computer vision, however, the problem of perception extends beyond that of classification, main idea is training a specified structure to perform the classifying a given pattern. In this work, we have proposed the application of pattern recognition techniques and neural networks with back propagation learning procedure for the autonomous robots navigation. The objective of this work is to achieve that a robot is capable of performing a path in an unknown environment, through pattern recognition identifying four classes that indicate what action to perform, and then, a dataset with 400 images that were randomly divided with 70% for the training process, 15% for validation and 15% for the test is generated to train by neural network with different configurations. This purpose ROS and robot TurtleBot 2 are used. The paper ends with a critical discussion of the experimental results.

Keywords: Autonomous robots · Artificial neural network · Pattern recognition · Neuro-controllers · ROS · TurtleBot 2

1 Introduction

Robotics, one of the most characteristic areas of Artificial Intelligence has been an amazing growth, has developed a lot of research regarding the autonomous mobile robots [1–3]. In the last years, advances in recent technologies in the area of robotics have made enormous contribution in different areas of application, robots have become a fundamental tool to produce, work and perform dangerous jobs on earth and beyond. Traditionally, applications of robotics [4] Were the

© Springer International Publishing Switzerland 2015
J.M. Ferrández Vicente et al. (Eds.): IWINAC 2015, Part II, LNCS 9108, pp. 320–329, 2015.
DOI: 10.1007/978-3-319-18833-1_34

focused mainly in the industrial sector, in the last two decades, the field of application of robotics has been extended to other sectors [5], for example, robots for construction [6, 7], domestic robots [8, 9], assistance robots [10–12], robots in medicine [13, 14], robots defense, rescue and security [15, 16], among others.

These investigations have been directed towards finding efficient and robust methods for controlling mobile robots. Today, can be considered a fully established scientific discipline, because they are emerging new areas of knowledge that attempt to improve the effectiveness, efficiency, performance and robustness in the autonomous robots navigation.

Computer vision has a multitude of applications in different areas, certainly, robotics is one of the main beneficiaries, because, there are many approaches to solve navigation of mobile robots, one of the most used techniques is the navigation based on image analysis. The computer vision field is formed by several techniques, and beside that, they are constantly combined with machine learning algorithms. In this research we have used a feedforward Scaled Conjugate gradient instead of Backpropagation, some researchers have used these techniques to solve other problems.

Khorrami et al. [17] have compared Continues Wavelet Transform (CWT) with the Discrete Cosine Transform (DCT) and Discrete Wavelet Transform (DWT) as a way to improve the performance of a Multi-Layered Perceptron (MLP) and a Support Vector Machine (SVM). The training or learning algorithms used in MLP and SVM are Backpropagation (BP) and KernelAdatron (KA), respectively.

Alwakeel and Shabaan [18] have proposed a new face recognition system based on Haar wavelet transform (HWT) and Principal Component Analysis (PCA) using Levenberg-Marquardt backpropagation (LMBP) on a neural network. For face detection haar wavelet is used to form the coefficient matrix and PCA is used for extracting features. These features are used to train the classifier based on artificial neural networks. A comparison between the proposed recognition system using DWT, PCA and Discrete Cosine Transform (DCT) is performed.

In the study by Nazir et al. have presented a gender classification technique more efficiently than existing using the Discrete Cosine Transform (DCT) for feature extraction and sorting the features with high variance [19].

Sridhar et al. [20] have proposed a new method by combining of Discrete Cosine Transform and Probabilistic Neural Network for brain tumor classification. According to the results, these algorithms are more efficient and fast compared with other classifiers.

2   Formal Description of the Problem

The autonomous robot navigation based on computer vision is a wonderful resource, because any other information that can be extracted by a camera can provide a great help in getting the robot motion.

The aim of this work is to find a safe way able to guide the robot, from a initial position to the final position o target, through obstacle-free path within a given environment either known or unknown, trying to move with minimal cost. The environment must indicate which is the starting position and the position to which you want to reach (target). For this purpose, we have taking into account four different images corresponding to the main classes that indicate what action to perform as presented in Fig. 1.

Fig. 1. Four classes that indicate the action to perform

3   Theoretical Foundations

In the last decades, pattern recognition has become an interesting research line in the area of robotics and computer vision. The classical techniques of pattern recognition more used are template matching, statistical classification, syntactic or structural matching and neural networks [21]. In this article, we have focused on the neural network approach to achieve an autonomous robot navigation using pattern recognition. The problem of perception extends beyond that of classification, main idea is training a specified structure to perform the classifying a given pattern.

Artificial neural networks (ANN's) have been studied since the 60's until today by various researchers in the scientific community to solve problems in many different application areas. Basically, the ANN's as its name indicates, are composed for a number of interconnected neurons, where its input parameters are the set of signals received from the environment, after, calculations are performed using an activation function and finally to obtain the output signal as shown in Fig.2. In this paper we have used supervised training using neural networks toolbox module of the mathematical software MATLAB. This process requires a set of training patterns, which are randomly propagated through the ANN to generate the output, using the backpropagation training algorithms.

3.1   Backpropagation Algorithm

As is known, the "knowledge" acquired by neural networks is obtained through a learning algorithm, in which the weights are adjusted by iterations until to

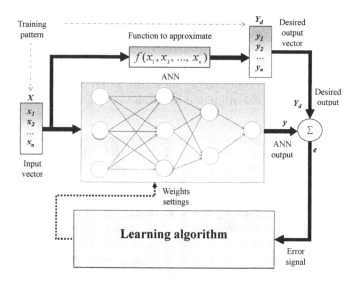

Fig. 2. General scheme of a neural network and representative diagram of supervised training

achieve desired outputs within the accuracy level established. The backpropagation algorithm is the most used method in the literature, however, over time they have developed new techniques that are more stable and allow convergence faster, such as: gradient descent with momentum, gradient descent with adaptive learning rate, gradient descent with momentum and adaptive learning rate, resilient backpropagation, conjugate gradient backpropagation with fletcher-reeves update, conjugate gradient backpropagation with powell-beale restarts, BFGS quasi-newton backpropagation, scaled conjugate gradient backpropagation, among other. A brief description of these algorithms is presented in [22].

In this work, we used training algorithm Scaled Conjugate Gradient to train the multilayer feedforward network. We are trying to solve a simple pattern recognition problem using a network with backpropagation learning procedure, our task is to teach the neural network to recognize 4 images.

4   Implementation Tools

The Robot Operating System (ROS) is a development platform open source for robotic systems. Provides a range of services and libraries that greatly simplify the creation of complex applications for robots. ROS allows the use of different programming languages. Officially supported Python, C++ and Lisp, besides many others. Currently, the library is dedicated to the Ubuntu operating system that is completely stable, although it is also adapting to other operating systems like Fedora, Mac OS X, Arch, OpenSUSE, Slackware, Debian and Microsoft

Windows. At present, there are many groups using ROS to power their robots, some examples are: Care-O-bot 3, iRobot Create, Aldebaran Nao and TurtleBot 2 [24, 25].

Turtlebot 2 is a mobile robot of differential kinematics is programmed with ROS and can be used for multiple applications, is an open robotics platform designed specifically for education and research (see Fig. 3).

Fig. 3. Turtlebot 2 equipped with a sensor in 3D, bumpers and Kinect Xbox 360, it can navigate in indoor and outdoor environments

5   Experimental Results

To carry out the development of this work we have performed in two stages, in order to obtain an effective solution to the problem proposed. The first stage, consisted of the acquisition of data or images obtained from TurtleBot 2 by controller Openni, in order to obtain the characteristics of the patterns to recognize. For best results in the acquisition of images, we performed a median filter and binarized the image to perform operations on it. Subsequently, we proceeded to reduce noise removing objects smaller than 300 pixels and we identified the total of objects found in the image using the integrated function in Matlab bwlabel.

Once we have recognized objects, in the second stage, we obtain the exact coordinates where the object is located in the image, in order to remove the object from the image and analyze it using the neural network. To carry out the training process of the neural network we have generated a dataset, obtaining 100 examples of each image with rotation variations and applying the technique of Discrete Cosine Transform (DCT) [23], where we get a vector that starts with the most significant values of the image and ends with the least significant.

At the end, we obtained a dataset with 400 images that were randomly divided with 70% for the training process, 15% for validation and 15% for the test. After that dataset is integrated, we have proceeded with the creation of the neural network using a structure optimized for pattern recognition. The neural network of the present design consists of three layers with 4 neurons each, 4 neurons in the hidden layer with a sigmoid activation function, 4 output neurons using SoftMax function and the algorithm scaled conjugate gradient backpropagation for training, and finally, there are 300 inputs to the network as shown in Fig. 4.

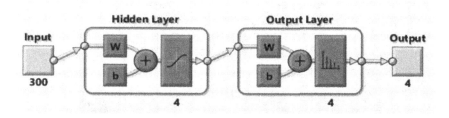

Fig. 4. Neural network structure

We have conducted several experiments to evaluate the performance index, we considered some variants for the neural network configuration, specifically, the number of neurons in the hidden layer was modified. In table 1 shows the training results for 3, 4 and 150 neurons in the hidden layer respectively.

Table 1. Training results for 3, 4 and 150 neurons in the hidden layer respectively

	Samples	CE	%E
Training	280	1.08021-0	6.78571e-0
Validation	60	2.76247e-0	10.0000e-0
Testing	60	2.75740e-0	10.0000e-0
Training	280	3.21172e-0	0
Validation	60	9.20532e-0	0
Testing	60	9.2148e-0	0
Training	280	3.09154e-0	0
Validation	60	9.33340e-0	0
Testing	60	7.11894e-0	0

Fig. 5 presents four confusion matrices: the training, validation and test. The fourth confusion matrix is obtained from the data of the three matrices previously mentioned. It can be seen that 100% correct classification for the entire dataset was obtained.

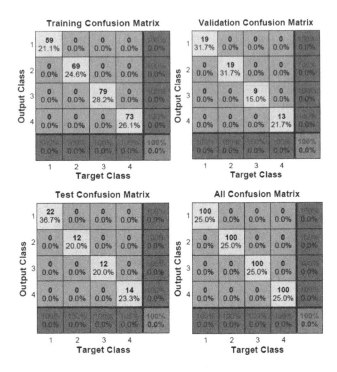

Fig. 5. Confusion matrix

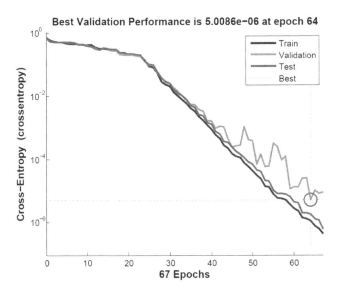

Fig. 6. Performance

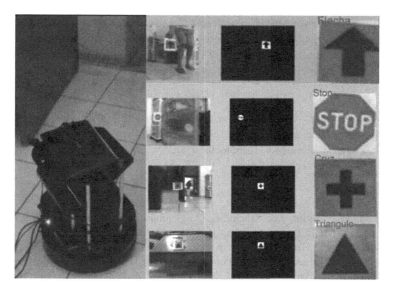

Fig. 7. Implementation in the turtlebot 2: the robot is able to recognize the four classes into the environment

Fig. 6 shows the evolution of the performance index obtained when the best performance in the validation was achieved, it is observed that not show any major mistake with training, this is, the validation curve is greater than the test. If the test curve had increased significantly before the validation curve, then it is possible that occurred overfitting.

Fig. 7 shows a sequence of images with the implementation in the Turtlebot 2, it can clearly see how the robot uses the data and characteristics of the environment to conduct a successful autonomous navigation, using pattern recognition techniques and neural networks.

6   Conclusions and Further Research Work

In this paper we have applied pattern recognition techniques and neural networks for autonomous navigation of robots. According to the results of the simulation on the performance index, we can conclude that a greater amount of data entering the network can generate an overfiting affecting the generalization of the network, because it tends to identify the images only when exactly replicate the conditions in which they were taken.

It has been verified that the algorithm is not affected by changes of scale and image rotation due to preprocessing performed, an important factor to obtain a better learning performance of the network is enlightenment, because when improving lighting conditions the coefficients generated by the DCT are more consistent.

Finally, experiments on the TurtleBot 2 were favorable, the robot was able to recognize patterns successfully. However, we considered it would be interesting to make a comparison with other training techniques in order to obtain a better learning performance, as well as the evaluation with different activation functions.

Acknowledgements. The authors would like to thank Universidad Autónoma de Sinaloa for supporting and financing this research project (PROFAPI2013/146) and Consejo Nacional de Ciencia y Tecnología.

References

1. Burgard, W., Moors, M., Stachniss, C., Schneider, F.: Coordinated multi-robot exploration. IEEE Transactions on Robotics 21(3), 376–386 (2005)
2. Chaimowicz, L., Grocholsky, B., Keller, J.F., Kumar, V., Taylor, C.J.: Experiments in multirobot air-ground coordination. In: IEEE International Conference on Robotics and Automation, vol. 4, pp. 4053–4058 (2004)
3. Howard, A., Parker, L.E., Sukhatme, G.S.: Experiments with a large heterogeneous mobile robot team: exploration, mapping, deployment and detection. The International Journal of Robotics Research 25(5-6), 431–447 (2006)
4. Parker, L.E.: Multiple Mobile Robot Systems. In: Bruno, S., Oussama, K. (eds.) Springer Handbook of Robotics (2008)
5. Braunl, T.: Embedded robotics: mobile robot design and applications with embedded systems. Springer, Heidelberg (2008)
6. Baeksuk, C., Kyungmo, J., Youngsu, C., Daehie, H., Myo-Taeg, L., Shinsuk, P., Yongkwun, L., Sung-Uk, L., Min, C.K., Kang, H.K.: Robotic automation system for steel beam assembly in building construction. In: IEEE 4th International Conference on Autonomous Robots and Agents, pp. 655–661 (2009)
7. Hanjong, J., ChiSu, S., Kyunghun, K., Kyunghwan, K., Jaejun, K.: A study on the advantages on high-rise building construction which the application of construction robots take. In: IEEE Control, Automation and Systems, pp. 1933–1936 (2007)
8. De Almeida, A.T., Fong, J.: Domestic service robots. IEEE Robotics and Automation Magazine 18(3), 18–20 (2011)
9. Sahin, H., Guvenc, L.: Household robotics: autonomous devices for vacuuming and lawn mowing. IEEE Control Systems Magazine 27(2), 20–90 (2007)
10. Linder, T., Tretyakov, V., Blumenthal, S., Molitor, P., Holz, D., Murphy, R., Tadokoro, S., Surmann, H.: Rescue robots at the collapse of the municipal archive of cologne city: a field report. In: International Workshop on Safety Security and Rescue Robotics, pp. 1–6 (2010)
11. Nagatani, K., Okada, Y., Tokunaga, N., Yoshida, K., Kiribayashi, S., Ohno, K., Takeuchi, E., Tadokoro, S., Akiyama, H., Noda, I., Yoshida, T., Koyanagi, E.: Multi-robot exploration for search and rescue missions: a report of map building in RoboCupRescue 2009. In: International Workshop on Safety Security and Rescue Robotics, pp. 1–6 (2009)
12. Santana, P., Barata, J., Cruz, H., Mestre, A., Lisboa, J., Flores, L.: A multi-robot system for landmine detection. In: IEEE Conference on Emerging Technologies and Factory Automation, vol. 1, pp. 721–728 (2005)
13. Guglielmelli, E., Johnson, M.J., Shibata, T.: Guest editorial special issue on rehabilitation robotics. IEEE Transactions on Robotics 25, 447–480 (2009)

14. Low, K.H.: Robot-assisted gait rehabilitation: from exoskeletons to gait systems. In: Defense Science Research Conference and Expo (DSR), pp. 1–10 (2011)
15. Okamura, A.M., Mataric, M.J., Christensen, H.I.: Medical and health-care robotics. IEEE Robotics and Automation Magazine 17(3), 26–37 (2010)
16. Reed, K., Majewicz, A., Kallem, V., Alterovitz, R., Goldberg, K., Cowan, N., Okamura, A.: Robot-assisted needle steering. IEEE Robotics and Automation Magazine 18(4), 35–46 (2011)
17. Khorrami, H., Moavenian, M.: A comparative study of DWT, CWT and DCT transformations in ECG arrhythmias classification. Expert Systems with Applications 37(8), 5751–5757 (2010)
18. Alwakeel, M., Shaaban, Z.: Face Recognition Based on Haar Wavelet Transform and Principal Component Analysis via Levenberg-Marquardt Backpropagation Neural Network. European Journal of Scientific Research 42(1), 25–31 (2010)
19. Nazir, M., Ishtiaq, M., Batool, A., Jaffar, M., Mirza, A.: Feature Selection for Efficient Gender Classification. Recent Advances in Neural Networks, Fuzzy Systems & Evolutionary Computing, 70–75 (2010)
20. Sridhar, D., Murali, K.: Brain Tumor Classification using Discrete Cosine Transform and Probabilistic Neural Network. In: IEEE International Conference on Signal Processing Image Processing & Pattern Recognition, pp. 92–96 (2013)
21. Jain, A.K., Duin, R.P., Mao, J.: Statistical pattern recognition: A review. IEEE Transactions on Pattern Analysis and Machine Intelligence 22(1), 4–37 (2000)
22. MATLAB Neural Network Toolbox. Users Guide, http://www.mathworks.com/help/nnet/index.html (accessed on March 2, 2015)
23. Bai, C., Kpalma, K., Ronsin, J.: Analysis of Histogram Descriptor for Image Retrieval in DCT Domain. In: Tsihrintzis, G.A., Virvou, M., Jain, L.C., Howlett, R.J. (eds.) IIMSS 2011. SIST, vol. 11, pp. 227–235. Springer, Heidelberg (2011)
24. Cousins, S., Gerkey, B., Conley, K., Garage, W.: Sharing software with ROS. IEEE Robotics & Automation Magazine 17(2), 12–14 (2010)
25. Araujo, A., Portugal, D., Couceiro, M., Rocha, R.: Integrating Arduino-Based Educational Mobile Robots in ROS. Journal of Intelligent & Robotic Systems 77(2), 281–298 (2014)

Collaborative Group Formation Using Genetic Algorithms

M. Angélica Pinninghoff[1]([✉]), Miguel Ramírez[1],
Ricardo Contreras Arriagada[1], and Pedro Salcedo Lagos[2]

[1] Department of Computer Science,
University of Concepción, Concepción, Chile
[2] Research and Educational Informatics Department
University of Concepción, Concepción, Chile
{mpinning,rcontrer,psalcedo}@udec.cl

Abstract. Collaborative learning is a process in which two or more individuals interact in order to learn or attempt to learn something. The success of the process depends heavily on the way in which the individuals are engaged in a community, i.e. where the process takes place. Individuals in the community are grouped into small clusters that i) posses homogeneous properties and ii) emphasize the diversity inside the group, allowing flourishing of diverse points of view. In this work we focus on the formation of groups of individuals. More specifically, we apply genetic algorithms in the formation process in order to deal with the high level of complexity. We developed a prototype to evaluate the approach and the results are discussed.

Keywords: Genetic algorithms · Collaborative learning

1 Introduction

Teaching strategies have evolved from the classic individual and competitive learning, to models that focuse on the group interaction, such as collaborative learning. Collaborative learning is the combination of individual efforts to accomplish a collective task [3]. Different approaches vary to some extent, but essentially they all promote the idea that young people's learning is best served when they have opportunities to learn with and from each other, and are shown how to do so effectively.

The responsible and individual work assigned to each member of the group is critical for the group, because learning is always regarded as *an individual process (non collective) influenced by external factors* [1]; including among these external factors the collaborative work influence and the learning supported by the classmates. This mutually supported learning establishes a key difference respect to the classic learning process, restricted to the teacher explanation or the reading of a text.

It is not difficult to understand the importance of collaborative learning and hence the need for an adequate strategy in the process of forming groups.

J.M. Ferrández Vicente et al. (Eds.): IWINAC 2015, Part II, LNCS 9108, pp. 330–338, 2015.
DOI: 10.1007/978-3-319-18833-1_35

The objective of this work is to design and implement a software for helping in the task of forming learning groups, by using artificial intelligence techniques, in particular genetic algorithms. Groups should be similar, i.e., groups should present similar characteristics, but members in a group should present heterogeneous features combined with different abilities. The proposed approach take into account similarities and differences that need to be mixed to get a potentially successful group. We intend this tool is a support for teachers, by considering different features students exhibit.

This article is structured as follows. The first section is made up of the present introduction; the second section describes the context of the problem and related work and the proposal; the third section presents a resume on genetic algorithms, while the fourth section shows our proposal. Results are shown in fifth section, and the final section is devoted to conclusions of the work.

2   The Problem Context

Collaborative groups are an essential part of many academic activities. Unfortunately, it doesn't exist a clear methodology that can address the steps for choosing the adequate group. The problem is, in this case, that different groups can present important differences among them, affecting the learning process and misusing the advantages of the group conformation. Typical signals of it are the non existence of a leader or groups of members with non compatible personalities; symptoms that affect the accomplishment of the goals established for the team [10].

When dealing with collaborative work, it is important that members belonging to a group can feel comfortable and that the academic level of each member allows a positive interaction. Some researchers argue that the group structure must emerge from inside the group, but there is no a general agreement about this issue. Nevertheless an important number of works highlight the fact that some elements must be always present, additionally to the roles that need to be present in every working group.

Belbin [2] proposes a theoretical frame for analysis and integration of working groups. The frame is supported by a solid experimental work and over 25 years of related work in the Management College of Henley, Cambridge. Group roles refer to the individual and personal way in which we behave, the way in which we contribute to the task and the way in which we conduct our relationships with other members of the group. These roles are different to the functional roles. The group roles are spontaneous, intuitive and emotional. These roles have been studied and classified into different categories.

Trying to put technology on the topic, an important number of diverse initiatives have been conducted, keeping in mind the idea of a platform supporting the learning process. One of these projects is Teamie a platform that combines social networking and Education in a safe and secure manner, to drive collaborative learning and student engagement. Teamie allows teachers to create and upload lessons, create quizzes and publish gradebook and see reports. In other words, it acts as a classic supporting platform that offer a set of facilities including the use of messages[1].

A closer approach to our proposal can be found in the work of Yeoh et al. [13]. In this work, authors propose an algorithm to conform balanced groups of students, between three and five students based on diversity elements such as gender, race, nationality among others, considering also average capabilities quantified by a single number such as the average Cumulative Grade Point Average (CGPA). They create groups by following an arbitrary schema, and then compare groups to exchange members for diminishing the probability of having a bad group (a group that has a single gender, having too few of a race, etc.). Besides the limits imposed to the group size, the class profile must be fixed, to avoid a new regrouping process during the academic period.

An interesting work is described in [14], where authors present a method for group formation using genetic algorithms. The method focuses on working group for TI projects, in particular, software development groups. Authors emphasize that one of the key factors to be considered is good programming skill. So, this paper take into account how to form groups with balanced programming skills to ensure that every group members can complete the software project successfully.

The work in [8] introduces an algorithm that doesn't limit the number of features of students for grouping and, in addition of using values of features, their priorities are also involved in the grouping process. Authors claim that one significant improvement is the fact that they consider, for evaluation, a inter-fitness and a intra-fitness function. This approach is similar to our proposal; but the difference is that we introduce a higher dynamic behavior, because values associated to features are obtained through a set of tests, directly from students.

Our proposal is aimed to present an artificial intelligence based tool, for the process of grouping students in a collaborative learning strategy. The key idea is that groups, seen as a set, should contain homogeneous elements and, at the same time, have to be heterogeneous when observed in a intra-relationship view. If we consider a set of features, with every feature presenting a set of different possible values, the problem turns into a combinatorial problem, hard to solve because it is not computationally bounded to deal with a high number of probable combinations.

We believe that there is a set of features a student can present, that are not easy to measure, but that can guide the process of forming groups, trying to take into account some variables related to learning processes in general. To characterize a student, four different tests were applied.

- Multiple intelligences The multiple intelligences test based on Gardner [5], considers seven characteristics that correspond to seven types of intelligence that work together inside each person's overall development and structure: linguistic, logical/mathematical, visual/spatial, bodily/kinesthetic, musical, interpersonal and intrapersonal.

---

[1] https://theteamie.com/

- Learning styles Learning styles were developed by Honey and Mumford [7], based upon the work of Kolb, and they identified four distinct learning styles or preferences: Activist, Theorist, Pragmatist and Reflector. These are the learning approaches that individuals naturally prefer.
- Leadership Lewin's styles of leadership [9] identify three sales of leadership of decision making: autocratic leadership, democratic leadership and laissez-faire or liberal leadership.
- Assertiveness Rathus [12], presented a schedule for measuring assertiveness in a student. The student is asked to rate each of the 30 items on a 6-point scale, ranging from extremely descriptive to extremely undescriptive.

Given the high number of different combinations of variables a student can present, it is necessary to use an approximate mechanism for finding a, probably, optimal solution. Genetic algorithms offer a typical optimization mechanism that helps to find acceptable solutions when it is not possible to check every possible alternative.

3   Genetic Algorithms

Genetic algorithms (GA) are a particular class of evolutionary algorithms, used for finding optimal or good solutions by examining only a small fraction of the possible space of solutions. GAs are inspired by Darwin's theory about evolution. The basic concept of GAs is designed to simulate processes in natural system necessary for evolution, specifically those that follow the principles of survival of the fittest. As such they represent an intelligent exploitation of a random search within a defined search space to solve a problem.

The structure of a genetic algorithm consists of a simple iterative procedure on a population of genetically different individuals. The phenotypes are evaluated according to a predefined fitness function, the genotypes of the best individuals are copied several times and modified by genetic operators, and the newly obtained genotypes are inserted in the population in place of the old ones. This procedure is continued until a *good enough* solution is found [4].

In this work, a chromosome represents a specific distribution of student of a classroom in groups. In genetic algorithm terminology, each student is a gen and the identification of a student represents the value (allele) that the gene has. A good fitness means that a particular distribution contains a set of groups of students, where all the groups are homogeneous and each group has students with different characteristics. These characteristics are measured through the set of tests described in section 2.

Genetic Operators: Different genetic operators were considered for this work. These genetic operators are briefly described bellow:

- <u>Selection</u>. Selection is accomplished by using the roulette wheel mechanism or tournament [4]. It means that individuals with a best fitness value will have a higher probability to be chosen as parents.
- <u>Cross-over</u>. Cross-over is used to exchange genetic material, allowing part of the genetic information that one individual has, to be combined with part of the genetic information of a different individual. It allows us to increase genetic variety, in order to search for better solutions. Due to the nature of the problem we are dealing with, specific cross-over operator is considered: OX [11].
- <u>Mutation</u>. By using this genetic operator, a slight variation is introduced into the population so that a new genetic material is created.

By using GA, it is possible to achieve adaptability, because in optimization problems we can use different types of constraints, and different functions, linear or non linear, including discrete or continuous variables. GA are robust in the sense that using evolving procedures encourage the search for global solutions (over local solutions). Additionally, GA are flexible enough to combine with other heuristics for efficient implementations when solving specific problems [6].

4   The Proposal

To implement a GA, it is necessary to define six components: a representation or coding for problem solutions; an evaluation function to measure and compare different candidates; a mechanism for randomly generating an initial population for encourage the diversity in individuals; the choosing of the genetic operators that are to be used for evolution; the survival criteria to be accomplished for individuals that are conserved into the next generation, on each iteration and; finally, to test a set of parameters. These components are considered in the formation of groups, and are described in the following.

A solution is a class in which students are arranged into different groups; ideally having the same size. A class is formed by **n** students, grouped into **m** groups and each group consists of **s** students. For coding a class as a set of chromosomes, every student is identified by a label (a number ranging from **1** to **n**. In doing so, a chromosome (an individual) contains an ordered set of size **n**, where **n** is the number of students. This ordered set is composed by **m** sections containing **s** numbers; every section represents a specific group. Figure 1 shows an example of a class having 16 students and four groups containing four students each.

Fig. 1. A class of 16 students

In a separate table, the results of the four tests is stored. Every student is linked to a set of four numbers, that correspond to the selected alternatives for each test. So, we can say, for example, that for student 10, the corresponding sequence is (5,2,1,6), that means that in the test of multiples intelligences he/she is classified as musical/rhythmical; reflexive learning in the test about learning styles, authority leadership in the leadership test, and extremely non descriptive in the assertiveness test.

The evaluation function allows to assign a value to each chromosome. In this case, the lower value the chromosome presents, the better value for the evaluation function. The value for the chromosome depends on two issues: first, it is necessary that every characteristic a student holds should be different to the characteristics their classmates present. Secondly, every group should be similar to the rest of the groups. The ideal fitness is zero, that indicates there are not repeated characteristics inside a group; in other word, every student has a different value for a specific characteristic. When a characteristic is repeated, this fact counts in the evaluation function. To reflect the fact that this condition is undesired, we take the resulting value for each group, and for making important the penalty, arbitrarily we rise the value to the power of three. It is arbitrary, but experiments validated this issue.

The final value, obtaining by adding the values for all the groups, is the final value for the chromosome.

The initial population must be generated randomly, to obtain a higher diversity. In this case, we seek for a feasible population, in other words a population in which every individual is a feasible solution.

We considered three genetic operators: selection, cross-over and mutation. The selection operator is tournament, that consists of organizing a tournament among a small subset of individuals in the population for every offspring to be generated. The procedure starts by randomly picking **k** (the tournament size) individuals from the population; the individual with the best fitness among the **k** individuals generate an offsping. All **k** individual are put back into the population and are eligible to participate in further tournaments [4]. In this work the value of **k** was set to 4.

The chromosome is an ordered set of students. Because all students must be present in every chromosome, it was necessary to consider a specific crossover operator, OX (Ordered Crossover) crossover operator [11]. Operator OX builds one offsprings by choosing a subsequence of one parent and preserving the relative order of the rest of sequence from the other parent. To build the second offspring it is the same procedure, but with the oposite parents. By using this strategy there is no losing of students when moving from one generation to the next.

The mutation operator, due to the same reason just explained, is Displacement. This operator selects a subsequence in the chromosome and inserts it in a random place [11].

The survival criteria is elitism. Elitism allows a small number of best evaluated individuals are passed unchanged to the next generation. It is a very simple strategy used for keeping in the population the best evaluated individuals, if we do not adopt this strategy, a good individual (solution) can be lost.

Parameters are tuned through experimentation, and correspond to size of the population, number of generations, cross-over probability and mutation probability.

5   Results

Experiments for establish the parameters value were carried out on three types of classes: a small class (20 students), a medium class (32 students) and a big class (48 students). These sizes correspond to typical sizes for chilean educational reality.

Table 1, 2 and 3, summarizes values of the parameters considered for testing in each type of course.

Table 1. Parameters used for testing in a small course

Item	Value(s)
Population	50, 100, 150, 200
Number of generations	200
Crossover	70%, 80%, 90%, 100%
Mutation	1-7%

Table 2. Parameters used for testing in a medium course

Item	Value(s)
Population	50, 100, 150, 200, 300, 400, 500
Number of generations	500
Crossover	70%, 80%, 90%, 100%
Mutation	1-7%

Table 3. Parameters used for testing in a big course

Item	Value(s)
Population	300, 500, 700, 800, 900, 1000
Number of generations	1000
Crossover	70%, 80%, 90%, 100%
Mutation	1-7%

Table 4. Results of testing

Item	small	medium	big
Population	200	300	900
Number of generations	50	460	260
Crossover	100%	100%	70%
Mutation	4%	5%	4%

Result of experiments for fixing parameters are shown in Table 4. For these values we obtained the best solutions for different classes size. Column names refer to the types of classes as explained above.

We develop the analysis by considering separately the different classes size.

In a small class, the algorithm behaves reaching rapidly an optimum value. Working with a bigger population size, convergence to the optimum value is faster than the case in which we work with a reduced population size. The execution time, as expected, is influenced more by the number of iterations than by the population size. We decided to use a bigger population, diminishing the number of iterations. Execution time takes only a few seconds. From the set of tests, the optimum value was reached in the 80% of experiments.

In medium size classes, by considering a population below 200 individuals, fitness never reaches the value zero. By using a population size above 300 individuals, convergence is reached approximately in generation number 500. However, when population is increased to 500 individuals, convergence is reached in approximately 300 iterations. In both cases, execution time takes less than two minutes. For this size of class, 60% of experiments reached the optimum value.

For a big size class, even if the optimum is not reached, a result can be accepted as useful when a local optimum close to zero is found. For this group of experiments there are three different behaviors, depending on the size of the population. The first case that considers a population below 500 individuals, causes the algorithm to execute a great number of iterations, with a rapid convergence but far from a good value for fitness. The second case corresponds to a population size between 600 and 800 individuals, finding acceptable solutions in 50 iterations but with a high execution time. In the third case, that considers a population between 900 and 1000 individuals, the algorithm reaches an acceptable solution in approximately 250 iterations, with an execution time of about three minutes. For this instance, if the execution time is extended to 5 minutes, the optimum solution is reached in approximately 20% of tests.

6   Conclusions

The approach was implemented as a web product that allows teachers form groups, for an efficient and effective collaborative learning. The success of this approach is supported by the multidisciplinary interaction of different professionals, including psychologist and education experts. In particular, understanding the way in which a particular test can help in detecting particularities in individuals is absolutely necessary to transfer a pedagogic strategy to a software product. Beyond results, it is clear that by incorporating different points of view, results may be improved.

Once again, genetic algorithms have shown a great capability for dealing with problems that offers a highly combinatorial feature; but the isolated consideration of this technique is not enough, a good students characterization is a key element for this type of problems.

Acknowledgements. This work is partially supported by the Chilean National Fund for Scientific and Technological Development, FONDECYT, through project number 1140457.

References

1. Ariza, A., Oliva, S.: Las nuevas tecnologías de la información y la comunicación y una propuesta para el trabajo colaborativo. V Congreso Iberoamericano de Informática Educativa, Chile (2000)
2. Belbin, M.: Management Teams: why they succed or fail, 3rd edn. Rouledge (2010)
3. Calzadilla, M.: Aprendizaje colaborativo y tecnologías de la información y la comunicación. Revista Iberoamericana de Educación (2002)
4. Floreano, D., Mattiussi, C.: Bio-Inspired Artificial Intelligence. Theories, Methods, and Technologies. The MIT Press (2008)
5. Gardner, H.: Frames of Mind: The Theory of Multiple Intelligences, 3rd edn. Basic Books (2011)
6. Goldberg, D.E.: Genetic Algorithms in Search, Optimization and Machine Learning. Addison-Wesley (1989)
7. Honey, P., Mumford, A.: Manual of Learning Styles. P. Honey, London (1982)
8. Mahdi Barati Jozan, M., Taghiyareh, F., Faili, H.:An Inversion-Based Genetic Algorithm for Grouping of Students. In: The 7th International Conference on Virtual Learning, ICVL 2012, Brasov, Romania (2012)
9. Lewin, K., Lippit, R., White, R.K.: Patterns of aggressive behavior in experimentally created social climates. Journal of Social Psychology 10, 271–301 (1939)
10. Lucero, M.: Entre el trabajo colaborativo y el aprendizaje colaborativo. Revista Iberoamericana de Educación (2003)
11. Michalewicz, Z., Fogel, D.B.: How to Solve It: Modern Heuristics. Springer (2000)
12. Rathus, S.: A 30-item schedule for assessing assertive behavior. Behavior Therapy 4(3), 398–406 (1973)
13. Yeoh, H.K., Mohamad Nor, M.I.: An Algorithm to Form Balanced and Diverse Groups of Students. Computer Applications in Engineering Education 19, 582–590 (2011)
14. Ani, Z.C., Yasin, A., Husin, M.Z., Hamid, Z.A.: A Method for Group Formation Using Genetic Algorithm. International Journal on Computer Science and Engineering 02(09), 3060–3064 (2010)

The CALIMACO Multimodal System: Providing Enhanced Library Services Using Mobile Devices

David Griol[✉], Miguel Ángel Patricio, and José Manuel Molina

Computer Science Department
Carlos III University of Madrid
Avda. de la Universidad, 30, 28911, Leganés, Spain
{david.griol,miguelangel.patricio,josemanuel.molina}@uc3m.es

Abstract. Multimodal conversational agents have became a strong alternative to enhance educational systems with intelligent communicative capabilities. The combination of these systems with smart mobile devices have promoted in recent years more sophisticated human-machine interfaces that provide a more engaging and human-like relationship between users and the system. Among the different educational application domains, this combination allows developing enhanced digital libraries containing new types of multimedia contents, providing new interaction possibilities, and adapting this interaction taking into account the specific users requirements and preferences. In this paper we propose the practical application of multimodal conversational agents to develop advanced University Digital Libraries (UDL) adapted to the new interaction scenarios of mobile devices. Our proposal integrates features of Android APIs on a modular architecture that emphasizes interaction management to create robust applications, easily updated and adapted to the user.

Keywords: Conversational Agents · Digital Libraries · Assistants · Multimodal Interaction · Spoken Interaction · Mobile Devices · Android

1 Introduction

With the growing maturity of conversational technologies, the possibilities for integrating conversation and discourse in e-learning are receiving greater attention [12,9]. Using natural language in educational software allows students to spend their cognitive resources on the learning task, and also develop more social-based agents [14].

Current possibilities to employ multimodal conversational agents for educative purposes include tutoring applications [13], question-answering [16], conversation practice for language learners [3], pedagogical agents and learning companions [1], and dialogs to promote reflection and metacognitive skills [8]. These agents may also be used as role-playing actors in immersive learning environments [5].

Systems developed to provide these functionalities typically rely on a variety of components, such as speech recognition and synthesis engines, natural language processing components, dialog management, databases management, and

© Springer International Publishing Switzerland 2015
J.M. Ferrández Vicente et al. (Eds.): IWINAC 2015, Part II, LNCS 9108, pp. 339–348, 2015.
DOI: 10.1007/978-3-319-18833-1_36

graphical user interfaces. Laboratory systems usually include specific modules of the research teams that build them, which make portability difficult. Thus, it is a challenge to package up these components so that they can be easily installed by novice users with limited engineering resources. In addition, due to this variability and the huge amount of factors that must be taken into account, these systems are difficult to develop and typically are developed ad-hoc, which usually implies a lack from scalability. Our work represents a step in this direction.

In addition, mobile devices have lead to a new paradigm in which they can collect information from the user pervasively. The fact that increasingly more individuals have always with them a device with numerous displays, sensors and connectivity possibilities, opens new interaction scenarios that demand more sophisticated interfaces, such as multimodal conversational agents.

In this paper we propose the practical application of multimodal conversational agents and mobile devices to develop advanced digital libraries, which extends the concept of Library 2.0. To do this, we have relied on the recommendations for the development of University Digital Libraries (UDL) [15] compiled by recent studies [11,2], which propose using new open-source technologies to build complex systems at less cost, increasing the efficiency and sharing of bibliographic records for all libraries, maximizing the use of these libraries, providing new types of multimedia contents and interaction possibilities, and transferring library efforts into higher-value activities.

At the end of 2014, 75% of smartphones and tablets operate with Android OS [9]. Also, there is an active community of developers who use the Android Open Source Project and have made possible to have more than one million applications currently available at the official Play Store, many of them completely free. For these reasons, the developed application makes use of different facilities integrated in Android-based devices.

2   The CALIMACO Multimodal Application

The developed Android application consists of a multimodal virtual assistant that provides information and services related to a University Digital Library. The application is called Calimaco in honor of Callimachus of Cyrene, director of the Library of Alexandria and creator of pinakes (first catalogs of books that served as tables of contents of the library). As a multimodal application, the application allows users to interact by means of the tactile mode, by voice or by combining both modes. The following subsections summarize the main technologies and functionalities of the application.

2.1   Technologies Used

A multimodal conversational agent providing spoken interaction integrates six main tasks to deal with user's inputs: automatic speech recognition (ASR), natural language understanding (NLU), dialog management (DM), natural language generation (NLG), visual generation (VG), and text-to-speech synthesis (TTS).

Speech recognition capabilities on Android devices have evolved rapidly since Android 2.1 with each new version of Android. Besides the recognition capabilities that are implemented within the Android operating systems, there is the possibility to build Android apps with speech input and output using the Google Speech API (package *android.speech*). With this API, speech recognition can be carried out by means of a *RecognizerIntent*, or by creating an instance of *SpeechRecognizer*. In both cases, the results are presented in the form of an N-best list with confidence scores.

Once the conversational agent has recognized what the user uttered, it is necessary to understand what he said. Natural language processing generally involves morphological, lexical, syntactical, semantic, discourse and pragmatical knowledge. We propose the use of grammars in order to perform the semantic interpretation of the user inputs.

The dialog manager decides the next action of the system, interpreting the incoming semantic representation of the user input in the context of the dialog. The developed application integrates a statistical methodology that combines multimodal fusion and dialog management functionalities [4]. This module applies presentation strategies that decompose the complex presentation goal into presentation tasks. It also decides whether an object description is to be uttered verbally or graphically. The result is a presentation script that is passed to the visual generation and natural language generation modules.

The visual generation module creates the visual arrangement of the content using dynamically created and filled graphical layout elements. Since many objects can be shown at the same time on the display, the manager re-arranges the objects on the screen and removes objects, if necessary. The visual structure of the user interface (UI) is defined in an Android-based multimodal application by means of layouts. Layouts can be defined by declaring UI elements in XML or instantiating layouts elements at runtime. Both alternatives can be combined in order to declare the application's default layouts in XML and add code that would modify the state of the screen objects at run time.

UI layouts can be quickly designed in the same way a web page is generated. Android provides a wide variety of controls that can be incorporated to the UI, such as buttons, text fields, checkboxes, radio buttons, toggle buttons, spinners, and pickers. The *View* class provides the means to capture the events from the specific control that the user interacts with. The user interactions with the UI are captured by means of event listeners. The default event behaviors for the different controls can also been extended using the class event handlers.

Natural language generation is the process of obtaining texts in natural language from the non-linguistic representation, internal representation of information handled by the dialog system. The simplest approach consists in using predefined text messages (e.g., error messages and warnings). Finally, a text-to-speech synthesizer is used to generate the voice signal that will be transmitted to the user. We propose the use of the Google TTS API to include the TTS functionality in an application.

The text-to-speech functionality has been available on Android devices since Android 1.6 (API Level 4). The *android.speech.tts* package includes the classes and interfaces required to integrate text-to-speech synthesis in an Android application. They allow the initialization of the TTS engine, a callback to return speech data synthesized by a TTS engine, and control the events related to completing and starting the synthesis of an utterance, among other functionalities. Every Android device incorporates a default TTS motor. In addition, Android allows the installation and personalization of several motors, like Pico TTS, IVONA TTS HQ, SVOX Classic TTS, Samsung TTS, or eSpeak TTS.

The application uses different databases as data repositories, several of them arranged in an external Web server in order to allow access to the application and user identification using mobile devices. A specific web service has been developed to transmit a request from the Android application to a web server, which accesses the MySQL database and returns the requested information to the Android application in JSON format (JavaScript Object Notation). A JSON element consists of two structures: a collection of pairs of name-value and an ordered list of values. The Android application extracts information from these elements using the *org.json.JSONObject* and *org.json.JSONArray* classes. The Android application also accesses an internal SQLite database to store locally specific user information.

The application also includes features that require extract information from different web pages (e.g., to access the catalog of books or consult the regulatory requirements in a library). To perform this query, the Android application uses the *JSoup* Java library to connect the web page and extract the HTML contents.

2.2   Main Functionalities

The main functionalities provided by the multimodal application can be classified into six main modules. The first module allows registering in the application (required to personalize the application by collecting the most common queries completed by the user) and configure a wide range of options related to the main functionalities of the application (see Figure 1, first and second images). Also, users can access a wide range of Web 2.0 tools provided by the Carlos III University of Madrid (website of the UC3M Library, Twitter account, Facebook, and institutional repository of news and weekly schedule).

The second module offers access to the library catalog stored in different databases. Users can provide their query orally or using the tactile elements. The application provides the completed information of the requested resource, their current state (available, borrowed or waiting list), and images with a visual description extracted from the web repository of the university library (see Figure 2).

The application can also guide the user through the UC3M Library at the Leganés Campus. To do this, the user can provide the name of a specific location, a service offered by the library, the title of a bibliographical resource, or subject in the catalog. The application uses images, photographs and synthesized speech

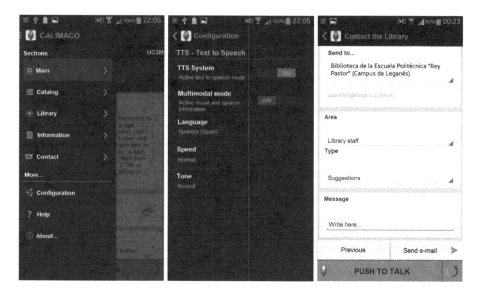

Fig. 1. Screenshots of the configuration and contact modules

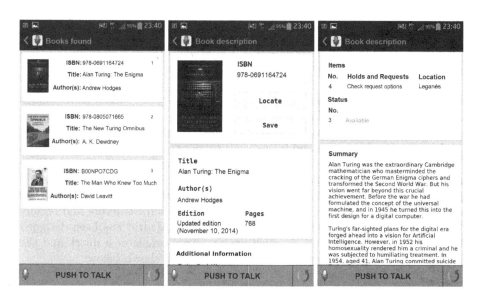

Fig. 2. Screenshots of the library catalog module

that convey specific step by step instructions to locate the required location starting from the main door of the library (see Figure 3, first and second images).

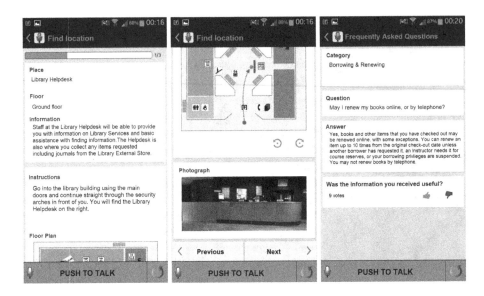

Fig. 3. Screenshots of the find location and FAQ modules

The fourth module of the application is related to the frequently asked questions (FAQ). The set of questions that the application is able to answer has been extracted from the lists provided by the university libraries of the University of León, University of Alcalá de Henares, University of Granada, Carlos III University of Madrid, University of Almería, University of Murcia, Autonomous University of Barcelona, University of Córdoba, and University of Alicante.

This set of questions has been classified into 13 categories (General, University Identity Card, Services and Infrastructure, Books, Borrowing and Renewing, Interlibrary Loan, Finding Information and Access to Resources, Search, Electronic journals, eBooks, Institutional Repository, Training courses, Impact and quotations, and Acquisitions), encompassing more than 250 FAQs asked by users of these libraries. A specific response adapted to the infrastructure and regulations of the libraries at the Carlos III University of Madrid has been designed for each question (see Figure 3, third image). An example of a query and the corresponding response provided by the application is shown below:

May I renew my books online, or by telephone?

Response: *Yes, books and other items that you have checked out may be renewed online, with some exceptions. To renew online, use the My Library Account link and log in using your currently valid University Network ID and password. You can renew an item up to 10 times from the original check-out date unless another borrower has requested it, an instructor needs it for course reserves, or your borrowing privileges are suspended. You may not renew books by telephone.*

The fifth module of the application provides information on how to contact the Library. This module provides access to a web form to request information directly to university staff, transmit suggestions, complaints, or compliments (see Figure 1, third image).

Finally, the Help & Quick Find module of the application provides a detailed help about the different functionalities of the application, as well as a shortcut to each of them. This information is provided by means of screenshots of the application, instructional videos, use cases, synthesized speech, and practical examples explaining how to use the previously described functionalities.

3 Evaluation

The methodology used to evaluate the multimodal interaction with the developed application is based on the work presented in [10], which points out that the usability assessment of a multimodal system requires the evaluation of the different interaction modalities. To do this, we have developed the assessment questionnaire shown in Table 1, which is based in standard questionnaires like AttrakDiff [6] and SASSI [7].

As it can be observed, the questionnaire consists of 9 questions. Each question has 5 possible answers from which only one is selected. The main aspects that are

Table 1. Questionnaire designed for the usability assessment of the application

Previous experience using multimodal interfaces	
Q1. Assess on a scale of 1-5 your previous experience using voice interfaces.	(1 = "Low", 5 = "High")
Q2. Assess on a scale of 1-5 previous experience using multimodal interfaces.	(1 = "Low", 5 = "High")
Understanding of user responses	
Q3. How well did the system understand you?	Never, Seldom, Sometimes, Usually, and Always
Understanding of system responses	
Q4. How well did you understand the system messages?	Never, Seldom, Sometimes, Usually, and Always
Interaction rate	
Q5. In your opinion, the interaction rate was...	Very slow, Slow, Suitable, Fast, Very fast
Difficulty level using the system	
Q6. Indicate the difficulty level of the system.	Very difficult, Difficult, Normal, Easy, Very easy
Presence of errors	
Q7. Have you noticed errors during the interaction?	Never, Seldom, Sometimes, Usually, and Always
Certainty of what to do at each moment	
Q8. Was it easy to decide what to do after each system turn?	Never, Seldom, Sometimes, Usually, and Always
Global satisfaction with the system	
Q9. In general, are you satisfied with the system?	Very dissatisfied; Dissatisfied; Satisfied; Quite satisfied; Very satisfied

Table 2. Results of the usability assessment (1=worst, 5=best evaluation)

Previous experience using multimodal interfaces			
Question	Avg. value	Max. value	Min. value	Std. deviation
Q1	3.36	5	1	1.26
Q2	3.24	5	1	1.34
User understood by the system			
Question	Avg. value	Max. value	Min. value	Std. deviation
Q3 (tactile mode)	4.80	4	5	0.40
Q3 (oral mode)	4.64	4	5	0.48
Q3 (multimodal mode)	4.84	4	5	0.37
System understood by the users			
Question	Avg. value	Max. value	Min. value	Std. deviation
Q4 (tactile mode)	4.92	4	5	0.27
Q4 (oral mode)	3.80	3	5	0.69
Q4 (multimodal mode)	4.48	4	5	0.50
Interaction rate			
Question	Avg. value	Max. value	Min. value	Std. deviation
Q5 (tactile mode)	3.64	3	5	0.69
Q5 (oral mode)	2.80	2	5	1.02
Q5 (multimodal mode)	4.76	3	5	0.51
Easiness of the interaction			
Question	Avg. value	Max. value	Min. value	Std. deviation
Q6 (tactile mode)	4.24	3	5	0.51
Q6 (oral mode)	3.60	3	5	0.69
Q6 (multimodal mode)	4.84	3	5	0.37
Absence of errors			
Question	Avg. value	Max. value	Min. value	Std. deviation
Q7 (tactile mode)	4.80	4	5	0.40
Q7 (oral mode)	3.80	2	5	0.98
Q7 (multimodal mode)	4.44	4	5	0.50
Certainty of what to do at each moment			
Question	Avg. value	Max. value	Min. value	Std. deviation
Q8 (tactile mode)	4.00	3	5	0.49
Q8 (oral mode)	4.40	3	5	0.63
Q8 (multimodal mode)	4.84	4	5	0.37
Global satisfaction with the system			
Question	Avg. value	Max. value	Min. value	Std. deviation
Q9 (tactile mode)	4.44	4	5	0.50
Q9 (oral mode)	3.64	3	5	0.74
Q9 (multimodal mode)	4.88	4	5	0.32

evaluated are users previous experience using multimodal interfaces, the degree to which the user finds that the system understood him and they understood the system, the perceived interaction rate, the perceived difficulty level of the interaction with the system, the presence of errors, the certainty of the user of what to do at each time, and the global level of satisfaction with the system. A total of 25 native Spanish users (12 men, 13 women, aged 22 to 54, avg. age 34.6) participated in the evaluation.

Table 2 shows the results of the subjective evaluation using the described questionnaire. We can observe that the participants' previous experience using multimodal interfaces is very varied, as our objective was to evaluate the system with users with different degrees of familiarity with these systems.

With respect to the extent to which the users feel that the system understood them, it is possible to see that the recognizer had a very good performance. Tactile mode was perceived as more accurate than oral as expected, but the oral mode was punctuated very high by the users, with 4.64 over 5 respectively. As can be observed, users felt that the system understood them better with the multimodal than with the tactile mode.

The users also felt that the system responses were comprehensible, specially with the tactile and multimodal modes. The results of the oral mode were lower probably due to the quality of the synthesized voice. We used the standard English voice provided by Android. The results were higher when using better synthesizers like the voices of IVONA TTS. Also with the library catalog functionality, sometimes the data of the bibliographical resources were in languages that differed from the one selected for TTS which reduced the intelligibility of the results in these particular cases.

Regarding the interaction rate, it was found adequate in most cases, though in some cases the participants reported they were expecting barge-in mechanisms in the oral mode. The multimodal mode was found very useful to for the users to interact with the system at a pace which is adequate for their needs, as they could switch between interaction modes. This is in consonance with the fact that the multimodal mode has reported the maximum perceived easiness of use.

Generally, the users have not perceived errors during their interactions in the tactile mode, while in the oral and multimodal modes more errors were detected though they did not imply a fail in the interaction and in every case they could complete the task. In particular, the participants reported that the multimodal mode was more useful than the oral mode for reporting errors to the users.

About the certainty of what to do at each moment of the interaction, participants felt more security in the multimodal mode, which also received the better punctuation in the overall satisfaction as it brings more flexibility to the user.

4 Conclusions and Future Work

Multimodal interactive systems offer users combinations of input and output modalities for interacting with their devices, taking advantage of the naturalness of speech. Different vendors offer APIs for the development of applications that use speech as a possible input and output modality, but developers have to design ad-hoc solutions to implement the interaction management. In this paper we have described how to develop multimodal conversational agents providing enhance library services that can be easily integrated in hand-held Android mobile devices.

We have completed an evaluation of the developed application to assess the benefits of the multimodal interaction. To do this, the users employed the system with visual only, voice only and multimodal modes. The results show that the maximum satisfaction rates were achieved by the multimodal mode, as the users were able to switch between modalities and found this flexibility very useful.

We are currently undergoing the next phases in the deployment of the application. First, we want to conduct a more comprehensive evaluation of the system's functionalities. We also want to extend user awareness using a register to store the previous user's interactions with the application and adapting its operation according to their preferences, mistakes and main uses. With the results of these activities, we will optimize the system, and make it available in Google Play.

Acknowledgements. This work was supported in part by Projects MINECO TEC2012-37832-C02-01, CICYT TEC2011-28626-C02-02, CAM CONTEXTS (S2009/TIC-1485).

References

1. Cavazza, M., de la Camara, R.S., Turunen, M.: How Was Your Day? A Companion ECA. In: Proc. AAMAS 2010, pp. 1629–1630 (2010)
2. Chen, H., Albee, B.: An open source library system and public library users: Finding and using library collections. Library and Information Science Research 34(3), 220–227 (2012)
3. Fryer, L., Carpenter, R.: Bots as Language Learning Tools. Language Learning and Technology 10(3), 8–14 (2006)
4. Griol, D., Callejas, Z., López-Cózar, R., Riccardi, G.: A domain-independent statistical methodology for dialog management in spoken dialog systems. Computer Speech and Language 28(3), 743–768 (2014)
5. Griol, D., Molina, J., Sanchis, A., Callejas, Z.: A Proposal to Create Learning Environments in Virtual Worlds Integrating Advanced Educative Resources. Journal of Universal Computer Science 18(18), 2516–2541 (2012)
6. Hassenzahl, M., Burmester, M., Koller, F.: AttrakDiff: A questionnaire for measuring perceived hedonic and pragmatic quality. In: Mensch & Computer 2003. Interaktion in Bewegung, pp. 187–196. Vieweg+Teubner Verlag (2003)
7. Hone, K., Graham, R.: Subjective assessment of speech-system interface usability. In: Proc. Eurospeech (2001)
8. Kerly, A., Ellis, R., Bull, S.: CALMsystem: A Dialog system for Learner Modelling. Knowledge Based Systems 21(3), 238–246 (2008)
9. McTear, M., Callejas, Z.: Voice Application Development for Android. Packt Publishing (2013)
10. Metze, F., Wechsung, I., Schaffer, S., Seebode, J., Möller, S.: Reliable evaluation of multimodal dialogue systems. In: Jacko, J.A. (ed.) HCI International 2009, Part II. LNCS, vol. 5611, pp. 75–83. Springer, Heidelberg (2009)
11. Moreno, M., Castao, B., Barrero, D., Helln, A.: Efficient Services Management in Libraries using AI and Wireless techniques. Expert Systems with Applications 41(17), 7904–7913 (2014)
12. Pieraccini, R.: The Voice in the Machine: Building Computers that Understand Speech. The MIT Press (2012)
13. Pon-Barry, H., Schultz, K., Bratt, E.O., Clark, B., Peters, S.: Responding to student uncertainty in spoken tutorial dialog systems. Int. Journal of Artificial Intelligence in Education 16, 171–194 (2006)
14. Roda, C., Angehrn, A., Nabeth, T.: Conversational Agents for Advanced Learning: Applications and Research. In: Proc. of BotShow 2001, pp. 1–7 (2001)
15. Vassilakaki, E., Garoufallou, E.: Multilingual Digital Libraries: A review of issues in system-centered and user-centered studies, information retrieval and user behavior. Information and Library Review 45(1-2), 3–19 (2013)
16. Wang, Y., Wang, W., Huang, C.: Enhanced Semantic Question Answering System for e-Learning Environment. In: Proc. AINAW 2007, pp. 1023–1028 (2007)

Agent Based Modelling to Build Serious Games: The Learn to Lead Game

Andrea Di Ferdinando[1], Massimiliano Schembri[2], Michela Ponticorvo[1(✉)], and Orazio Miglino[1]

[1] Department of Humanistic Studies, University of Naples "Federico II", Naples, Italy
michela.ponticorvo@unina.it
[2] Department of Psychology, Sapienza, University of Rome, Rome, Italy

Abstract. In the present paper we describe how to build up digital Serious Games (SG) adopting an Agent Based Modelling (ABM) approach. In particular the design of an Agent Based digital Serious Game must address two different levels: the shell and the core. To better explain this design pathway we will introduce the Learn to Lead project that developed an useful educational/training tool about leadership.

Keywords: Agent based modelling · Serious games · Leadership · Game design

1 Introduction

Games can engender motivation. Games are hugely popular. Games require players to learn. These are the premises for creating games that convey useful learning experiences, in other words educational Serious Games (SG) [1,9].

Which features SG have? First of all educational games, as mainstream games, increase learner motivation, time-on-task and, consequently, learning outcomes. In other words learning achievements are improved. Game based learning appears to contribute to the achievement of high learning outcomes. For people it is easy to learn with games. They appear to learn the knowledge embedded in the game, and to remember it well as their activity was engaging. In a quite recent review by Vandercruysse and colleagues [13] it is shown that many studies report a positive effect of using games on learning and motivation, even if this effect is moderated by different learner variables and depends on different context variables.

The second relevant feature is that SG are engaging. Players feel more actively involved in the game based learning activity than in a lecture, and define games as a new and pleasant way to learn. Players are engaged by design [5] as game designers incorporate a number of strategies and tactics for engaging players in gameplay that can be integrated into the framework of engaged learning.

The third feature is that games stimulate self-regulation and learning by doing. People involved in educational games are highly autonomous during game sessions and manage to interact successfully with the computer and with the

© Springer International Publishing Switzerland 2015
J.M. Ferrández Vicente et al. (Eds.): IWINAC 2015, Part II, LNCS 9108, pp. 349–358, 2015.
DOI: 10.1007/978-3-319-18833-1_37

game interface. In literature this has been observed in many cases, consider, as an example, Kim and colleagues [7] who observed that players used different meta-cognitive strategies: self-recording, modelling and thinking aloud in managing game activity.

How can these features be included in SG? With the appropriate design, of course. SG, as many other kind of games, are expressed trough a narrative metaphor; it is therefore important defining the plot, the scenario, the characters, in one word, the setting. According to Barthes [3] narrative can be found in myth, legend, fairy tale, novel, epic, history, news, cinema, comics, in every place and every society, and obviously, in games. Designing learning environment that are based on micro-worlds [12] links together simulation in a digital environment and narration, thus using narrative structures to marry these innovative tools with motivational and emotional factors to enhance learning. The appropriate setting allows to attract the player, to immerge him/her in a completely different environment, that is relevant in an educational context.This is what we call the shell level.

On the shell level there are players immersed in the setting, defined by the narrative: for example the characters can be the prince charming and the princess in the fairy tale, the leader and the employee in a company, two armies in the chess game or everything the narrative metaphor suggests. This level holds an hidden level with specific operations, the core. This latter level can be conceptualized as an Agent Based simulation. This requires to represent the player and the other roles, explicitly foreseen and functional to the educationa/training goals such as teacher, tutor, supervisor, facilitator, guide, assessment responsible as agents. These figures implement learning analytics in SG.

In particular, we propose to adopt the Agent Based Modelling (ABM) approach. ABM [6,4] is a class of computational models used to simulate phenomena belonging to various domain ranging from biology to psychology and sociology starting from the action and interaction of simple agents. These agents are autonomous and can represent individual or collective entities such as groups. We adopt a wide definition of ABM: in SG, ABM is not used to understand collective behaviours starting from simple rules, but it aims at representing in detail agents interaction and the agent itself. In other words, a great effort is devoted to modelling agents too, in this respect resembling multi-agents systems approach [13] where agents can be very complex.

In the core level, every agent is defined in function of its sensory features, what it sees, hears, smells, touches in the setting and about the core, and action endowment, what it can do to affect the core state. For example, the agent could move to reach another agent or express an opinion trough a text typing in the surface: these actions result in a change in the core state space, in chess terms, they are moves. These moves must follow game rules that are defined both by setting constraints and by agents action chances and that reside in the core level. As we are in the domain of digital SG, agents can also be artificial agents: in this case the agents is not human, but a bot.

In what follows we will introduce a digital SG to train leadership based on ABM, following the design approach we have just described.

2   Learn to Lead Game

In the context of the European project Learn to Lead, L2L, our research team aimed to design, implement, and test a novel, online approach to train in team leadership, suitable for use in SMEs, small government offices, NGOs etc. The training provided by Learn2Lead is based on an online game. In the game, each learner manages a simulated team of employees (e.g. a team of workers in a bank agency, a post-office or a local government office) which competes against other teams to maximize its objectives (e.g. profit, volume of services delivered, customer satisfaction). Learn2Lead research project was funded by the European Agency for Education, Culture and Audiovisual.

In next section we will describe the design process addressing the shell level and the core level.

2.1   The Shell Level

In designing L2L game we started from the shell level, the game narrative that must enclose the player. L2L is a digital virtual laboratory, an online serious game where an user, acting as a leader, learns by governing a team of artificial agents, the followers. It is a point and click online game in a 2D environment which takes place in a firms office, what can be inferred from some decorative elements: desks, chairs, PCs and mobiles, stacks of paper, etc. (Figure 1). This physical setting varies across all game levels looking nicer and nicer as the player advances in his/her career.

In L2Lwe have the player acting as a leader and the followers, artificial agents in hierarchical interaction. One human player is involved in the game at a time with artificial agents.

Each learner manages a simulated team of agents that represent for example a team of workers in a bank agency, a post-office or a local government office. This simulated team competes against other teams to maximize its objectives (e.g. profit, volume of services delivered, customer satisfaction).

The narrative is the following: the player is hired to work in a large corporation. The CEO has picked the player out as a future leader and has organized for him/her to follow a programme of on-the-job training, where he/she will learn all what is necessary to become a great leader. The game is played across a number of levels. Across those levels the player will lead teams in a number of different corporation departments, from the catering department to the research and development one. During the game, the player has two goals:

1. to ensure that the company runs efficiently and productively;
2. to ensure that his/her followers develop appropriately, as outlined by the Full-Range Leadership (FRL) model (see below).

The day-to-day running of the department involves dealing with jobs that have specified deadlines and workloads, and assigning staff to work on those jobs. The basic challenge is to ensure that followers finish all jobs in time.

Fig. 1. The firm office where the game takes place at level 1

The leader is in control of assigning followers to work on jobs. With a smart management, it is possible to finish a respectable number of jobs within their deadlines. However, leadership involves more than management, and if the player uses a strategy for developing his/her followers he/she will get an effective advantage in the game. In fact each follower has an ability level and a motivation level: these two variables combined define the amount of work the follower can do in any given day. The player can consider his/her followers workload, ability, stress levels and personality in assigning players to jobs, a typical management task, and this will improve performance with respect to a leader that limits to standard managing. In this optic, the player has the option of running workshops, organizing team-building events, performing one-to-one coaching, getting involved in day-to-day work, sending memos, proposing training course, giving lectures about poor performance, deliver evocative speeches at staff meetings, etc. thus helping his/her followers development in ability and intrinsic motivation. More developed followers can complete more jobs, thus, spending time on developing staff helps the player to reach both efficiency and development, the goals that players should be aiming for, according to FRL model. Followers experience stress, which affects their ability to work. The work has also a competitive side: the player must compete with other groups to achieve goals: manage workloads, stress and employees motivations in order to increase profit, create new services and improve customers satisfaction. The system allows users to experiment different approaches before competing with other players. At the beginning of the game the player is explicitly introduced in the narrative with a specific description of what the game will be like. The player has different learning chances during the game, it is possible to follow tutor suggestions or study the provided materials beside playing the game.

L2L is divided into 5 levels, more and more complex and articulated. At the beginning of each level reading material, tutorials and mini-games are provided, regarding some theories and principles of Leadership such as the Path Goal Theory, the Full Range Leadership (see core section for details). After the player goes through the theoretical part, he/she is able to practice it in a safe environment (the player is not fired if he/she fails), where the level of acquired knowledge is evaluated. In order to surpass a level it is needed to study Leadership theory and apply it to the virtual environment as the game requires. Each level has a specified length in time; for example, a week. Each hour in that week, the player will have the opportunity to view the performance and progress of their team, make any change to job assignments and perform any leadership tasks they feel necessary. Once satisfied, the player will press a button to advance the game one hour, whereupon the game engine will calculate what happened in that hour. In order to recreate the dynamics that occur in real teams, we simulated the way in which interactions among team members and managerial actions combine to create the outcomes (e.g. profit, production, customer satisfaction), used to score players performance. The score obtained is based not only on the number of jobs finished, but also on the development of staff ability and motivation.

In the first tutorial level, the player covers a non- prestigious role. After reading the first instructions which represent the first theoretical part regarding staff and workload management, the first task is assigned: managing one single shipment with a time limit, and with only one employee. This first step is definitely simple and it is needed to introduce the player to the games mechanics. As the game proceeds, levels become more difficult, roles more complex and ambitious, the workload increases and personnel is larger and characterized by subjective and distinctive peculiarities: each single employee has his/her own personality, strengths and weaknesses, and the leader has to take these features into account to efficiently monitor the task and achieve it in the pre-established time limit. The team is formed by brilliant and motivated employees as well as by agents who need to be motivated. In order to achieve newer and more complex goals it is therefore needed to manage employees workload and tasks, without ignoring their stress and satisfaction levels. Verifications or weekly checks must be run to support and provide feedback, rewarding or punishing for the job done.

Players who complete the game have performed an effective training: they learnt about leadership theories on the theoretical side and about useful psychosocial dynamics to manage teams, motivational strategies, the importance of human resource management, of workloads and respecting deadlines on an practical side. These strategies are an important expertise for the manager as well as for the employee.

2.2   The Core Level

The L2L core layer encloses the game engine that defines the game operation. L2L is a logical structure where an asynchronous interaction takes place between the leader and the follower. A turn-based structure to play is implemented, so that players always have an unlimited amount of time to carefully consider their actions, and consult reference material about FRL if necessary before making a decision, before each move. The player acts on the work environment and team dynamics by setting the working plan of each follower. There are three main types of interaction between the leader and the followers: stimulation: the leader stimulates a follower, by talking to him/her or in some other way; punishment: the leader gives a punishment to a follower; reward: the leader gives a reward to a follower. Leader choices influence followers psychology, in particular their motivation, and thus their contribution to the team. These are the moves that change the core state. The player interacts with followers, artificial agents with their own psychological state that the player has to monitor carefully, if he/she wants to successfully advance in the game.

The followers (artificial) psychology is composed by three variables taken from McClelland [8] theory:

a. Achievement: followers have a disposition for excellence in performance, a continuing concern for doing better all the time. This motive concerns achieving excellence through individual efforts;
b. Affiliation: followers have a concern for establishing, maintaining, and restoring close personal relationships with others;
c. Power: followers have a concern for acquiring status and having an impact on others. High power motivation induces highly competitive behaviour.

It is important to stress that followers with different personalities should be managed in a different way to obtain optimal results. Figure 2 represents the tools the leader can exploit to monitor followers' stress and motivation.

Each follower has to accomplish a task, allocated by the leader, in different kinds of environments. The maximum workload is determined by two variables: ability and motivation. The player/leader can vary those variables through a series of possible actions (for example, by sending the follower to a training course, or by stressing them). Leaders get a score on the basis of the motivational and skill development of the followers. At the end, a leadership profile is created.

The idea underlying this general framework is that in some given conditions, the leader (i.e., the player) has to take some decision about one or more followers (moves). The player must feel that his/her actions do have reliable and realistic consequences on followers behaviour and development The teamwork is constituted by a set of artificial agents, as already said, and each agent is controlled by a neural network implementing a specific theory about leadership. The theory that is simulated in the game engine is Full-Range Leadership. The Full-Range Leadership (FRL) Theory, proposed by Avolio and Bass [2] recognizes three typologies of leadership behaviour: transformational, transactional, and non-transactional laissez-faire leadership. This theory has been extensively

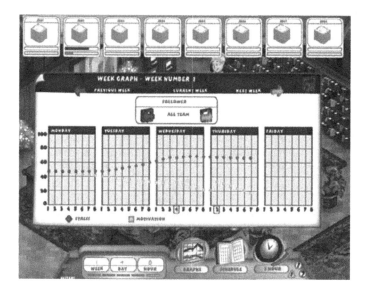

Fig. 2. Graphs representing the varying levels of stress and motivation of the team members

applied to dynamics in small groups. Transformational leaders are proactive, raise follower awareness in order to achieve better performance level ant to reach collective interests, thus helping followers to achieve extraordinary goals. Transactional leaders focus on standard organizational goals fulfilment and on setting objectives while monitoring outcomes by followers. Transformational leadership does not replace transactional leadership, but it augments it in achieving leader, associate, group, and organization objectives. Laissez-faire leader avoids making decisions, abdicates responsibility, and does not use his/her authority, he/she is therefore considered the most passive and ineffective form of leadership because he/she does not respond to situations and problems systematically, avoids specifying agreements, clarifying expectations, and providing goals and standards to be achieved by followers who does not give any feedback or support to. This leadership style has a negative effect on desired outcomes. Within the FRL model three important outcomes can be achieved as an effect of leadership: extra effort, group effectiveness and satisfaction. Moreover this model can be easily integrated with other models or theories pointing to personal characteristics of the leader, such as McClelland motives as well as to situational factors and their interaction with leaders and followers characteristics. A key point in FRL model is that every leader displays each style to some amount in different moments and context and has to govern changing within team dynamics. Effective team leaders can manage these dynamics in ways that help the team to meet its objectives.

When the leader agent (i.e. the user) takes some decision about one or more followers (the moves), these decisions are encoded in the followers network as some combination of inputs. Based on the inputs received from the leader, and the ones coming from the environment, the agents internal states will change and influence his/her contribution to the team job. The followers model represents the way the FRL theory is implemented inside the Learn2Lead game. Figure 3 represents agent model.

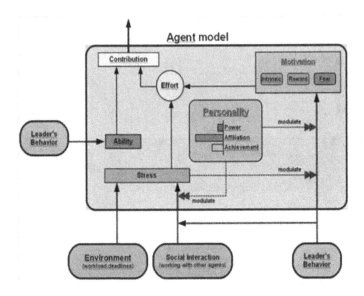

Fig. 3. L2L agent model. According to the FRL theory each agent behavior is guided by a set of external variables such as leaders behavior, workload, etc. and a set of internal variables related to personality, motivation, ability, and stress. This model in L2L has been implemented exploiting neural nets.

3   Conclusions

In this paper we have described how to design a Serious Game adopting an Agent Based approach. In the design process, we have identified two levels to work on: the shell level with the narrative and the core level where the agent model resides.

In particular the FRL theory has been translated in an agent based simulation allowing to create an infinite number of specific contexts in which the knowledge can be experienced and practiced. Artificial agents are essentially input-output systems with an internal state that changes over time depending on the external input and some internal variables. Every follower has some internal variables that affect the final contribution in getting through the jobs assigned. The most relevant to the FRL theory is the motivation. The motivation level is affected by three subcomponents: intrinsic, reward and fear. The intrinsic component models the dynamics of intrinsic motivation and it is related to the transformational leadership style, while the reward and fear components model extrinsic motivation and are related to the transactional leadership style. What differs among the three is the time dynamics, and specifically their decay. For example, the intrinsic component has a slower decay than the reward and fear, but can be activated only by appropriate leader behaviours (typically pertaining transformational style). The stress variable is linked to some external inputs like social interaction, workload and deadlines. It affects the contribution and is an important aspect to keep under control during the game. Stress also has a modulator effect on the leader motivation oriented behaviours. Personality and ability try to capture what the FRL theory says about individual consideration. Ability level is linked to follower performance. Personality is conceived as a modulator for the leader behav-iour so that the same leader action may have a different impact on followers with different personality. On the contrary the leader that aims at raising the motivation of the team as high as possible needs to perform some individualized consideration.

As L2L core is modelled according to a specific leadership theory, FRLT, the user while training about leadership, also learns about the theory itself. The game becomes a virtual lab the user can use to test what he/she has learnt in theory, intervening upon specific variables. In fact, the human behaviour is properly modelled in this agents simulator and learners can observe how agents behaviour change varying a set of psychological variables. In other words, observation takes place in an artificial lab scenario rather than in a natural context.

Leaders, along with technical skills related to their area of business, must have competencies in people management, resource management and organization. Effective leadership is obviously strongly demanded in organizational context but becoming a good leader can be a complex and expensive task. L2L offers a clear advantage instead: programmes about leadership are very expensive and time demanding whereas playing L2L can constitute an effective training much less expensive and that can be managed individually and autonomously. Although Learn2Lead could not substitute a long and professional training, it is a psychological theory implementation in an agents simulator and can therefore be used as a powerful supporting tool in teaching how to successfully manage a group of followers [10,11]. L2L particular game design facilitates the adaptation of the game to different work environments. It is fit for training in many different context, from SMEs, to no-profit organization up to public administration. Moreover, using the game results effective either in autodidact and tutored conditions, thus making it a very versatile tool.

The first L2L beta version runs on project website (www.learn2lead.unina.it) and it has been tested in three trial sessions in France, Italy and Spain. In the 2012 a freeware version and a commercial version have been released. The commercial version belongs to the Learn to Lead partnership that will use the game to offer tutor-supported training to their customers, but the free version is available to everyone who is interested in this training tool.

Acknowledgments. The Learn to Lead project has been funded by the European Agency for Education, Culture and Audiovisual under Leonardo Lifelong Learning Programme.

References

1. Abt, C.C.: Serious games. University Press of America (1987)
2. Avolio, B.J., Bass, B.M.: Full-range training of leadership. Manual. Bass/Avolio and Associates, Binghamton (1991)
3. Barthes, R.: Introduction to the structural analysis of narratives. Fontana, London (1977)
4. Bonabeau, E.: Agent-based modeling: Methods and techniques for simulating human systems. Proceedings of the National Academy of Sciences 99(suppl. 3), 7280–7287 (2002)
5. Dickey, M.D.: Engaging by design: How engagement strategies in popular computer and video games can inform instructional design. Educational Technology Research and Development 53(2), 67–83 (2005)
6. Helbing, D.: Agent-based modeling. In: Social Self-organization, pp. 25–70. Springer, Heidelberg (2012)
7. Kim, B., Park, H., Baek, Y.: Not just fun, but serious strategies: Using metacognitive strategies in game-based learning. Computers and Education 52(4), 800–810 (2009)
8. McClelland, D.C.: Human motivation. CUP Archive (1987)
9. Michael, D.R., Chen, S.L.: Serious games: Games that educate, train, and inform. Muska and Lipman/Premier-Trade (2005)
10. Miglino, O., Di Ferdinando, A., Schembri, M., Caretti, M., Rega, A., Ricci, C.: STELT (Smart Technologies to Enhance Learning and Teaching): una piattaforma per realizzare ambienti di realt aumentata per apprendere, insegnare e giocare. Sistemi Intelligenti 25(2), 397–404 (2013)
11. Miglino, O., Gigliotta, O., Ponticorvo, M., Nolfi, S.: Breedbot: an evolutionary robotics application in digital content. The Electronic Library 26(3), 363–373 (2008)
12. Papert, S.: Microworlds: transforming education. Artificial Intelligence and Education 1, 79–94 (1987)
13. Vandercruysse, S., Vandewaetere, M., Clarebout, G.: Game-based learning: A review on the effectiveness of educational games. In: Handbook of Research on Serious Games as Educational, Business, and Research Tools, pp. 628–647. IGI Global, Hershey (2012)

An Augmented Reality Application for Learning Anatomy

C. Rodriguez-Pardo, S. Hernandez, Miguel Ángel Patricio(✉), A. Berlanga, and José Manuel Molina

Applied Artificial Intelligence Group, University Carlos III of Madrid, Madrid, Spain
{carlos.rodriguez,sergio.hernandez,miguelangel.patricio,
antonio.berlanga,josemanuel.molina}@uc3m.es

Abstract. This work has been developed within the project of Applied Artificial Intelligence Group at UC3M, called "Augmented Science"[1], in order to disseminate and support education in science through Augmented Reality (AR) tools. The project is part of new developments provided with AR systems on mobile devices and their application in promoting science teaching. Generically, we understand that technology Augmented Reality as supplementing the perception and interaction with the real world and allows the user to be in a real environment augmented with additional information generated from a computerized device. In education the AR is an especially effective technology platform in everything related to how students perceive physical reality, since it allows break it down into its various dimensions, to facilitate the uptake of its various peculiarities, sometimes imperceptible to the senses. So with the AR is feasible to generate multidimensional models that simplify the complexity of the world, which, from the perspective of popularizing science, brings wholeness to any learning experience. In this context, we explore the suitability of the use of Artificial Intelligence techniques for the development of AR applications. As a use case, the development of an application for Android devices, which, through techniques of AR overlays a bony hand model is described. The application allows user interaction in order to discover the name of the bones of the hand. The article conducts an assessment of the application to analyze their educational impact.

1 Introduction

The term Augmented Reality (AR) has been given different meanings in specialized literature. We will focus on AR as a concept, whose definition can be more durable and useful tan defining AR as a technology. AR can be, therefore, defined from diverse perspectives,[1] Azuma defines AR "as a system that fulfills three basic features: a combination of real and virtual worlds, real-time interaction, and accurate 3D registration of virtual and real objects". [2] Klopfer and Squire

---

[1] http://cienciaaumentada.uc3m.es

© Springer International Publishing Switzerland 2015
J.M. Ferrández Vicente et al. (Eds.): IWINAC 2015, Part II, LNCS 9108, pp. 359–368, 2015.
DOI: 10.1007/978-3-319-18833-1_38

give a wider definition: "A situation in which a real world context is dynamically overlaid with coherent location or context sensitive virtual information". AR is, in short, a "situation or technology that blends real and virtual information in a meaningful way".[3]

AR is an excellent tool to improve the educational process, allowing interaction in both the real and virtual world. AR is based on a set of techniques and tools that allow user interaction between these two worlds. Artificial Intelligence techniques found in the AR one of their challenges:

1. First, it is necessary to contextualize and recognize the real world on which the virtual information is projected. Thus, different pattern recognition techniques are used, which, from the available sensors recognize the real world.
2. Once recognized the real world, rendering techniques that improve the overlay the virtual to the real world are needed. Usually, these will be 3D models, but also can be any multimedia information.
3. Finally, it is intended that the user interaction involves a rich experience. In this case, it would be of interest to research in intelligent behaviors of virtual objects created and overlaid on the real world.

The possible applications of Artificial Intelligence techniques to AR are broad. In this work we will conduct a proof of concept of the implications of AI in AR. Above all we will see how recognition techniques are applied using Vuforia SDK [4] as well as the use of 3D models for representation using Unity3D [5]. All with the aim of developing a mobile application for learning the bones of the hand using AR techniques.

2 Related Work

2-D media in education, such as paper and blackboards, is "very convenient, familiar, flexible, portable and inexpensive"[6], and is the most used option in educational institutions worldwide. Nevertheless, those classical teaching methods don't allow dynamic content, and, mainly, don't give students the opportunity to be immersed and interact with real world situations. Although Virtual Reality can solve this problem, it can make students "become divorced from the real environment"[6]. AR, however, gives students the opportunity to be immersed into virtualised real world situations which are mixed with their real context. For instance, AR could be used to teach open heart surgery to medicine students, who could try to conduct the surgical procedure by themselves. This hypothetical AR application is just one example, but it illustrate the real potential of AR in education, which is giving students the opportunity to visualize and interact with real life situations which aren't usually accessible or feasible to interact with.

Iulian Radu made an extensive analysis on Augmented Reality in education in his paper *"Augmented reality in education: a meta-review and cross-media analysis"* [7]. He analyzed 26 publications that compared learning in AR versus non-AR applications. By doing so, he identified "benefits and negative impacts of AR experience learning and highlight factors that are potentially underlying these effects". We will now offer a brief summary of his research.

Iulian Radu's research found that "A large propotion of the surveyed papers indicate that for certain topics, AR is more effective at teaching students than compared to other media such as books, videos, or PC desktop experiences".[7]

AR can be, if used properly, beneficial in many ways within the educational experience. It can increase long-term memory, group collaboration, motivation and content understanding. However,it should not be used as a substitute of conventional education, but rather as a complement to it.

We've already stated that AR can be beneficial for teaching and learning. Nevertheless, there are many ways to utilize this technology, both in hardware and software aspects. Hardware-wise, there are three popular options: Head-Mounted Displays, Spatial Augmented Reality and Handheld Devices. Handheld devices, such as smart-phones and tablets, are, for many reasons, the best option for modest AR experiences that have the purpose of enhancing the educational experience. They are inexpensive, and they use a great amount of sensors, which are critical for a useful AR system. Those sensors, like GPS, accelerometers, gyroscopes, digital compasses and, last but not least, cameras, can be used to correctly identify what and where the user is doing at any given time. They also count with the ability to connect fluently with the internet, and have enough computing power. However, the main advantage that smart-phones and tablets have is their penetration in the market. [8] For instance, in Spain, the smart-phone penetration in the market, as of 2013, was about 80% population-wide, number that grows to 91% for people younger than 25. It is clear that a platform like smart-phones is much more powerful for this purpose than the others we mentioned.

Many attempts have been made in the mentioned platform,and thus there are many AR projects that focus on teaching and learning. For instance, Google created Sky Map, an AR application for both iOS and Android handheld devices; that uses GPS location, real time sky tracking, and user's device's sensors like the gyroscope and digital compasses to show in the device's display a real-time projection of the night sky. This is just an example of astronomy teaching using AR. This technology is also used to teach and learn architecture, history, art, space exploration, chemistry, physics or even maths.

However, many of this, on the other hand, valuable applications, lack a real user-machine interaction, with which the student can feel immersed into the augmented world that is being projected. Artificial Intelligence (AI) developments can be used to correct these deficiencies. Techniques such as computer vision and real-time pattern recognition are helpful for enhancing the AR experiences, and advances have been made on the area. [9]

After an analysis of what is being done and what could be done, we discovered a lack of use of computer vision techniques on AR software made for educational purposes, and that this software (specially software for handheld devices) don't offer the user a real immersion and the opportunity to interact more strongly with the augmented world. Our proposal, which we will deeply explain now, combines computer vision, 3-D CGIs, and real time user-machine interaction to give students a new opportunity in anatomy learning. We will propose a general scheme and infrastructure for AR applications for handheld devices, which uses AI techniques, and give an example of how this scheme can be used.

3   A Proposed Architecture for AR Applications in Education

Given the undoubted growth of mobile devices, along with the increasing availability of wireless internet connection, our work will be based on mobile devices (Smart-phone or Tablet). Theses devices support a large number of sensors, of whom the most important are accelerometers, GPS, camera and compass, that enable the device to recognize its surroundings. Our proposed architecture is divided into three principal layers. First, the Situation Assessment module, whose function is contextualizing and recognizing the real world. In this layer, the device must be able to recognize its surrounding environment. The second layer is intended to complement real-world objects through content generated by a computer, which is the "Augmented layer". Finally, the Interaction layer, which is responsible of keeping track of stimuli received from the user, and modifying the behavior of the augmented layer.

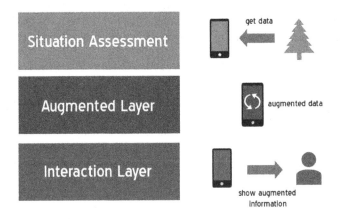

Fig. 1. Our architecture is divided in three principals layers

Situation Assessment. There are different ways to recognize where we are. In this case, we are interested in the use of the camera device, because our goal is to get an "imaged resource". A useful tool to do this work is OpenCV [10], a open source resource focused on real-time applications. Based on BSD license, it allows to use it with academic and commercial purpose. However, there is a problem with its efficiency, as it is not able to process a high mount of images in a powerless device like a mobile phone. Seeking efficiency, for our proposal we selected Vuforia SDK [4] as a system of recognition and object tracking for mobile devices.

Vuforia is a SDK developed by Qualcomm, whose focus is Augmented Reality technology for Android, iOS and Windows. This technology is able to recognize planar images and simple 3D objects, like cylinders and boxes, and use them as a "target", to get a point of reference. This point of reference allows the program to locate and know where it is. Vuforia tracks the target perspective, and takes it as the perspective of all the figures that the program will generate later, so if the user moves around the target, and takes another perspective, the application will show another view of the figures, like if there was an object over the target.

Vuforia offers some extra features, like an online target recognition, which allows to recognize any target in Vuforia's database, so the program is able to detect different targets and do something in function of the target that it has recognized. Figure 2 shows the possibility of representing different objects depending of the target. If the number of targets is small, their respective planar images could be stored in the device's local data store. Another useful feature are virtual buttons, which allow the program to detect if any part of the target has been covered, and to behave as in these parts there were a button. In the Figure 3 we can appreciate how an user can place his finger over the specified part, being detected by the program. This is used, for example, if it is needed a special animation when the user has to interact with the virtual object, or simply when you want to change the object. This can be specially useful when a real-time interaction with the user is needed.

Fig. 2. Possibility to track two different target by the same program

An advantage of using Vuforia is that you don't need to use a traditional marker, like a QR code. Vuforia allows to use any type of image, because it uses sophisticated algorithms to track the image and recognize it as a defined target. Vuforia has its own images rating, to evaluate how well a image could be tracked. It is recommended that a image has the best rating that it's possible, because

Fig. 3. Effect of pushing a virtual button

it will allow the device to detect the image easier, offering more efficiency and better user experience. Contrast is the other variable that we need to keep in mind. A high contrast image will have more features than a less contrast one. This also proves that a higher resolution image will be better, because the density of pixels indicates the precision of the image details. The features distribution influence too the target quality. A image whose features are only located in a small area of the image will have a bad rating. Accordingly, the features must be uniformly distributed around the image.

When the device has tracked the image, it has to render and transform the real world coordinates, to other coordinates that can be understood by the device using the equation $y = Tx$, where X are the coordinates of the real world, T are the algorithms needed to transform the real coordinates X, to a projection of them (Y)[2]. First of all, in order to test the target quality, Vuforia transforms the image to its gray scale representation. The image will get a bad rating if it doesn't have sufficient contrast and its histogram is narrow and spiky.

We know that in order to work with Vuforia it is necessary to have a target that is as well rated as possible, so the target needs a no repetitive pattern with high definition. Although this is not enough to have a real-time interaction with the user, there is where we need something to identify the user's hand, but the problem is that Vuforia technology is not able to detect complex objects like a hand. Here is where we could use virtual buttons, arranging the buttons like a hand shape, and when a user put his hand over the button's target, we are able to detect that this is the user's hand, and dispose the specified function. In this case, we want to show the hand's skeleton to the user. We will add instead constraints to detect if the object is a hand.

With this idea, the user only has to track the target, and put his hand over the target, "pushing" on all the virtual buttons. In this moment, the device will display the hand skeleton, with clickeable bones, that will let the user interact with the application and its augmented information of the hand skeleton.

Augmented Layer. When the device has recognized the hand, and has a reference point, our program has to transform this data to another which the device is able to use. First of all, the system needs to be secure that the object that is on the target is a hand. For solving that, we have created a series of constraints that lets the device to identify a hand form. Besides, we have to take care of the hand's size, because this program can be used by children and teachers, and thus we have established functions that are able to detect the user's hand size. The other data the device has got is the tracker, which informs us where the user's hand is in the space, allowing the device to know where exactly the skeleton hand will be represented. Besides, we have to take care to the user's movements, because we can't keep a static image in the same place if the user moves his hand. The program has to manage all of this constraints, transforming all of the data the device has tracked, on information that will be displayed to the user in the Interaction Layer.

In order to display the bones and the camera view concurrently, the device would need a high optimization of the functions that it use. This is the reason we preferred to use an automated optimization system to do this, Unity3D [5].

Unity is a game engine developed by Unity Technology focused in the design of video games for a lot of platforms. This environment is able to optimize the render of both constraints, allowing the application to show the hand bones properly to the user without lag. This platform allows programming with both Javascript and C#, and is totally compatible with android mobile devices.

Interaction Layer. A User Interface must be enjoyable and easy to use by the user, especially when our target market are children. This is the reason we have used colors to identify the different types of bones (red to distal phalanges, blue to intermediate phalanges, green to proximal phalanges, yellow to metacarpals and purple carpals). This allows the student to recognize and memorize easily the hand's bonds. If the user has any doubt, the application has a simple tutorial that appears at the beginning of the app, which can be skipped if the user is not interested in it. The interface has a interactive functionality that allows the user to click on each bone to get more information about them that the application shows by default. This functionality will let us to add an "game-mode" with the objective of memorizing the bones.

4   Evaluation

In order to assess whether this implementation is a useful tool to learn anatomy, we had to conduct a quantitative research. Based on the topic's literature, we based our test on Radu's work [7], and Di Serio's paper "Impact of an augmented reality system on students' motivation for a visual art course" [11]. They both propose some questions that can be useful to determine how successful an AR application for education has been. We will now list the questions we made, and detail afterwards the results our app had.

Q1 Have you had any previous contact with an augmented reality application? If so, please rate your experience from 1 to 5.
Q2 Have you had prior contact with an educational application? If so, please rate your experience from 1 to 5.

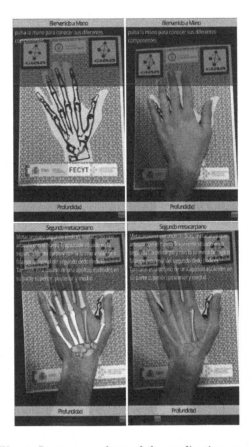

Fig. 4. Some screenshots of the application usage

Q3 Please rate the level of interaction with the user of the application "Hand's anatomy".

Q4 Please rate the difficulty level of use of the application "Hand's anatomy".

Q5 In general, are you satisfied with the application "Hand's anatomy"?

Q6 Do you think that "Hand's anatomy" is capable of doing learning anatomy easier?

Q7 Do you think that "Hand's anatomy" draws attention to the fundamentals of learning anatomy?

Q8 Please show your level of agreement / disagreement with the following statement: ""Hand's anatomy" allows the user to feel immersed in learning anatomy"

Q9 Please show your level of agreement / disagreement with the following statement: ""Hand's anatomy" allows the user to interact with spatially complex problems"

We gave smart-phones running the application to volunteers. They used it and interacted with the application for a few minutes, and then they answered the 9 questions we indicated before, and personal questions such as genre and age.

In this study, thirty-seven people aged 14-61 from different backgrounds were surveyed. They took the test just after testing the app, and answered the questions directly to the interviewer, who took note of their answers using Google Forms, which is a web-based application that allows conducting surveys. 17 out of 37 were males, 20 were females, and the average age was 24.

In Table 1, we sum up the results the survey showed, which we will comment later.

Table 1. The study results

Question	Average	Standard deviation
Q1	3.054	1.268
Q2	3.378	1.163
Q3	3.864	0.696
Q4	4.486	0.692
Q5	3.837	0.833
Q6	4.135	0.855
Q7	3.594	0.864
Q8	3.837	0.799
Q9	2.622	1.001

The table shows that the previous contact the respondents had with both AR and educational applications gave average results, with a lot of variability. However, when they were asked about our application, the results were much better, and with less deviation. Furthermore, the respondents agreed that our app is really easy to use and interaction was also well rated. In the educational aspect, respondents agreed that our implementation is useful for learning anatomy, and it makes the user feel immersed in the educational experience. Finally, our implementation was not as good as in the other areas in the "spacial complexity" area, probably due to the fact that this application doesn't focus on areas that can be spatially complex, such as geometry teaching.

5   Conclusions and Future Work

In this paper, we have demonstrated the ability of Augmented Reality techniques for use in learning. Moreover, we have seen the need to research on techniques and tools of Artificial Intelligence in order to improve the user experience in AR applications. The possibilities of AR, regarding the development of training materials and learning activities are multiple and heterogeneous in virtually all university disciplines primarily on scientific and technological disciplines.

Soaring performance of Smartphone and Tablet have turned these devices into useful tools for using AR technology. Through its new capabilities and its increasingly sensitive cameras, faster processors and complex functionalities allow to handle 3D graphics and incorporating sensors (accelerometers, compasses, gyroscopes, GPS, etc.) undoubtedly facilitate their adaptability to the requirements of AR systems. In the future will be required Artificial Intelligence techniques that use this set of sensors in order to improve the recognition of the environment of a user, and therefore create applications that enhance user experience. Society is changing, people are demanding more interaction, innovation and participation with content. And the AR in mobile environments meets all these requirements.

Acknowledgement. This work was supported in part by Projects MINECO TEC2012-37832-C02-01, CICYT TEC2011-28626-C02-02, CAM CONTEXTS (S2009/TIC-1485).

References

1. Azuma, R.T., et al.: A survey of augmented reality. Presence 6(4), 355–385 (1997)
2. Klopfer, E., Squire, K.: Environmental detectives the development of an augmented reality platform for environmental simulations. Educational Technology Research and Development 56(2), 203–228 (2008)
3. Milgram, P., Kishino, F.: A taxonomy of mixed reality visual displays. IEICE Transactions on Information and Systems 77(12), 1321–1329 (1994)
4. Qualcomm: Vuforia developer (2015), https://developer.vuforia.com/
5. Technologies, U.: Unity3d (2015), http://www.unity3d.com
6. Kesim, M., Ozarslan, Y.: Augmented reality in education: current technologies and the potential for education. Procedia-Social and Behavioral Sciences 47, 297–302 (2012)
7. Radu, I.: Augmented reality in education: a meta-review and cross-media analysis. Personal and Ubiquitous Computing 18(6), 1533–1543 (2014)
8. Nájera Aragón, F.: External analysis of the smartphone industry in spain (2014)
9. Zhou, Y.: AR Physics: Transforming Physics Diagrammatic Representations on Paper into Interactive Simulations. PhD thesis, University of Central Florida Orlando, Florida (2014)
10. Bradski, G.: Dr. Dobb's Journal of Software Tools
11. Di Serio, A., Ibáñez, M.B., Kloos, C.D.: Impact of an augmented reality system on students' motivation for a visual art course. Computers & Education 68, 586–596 (2013)

Data Cleansing Meets Feature Selection: A Supervised Machine Learning Approach

Antonio J. Tallón-Ballesteros[(✉)] and José C. Riquelme

Department of Languages and Computer Systems,
University of Seville, Seville, Spain
`atallon@us.es`

Abstract. This paper presents a novel procedure to apply in a sequential way two data preparation techniques from a different nature such as data cleansing and feature selection. For the former we have experienced with a partial removal of outliers via inter-quartile range whereas for the latter we have chosen relevant attributes with two widespread feature subset selectors like CFS (Correlation-based Feature Selection) and CNS (Consistency-based Feature Selection), which are founded on correlation and consistency measures, respectively. Empirical results on seven difficult binary and multi-class data sets, that is, with a test error rate of at least a 10%, according to accuracy, with C4.5 or 1-nearest neighbour classifiers without any kind of prior data pre-processing are outlined. Non-parametric statistical tests assert that the meeting of the aforementioned two data preparation strategies using a correlation measure for feature selection with C4.5 algorithm is significant better, measured with roc measure, than the single application of the data cleansing approach. Last but not least, a weak and not very powerful learner like PART achieved promising results with the new proposal based on a consistency measure and is able to compete with the best configuration of C4.5. To sum up, bearing in mind the new approach, for roc measure PART classifier with a consistency metric behaves slightly better than C4.5 and a correlation measure.

Keywords: Data cleansing · Feature selection · Classification · Outlier detection · Inter-quartile range

1 Introduction

Knowledge Discovery in Databases (KDD)[9] is a multidisciplinary paradigm of computer science comprising challenging tasks to transform a problem into useful models for prediction such as dealing with raw data, analysing the problem, data preparation [25] and data mining [17].

Several machine learning approaches have tackled the classification problem. Roughly speaking, algorithms may be divided into strong and weak algorithms according to inner complexity of the classifier. There is a good number of families to get classification models like those based on decision trees, rules, nearest neighbours, support vectors and neural networks.

© Springer International Publishing Switzerland 2015
J.M. Ferrández Vicente et al. (Eds.): IWINAC 2015, Part II, LNCS 9108, pp. 369–378, 2015.
DOI: 10.1007/978-3-319-18833-1_39

Data pre-processing is crucial due to some issues: a) real-world problems may be incomplete or noisy (with errors or outliers); b) the discovery of useful patterns depends on the starting data quality. Data cleansing [7] and feature selection [18] are two samples of data preparation approaches. The former aims to correct errors, to detect and analyse outliers, thus to purify data. The latter pursues to pick up the most important features in order to simplify the model and predict more accurately.

This paper goals to assess the potential usefulness of the ordered application in supervised machine learning problems of two very different data-preprocessing techniques like data cleansing and feature selection. Another additional aim is to improve the performance of the classification models.

The rest of this article is organized as follows: Sect. 2 describes some concepts about data cleansing and feature selection; Sect. 3 presents our proposal; Sect. 4 details the experimentation; then Sect. 5 shows and analyzes statistically the results obtained; finally, Sect. 6 states the concluding remarks.

2   Related Work

2.1   Data Cleansing

An outlier may be defined as a data point which is very different from the rest of the data based on some measure [1], or alternatively as a case that does not follow the same model as the remaining data and appears as though it comes from a different probability distribution [26].

The core question about outliers is to delete or not to delete them. The answer is unclear because there are contributions from both sides. On the one hand, some authors claim either that the deletion of outliers did not significantly change the overall error distribution, accuracy, ... [14], or that the elimination of instances which contain attribute noise is not a good idea, because many other attributes of the instance may still contain valuable information [26]. On the other hand, some works showed that dropping the outlier in the training set may be a beneficial action for the classifier [20].

2.2   Feature Selection

It can be defined as the problem of choosing a small subset of features that ideally is necessary and sufficient to describe the target concept. The different approaches for feature selection (FS) can be divided into two broad categories (i.e., filter and wrapper) based on their dependence on the inductive algorithm that will finally use the selected subset [16]. Filter methods are independent of the inductive algorithm, whereas wrapper methods use the inductive algorithm as the evaluation function. FS involves two phases: a) to obtain a list of attributes according to an attribute evaluator and b) to perform a search on the initial list. All candidate lists would be evaluated using a measure evaluation and the best one will be returned. Correlation-based Feature Selection (CFS) [12] and

Consistency-based feature selection (CNS) [6] are two of the most widespread feature subset selectors (FSS) and both work together with a search method such as Greedy Search, Best First or Exhaustive Search.

3 Proposal

A statistical outlier detection method based on the partial removal of outliers according the inter-quartile range of all the instances with the same class label was introduced in [22] with the name OUTLIERoutP. The framework can be reviewed in Figure 2 of the aforementioned work.

The current paper proposes to complete an additional data preparation stage after the data cleansing from a different perspective such as feature selection [16]. Typically, the application of feature selection has become a real prerequisite for model building in due to the multidimensional nature of many modeling task in some fields [21]. Figure 1 overviews the proposal. It is a generic methodology in the sense that there is no restriction in the kind of feature selection method or the number of classes that the classifier is able to operate with. As usually in data mining field, the data preparation techniques act on the training set and the test set stands unaltered and is evaluated by the first time once the classifier is trained. To the best of our knowledge, the main novelty of this work is to do a further data pre-processing phase by means of feature selection after the application of the data cleansing stage via an outlier detection method. According to the literature, the researches tackle data cleansing or feature selection in an isolated way.

4 Experimentation

Table 1 describes the data sets utilised together with the outlier level according to the taxonomy proposed in [22]. Most of them are publicly available in the UCI (University of California at Irvine) repository [4]. They come from real-world applications of different domains such as Finances, Physics, Life, Environment, and Analysis of Olive Oils. The following seven have been used: Cardiotocography (*CTG*), Statlog (*German* credit), MAGIC Gamma Telescope (*Magic*), Olive Oil (*Olitos*), Pima Indians Diabetes (*Pima*), *Tokyo* and *Water* Treatment Plant. Olitos problem is deeply explained in [3]. The size of the problems ranges from one hundred twenty to more than nineteen thousands. The number of features varies between eight and sixty one, while the number of classes is between two and four. The missing values have been replaced in the case of nominal variables by the mode or, when concerning continuous variables, by the mean, bearing in mind the full data set. The outlier level is computed once the imputation of the missing values has been carried out. These data sets contain a number of outliers that is between a low percentage (up to a 10%) and a moderate-high one (in the range 30-40%, for the *Tokyo* problem). It is important to stress that these outliers are originally present in the data set and we have not performed any artificial way to add them. The taxonomy of the problems depending on the

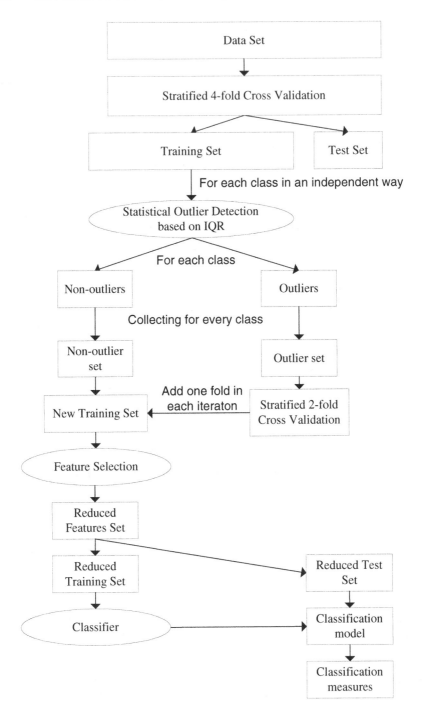

Fig. 1. Methodology OUTLIERoutP+FS

outlier level is based on the inter-quartile range (IQR) by classes. The common point of these data sets is that their important error rate in test phase without any kind of data pre-processing is about 10% or above with reference and robust classifiers such as 1-nearest neighbour (1-NN) [2] [5] or C4.5 [19]. In relation to the experimental design we have followed a stratified 4-fold cross validation [13], whereby the data set is divided into four parts and subsequently a partition is the test set and the three remaining ones are pooled as the training data. On the other hand, for the assessment of the classification models we have chosen the accuracy [15] and roc [8] measures. The former can be defined as the probability of correctly classifying a randomly select pattern. Sometimes, it is called as the number of successful hits [24]. The latter stands for receiver operating characteristic (ROC) and is the area under the ROC curve. We report both measures for the test set that is the performance with unseen data during the generalisation stage.

Table 1. Summary of the data sets

Data set	Size	Features	Classes	Outlier level
CTG	2126	23	2	II
German	1000	61	2	I
Magic	19020	11	2	I
Olitos	120	26	4	II
Pima	768	8	2	I
Tokyo	639	44	2	IV
Water	527	38	3	II
Average	3457.1	30.1	2.4	I − II

Table 2 depicts the data preparation methods concerning the different experiments that were conducted. The first one only includes the data cleansing, whereas the last two ones comprise the execution of the methodology OUT-LIERoutP followed by a feature subset selection evaluated with a correlation or consistency measure. Last column defines an abbreviated name for each of them that often will be referred in next sections. We have chosen CFS and CNS as representative feature subset selection methods, because they are based on different kind of measures, have few parameters and have provided a good performance inside the supervised machine learning area. Often, BestFirst search is the preferable option by the researchers for both FSS algorithms. CFS is likely the most used FSS in data mining. CNS is also powerful, however the quantity of published works is more reduced [21]. As classification algorithms we have experienced with C4.5 [19] and PART [10]. For the two previous FSS and classifiers we have used the implementations provided by WEKA tool [11] with default parameters that are those suggested by the own authors of the algorithms.

Table 3 overviews the properties of the data sets with a stratified 4-fold cross validation in three moments: i) in the initial situation (see all the columns containing *Or.* word), ii) after the data pre-processing stage (refer to columns

Table 2. List of data preparation methods for the experimentation

Data cleansing	Attribute evaluator	Search method	Feature selector name	Abb. Name
$OUTLIERoutP$ $-$	$-$	$-$		$OutP$
$OUTLIERoutP$ CFS	$BestFirst$	$CFS\_BestFirst$		$OutP + FS1$
$OUTLIERoutP$ CNS	$BestFirst$	$CNS\_BestFirst$		$OutP + FS2$

Table 3. Number of instances and features with the data preparation methods with a 4-fold cross validation

Data set	Or. Av. Tr. Sz.	Av. OutP Tr. Sz.	♯ Or. F.	♯ Av. F. FS1	♯ Av. F. FS2
CTG	1594.5	1490.6	23	6.8	7.8
$german$	750.0	744.6	61	7.8	19.9
$magic$	14265.0	14058.5	10	4.0	10.0
$olitos$	90.0	80.4	25	12.3	12.3
$pima$	576.0	556.1	8	3.9	7.3
$tokyo$	719.3	619.6	44	11.0	11.4
$water$	390.8	362.9	38	10.8	11.0
$Average$	2626.5	2559.0	29.9	8.1	11.4

$Or. = Original$ $Av. = Average$ $Tr. = Training$ $Sz. = Size$ $F. = Features$

labelled with $OutP$) and iii) once the two data preparation approaches have been carried out (see two last columns).

5   Results

Tables 4 reports the accuracy and roc test results averaged with an outer stratified 4-fold cross validation over the original data set followed by an inner stratified 2-fold cross validation adding one fold of the outlier set into the training set
in each iteration (see Fig. 1). Also, we have included the number of times that
$OutP + FS1$ or $OutP + FS2$ result is better than the $OutP$ one. Finally, last
row shows the average for each method with all the data sets. It is clear that
the performance measured with accuracy for $OutP + FS1$ and $OutP + FS2$ is
very similar to $OutP$ because the number of wins is lower than the half of the
number of data sets. On the other hand, the results with roc measure require to
be submitted to a deep analysis.

Table 5 shows the results of the non-parametric statistical analysis of OutP
(baseline approach) versus $Out + FS1$ or $Out + FS2$. We have represented the
average roc value for each method, their difference with the baseline case and its
ranking. According to Wilcoxon signed-ranks test, since there are 7 data sets,
the T value at $\alpha = 0.05$ should be less or equal than 2 (the critical value) to
reject the null hypothesis. On the one hand, $OutP + FS1$ is significantly better
than $OutP$. On the other hand, for $OutP + FS2$ the results are statistically
in the line of $OutP$ but the $R-$ value is three times the $R-$ value, thus the
performance of $OutP + FS2$ is promising in most of the cases.

Table 4. Classifier C4.5: Accuracy and roc test results averaged with a 4-fold cross validation

Data set				C4.5		
		Accuracy			Roc	
	$OutP$	$OutP + FS1$	$OutP + FS2$	$OutP$	$OutP + FS1$	$OutP + FS2$
CTG	89.49	86.50	88.88	0.8276	0.8379	0.8236
german	70.95	71.50	72.35	0.6170	0.6998	0.7168
magic	85.22	82.80	85.22	0.8691	0.8639	0.8691
olitos	65.42	68.75	68.75	0.7480	0.7730	0.7730
pima	74.86	74.60	74.66	0.7553	0.7645	0.7488
tokyo	90.62	91.71	91.03	0.9070	0.9290	0.9095
water	84.07	83.11	83.39	0.8035	0.8048	0.8199
Wins	3	3			6	4
Average	80.09	79.85	80.61	0.7896	0.8104	0.8087

Table 5. C4.5: Statistical tests for roc measure

Data set	$OutP$	$OutP + FS1$	$Difference$	$Ranking$	$OutP + FS2$	$Difference$	$Ranking$
CTG	0.8276	0.8379	0.0103	4	0.8236	−0.0040	3
german	0.6170	0.6998	0.0828	7	0.7168	0.0998	7
magic	0.8691	0.8639	−0.0052	2	0.8691	0.0000	1
olitos	0.7480	0.7730	0.0250	6	0.7730	0.0250	6
pima	0.7553	0.7645	0.0092	3	0.7488	−0.0065	4
tokyo	0.9070	0.9290	0.0220	5	0.9095	0.0025	2
water	0.8035	0.8048	0.0012	1	0.8199	0.0164	5
		$T = min\{26, 2\} = 2\ (*)$			$T = min\{21, 7\} = 7$		

5.1   Application of the New Proposal to Classifier PART

Once the new approach has been validated according to non-parametric statistical tests we extended it to the PART classifier. Generally speaking, CFS and CNS exhibited an intermediate performance as feature selectors operating directly on the original data [23]. In the current paper we do a previous data cleansing and after that the feature selection phase to evaluate the convenience or not to apply both data pre-processing strategies.

Tables 6 depicts the accuracy and roc test results for PART algorithm averaged with an outer stratified 4-fold cross validation and an inner stratified 2-fold cross validation as explained in the previous section. We have represented the number of times that $OutP + FS1$ is better than $OutP + FS2$ and the average for each method with all the data sets. We should remark that $OutP + FS2$ has an excellent performance for roc measure and wins 5 out of 7 times to $OutP + FS1$.

5.2   Statistical Comparison of the Two Best Classifiers with their Suitable Data Preparation Approach

This subsection compares the two best achievements that have been reported so far in the paper. Table 7 includes the results of the two best options and a

Table 6. Classifier PART: Accuracy and roc test results with OUTLIERoutP+FS for a 4-fold cross validation

Data set	PART			
	Accuracy		Roc	
	$OutP + FS1$	$OutP + FS2$	$OutP + FS1$	$OutP + FS2$
CTG	85.89	84.22	0.8210	0.8349
german	72.00	70.50	0.7058	0.6755
magic	82.90	84.71	0.8731	0.8980
olitos	67.50	67.50	0.7858	0.7858
pima	73.63	73.44	0.7571	0.7606
tokyo	91.71	91.29	0.9359	0.9543
water	83.30	83.49	0.8166	0.8443
Wins by pairs	4	2	1	5
Average	79.56	79.31	0.8136	0.8219

Table 7. Statistical comparison between C4.5 with OutP+FS1 and PART with OutP+FS2 for roc measure

Data set	C4.5	PART		
	$OutP + FS1$	$OutP + FS2$	Difference	Ranking
$CTG2$	0.8379	0.8349	−0.0030	1
german	0.6998	0.6755	−0.0243	4
magic	0.8639	0.8980	0.0341	6
olitos	0.7730	0.7858	0.0127	3
pima	0.7645	0.7606	−0.0039	2
tokyo	0.9290	0.9543	0.0253	5
water	0.8048	0.8443	0.0395	7
Wins	3	4		
	$T = min\{21, 7\} = 7$			

non-parametric statistical analysis via a Wilcoxon signed-ranks test of C4.5+ $OutP + FS1$ versus PART+$OutP + FS2$. Since the T value at $\alpha = 0.05$ is not less or equal than 2 the null hypothesis is accepted. Hence, both algorithms performs equally without significant differences. The good new is that PART behaves better according to the rankings; $R+$ value is three times the $R-$ value, thus PART with the proposed methodology should be consider as an interesting alternative to C4.5 classifier with our new approach.

6   Conclusions

An innovative methodology that performs two data preparation phases such as data cleansing via outlier detection and feature selection, in this order, was introduced. An empirical study on seven binary and multi-class classification problems with a test error rate of a 10% or above measured in accuracy was conducted. The experimentation shed light on that the roc measure is improved in global terms and the accuracy is increased in punctual cases. According to the non-parametric statistical tests, C4.5 with the new approach (OUTLIER-outP+FS) using a correlation measure overcame significantly the results for roc versus the framework OUTLIERoutP that was previously proposed. Moreover, C4.5 and a feature selector with a consistency measure, after the data cleansing stage, achieved better results than OUTLIERoutP in most of the data sets. Finally, the behaviour of a not very powerful classifier such as PART became excellent with the new approach until the extent that PART with a consistency measure reached slightly better results than the best setting of OUTLIERoutP+FS (significantly better than OUTLIERoutP) with C4.5 classifier.

Acknowledgments. This work has been partially subsidized by TIN2007-68084-C02-02 and TIN2011-28956-C02-02 projects of the Spanish Inter-Ministerial Commission of Science and Technology (MICYT), FEDER funds and P11-TIC-7528 project of the "Junta de Andalucía" (Spain).

References

1. Aggarwal, C.C., Yu, P.S.: Outlier detection for high dimensional data. In: ACM Sigmod Record, vol. 30, pp. 37–46. ACM (2001)
2. Aha, D.W., Kibler, D., Albert, M.K.: Instance-based learning algorithms. Machine Learning 6(1), 37–66 (1991)
3. Armanino, C., Leardi, R., Lanteri, S., Modi, G.: Chemometric analysis of tuscan olive oils. Chemometrics and Intelligent Laboratory Systems 5(4), 343–354 (1989)
4. Bache, K., Lichman, M.: UCI machine learning repository (2013)
5. Cover, T., Hart, P.: Nearest neighbor pattern classification. IEEE Transactions on Information Theory 13(1), 21–27 (1967)
6. Dash, M., Liu, H.: Consistency-based search in feature selection. Artificial Intelligence 151(1), 155–176 (2003)
7. Dasu, T., Johnson, T.: Exploratory data mining and data cleaning, vol. 479. John Wiley & Sons (2003)
8. Fawcett, T.: An introduction to roc analysis. Pattern Recognition Letters 27(8), 861–874 (2006)
9. Fayyad, U., Piatetsky-Shapiro, G., Smyth, P.: From data mining to knowledge discovery in databases. AI Magazine 17(3), 37 (1996)
10. Frank, E., Witten, I.H.: Generating accurate rule sets without global optimization (1998)
11. Hall, M., Frank, E., Holmes, G., Pfahringer, B., Reutemann, P., Witten, I.H.: The weka data mining software: an update. ACM SIGKDD Explorations Newsletter 11(1), 10–18 (2009)
12. Hall, M.A.: Correlation-based feature selection for machine learning. PhD thesis, The University of Waikato (1999)
13. Hjorth, J.S.U.: Computer intensive statistical methods: Validation, model selection, and bootstrap. CRC Press (1993)
14. Klawikowski, S.J., Zeringue, C., Wootton, L.S., Ibbott, G.S., Beddar, S.: Preliminary evaluation of the dosimetric accuracy of the in vivo plastic scintillation detector oartrac system for prostate cancer treatments. Physics in Medicine and Biology 59(9), N27 (2014)

15. Kohavi, R., et al.: A study of cross-validation and bootstrap for accuracy estimation and model selection. IJCAI 14, 1137–1145 (1995)
16. Langley, P.: Selection of relevant features in machine learning. Defense Technical Information Center (1994)
17. Larose, D.T.: Discovering knowledge in data: an introduction to data mining. John Wiley & Sons (2014)
18. Liu, H., Motoda, H.: Computational methods of feature selection. CRC Press (2007)
19. Quinlan, J.R.: C4. 5: Programming for machine learning. Morgan Kauffmann (1993)
20. Shin, K., Abraham, A., Han, S.-Y.: Improving kNN text categorization by removing outliers from training set. In: Gelbukh, A. (ed.) CICLing 2006. LNCS, vol. 3878, pp. 563–566. Springer, Heidelberg (2006)
21. Tallón-Ballesteros, A.J., Hervás-Martínez, C., Riquelme, J.C., Ruiz, R.: Improving the accuracy of a two-stage algorithm in evolutionary product unit neural networks for classification by means of feature selection. In: Ferrández, J.M., Álvarez Sánchez, J.R., de la Paz, F., Toledo, F.J. (eds.) IWINAC 2011, Part II. LNCS, vol. 6687, pp. 381–390. Springer, Heidelberg (2011)
22. Tallón-Ballesteros, A.J., Riquelme, J.C.: Deleting or keeping outliers for classifier training? In: 2014 Sixth World Congress on Nature and Biologically Inspired Computing, NaBIC 2014, Porto, Portugal, July 30 - August 1, pp. 281–286 (2014)
23. Tallón-Ballesteros, A.J., Riquelme, J.C.: Tackling ant colony optimization metaheuristic as search method in feature subset selection based on correlation or consistency measures. In: Corchado, E., Lozano, J.A., Quintián, H., Yin, H. (eds.) IDEAL 2014. LNCS, vol. 8669, pp. 386–393. Springer, Heidelberg (2014)
24. Witten, I.H., Frank, E., Mark, A.: Data mining: Practical machine learning tools and techniques (2011)
25. Zhang, S., Zhang, C., Yang, Q.: Data preparation for data mining. Applied Artificial Intelligence 17(5-6), 375–381 (2003)
26. Zhu, X., Wu, X.: Class noise vs. attribute noise: A quantitative study. Artificial Intelligence Review 22(3), 177–210 (2004)

Multicriterion Segmentation of Demand Markets to Increase Forecasting Accuracy of Analogous Time Series: A First Investigation

Emiao Lu and Julia Handl[(✉)]

Decision and Cognitive Sciences Research Centre,
University of Manchester, Manchester, UK
julia.handl@mbs.ac.uk

Abstract. Forecasting techniques for analogous time series require the identification of related time series as a fundamental first step of the analysis. The selection of meaningful groupings at this stage is known to be a major factor for final forecasting performance. In this context, a segmentation is typically obtained using clustering of the actual time series data, clustering of suspected causal factors (associated with each time series), or, alternatively, by taking into account expert opinion. Here, we consider the potential of multicriterion segmentation techniques to allow for the simultaneous consideration of multiple types of information sources at this segmentation stage. We use experiments on synthetic data to illustrate the potential this has in feeding forward into the analytics pipeline and, thus, improving the final robustness and accuracy of forecasting results. We discuss the potential of this approach in a real-world context and highlight the role that evolutionary computation can play in contributing to further developments in this area.

Keywords: Market segmentation · Multicriterion clustering · Analogous time series · Bayesian pooling · C_MSKF · Demand forecasting

1 Introduction

Market segmentations aim to identify meaningful sub-groups within the market, which can be useful in terms of identifying, understanding and targeting specific parts of the market. In the transport sector, segmentations are often based on a set of hypothesised demand drivers, such as ticket types, flow types and geographic areas. Segmentations obtained from these variables do not necessarily lead to segments that may be clearly differentiated in terms of their demand patterns, as the demand levels observed are not accounted for during this analysis. Nevertheless, the segments identified during market segmentation often feed forward into further analysis, including the use of these groups to develop segment-specific forecasting strategies for demand management. In this context, it is often ignored that the hypothesised market segments may be suboptimal for a forecasting context, as the nature of the segments may offer no or little opportunity to identify shared patterns between time series or to draw

J.M. Ferrández Vicente et al. (Eds.): IWINAC 2015, Part II, LNCS 9108, pp. 379–388, 2015.
DOI: 10.1007/978-3-319-18833-1_40

upon specialised approaches for better forecasting "analogous" time series, i.e. time series that share similarities in terms of both their demand patterns and their demand drivers.

Here, we explore the potential of improving performance by ensuring the generation and use of segments that are suitable from the perspective of demand forecasting and, specifically, demand forecasting in the presence of analogous time series. For this purpose, we describe a multicriterion segmentation framework that ensures the simultaneous consideration of the variable sets usually accounted for at two different stages of the analytical process. Specifically, we aim to ensure the identification of segments that are interpretable at a market level (as represented by similarities in the values of a set of shared potential demand drivers) but, simultaneously, show similarities in their time-based demand patterns. We illustrate that segments of this type have significant potential in the context of forecasting strategies that draw prediction power from the concurrent consideration of multiple analogous time series.

The remainder of this paper is structured as follows. Section 2 briefly surveys the background to this work, including standard segmentation approaches, multicriterion clustering and research around the forecasting of analogous time series. Section 3 describes the core methodology employed in this paper, as well as the data used. Results for our experiments are presented in Section 4 and, finally, Section 5 concludes.

2   Background

Segmentations play a central role in many analytics pipelines, and are often applied as the first step to precede further, extensive analysis. In consequence, the accuracy of subsequent analytical steps may depend heavily on the results of the initial segmentation stage.

2.1   Limitations of Multi-stage Segmentation

A similar dependency can be observed in settings where segmentations attempt to account for multiple types of information that are incommensurable and thus difficult to combine in a single feature space. In this situation, the different data sources are usually integrated in a multi-stage manner. For example, in [1], the rail demand market was initially segmented through a clustering of the feature space defined by a set of demand drivers. Following on from this first-stage segmentation, customers were then grouped by their ticket types, as source of information that serves as a proxy for the customers' journey purpose. Compared to a one-stage approach, multi-stage segmentations are capable of producing segments with a richer interpretation (in terms of the range of factors considered) but they are inherently suboptimal as information found in one stage is shared with other stages in a sequential, suboptimal manner.

In the specific context of demand forecasting, information that is available often comprises data regarding demand patterns over time, as well as the values of a set of potential (hypothesised) demand drivers. Segments that share similarity in terms of their demand patterns are valuable, as they open up opportunities to improve forecasting by exploiting information from sets of similar time series. On the other hand, segments that are recognisably similar in terms of the values of certain demand drivers are useful as they allow for an immediate interpretation of the demand patterns found. If both aspects of the problem are to be considered, traditional multi-stage segmentation approaches are likely to be sub-optimal, as discussed above. On the other hand, the simultaneous consideration of both information sources in a single segmentation stage can be difficult, as it would require the representation of the information in a single feature space. Alternatively, overall distance can be calculated as a weighted sum of the distance in different feature spaces: this approach is more promising, but the identification of suitable weights can still be difficult, as the relative importance and reliability of different information sources is typically unknown and may be highly application-dependent.

2.2   Multicriterion Segmentation

A suitable approach in this setting is provided by multicriterion clustering techniques, as described e.g. in [3,8,10]. These techniques approach the segmentation problem using Pareto Optimisation, i.e. they allow for the identification of a set of solutions that provide optimal trade-offs regarding a set of possible criteria. In the context of market segmentation, this allows for the incorporation of multiple information sources into the segmentation process, without the need for prior information regarding the information (or reliability) of each information source.

Some of the first literature on multicriterion clustering focused on the consideration of different clustering criteria to combat the limitations of individual clustering criteria [5,10]. E.g., an exact approach to bicriterion data clustering was first suggested in [5], but it was specific to a particular pair of clustering criteria, and its usability was restricted by its computational expense. Later work extended this approach to account for clustering criteria specific to different information sources [8], and this is the setting that is required in the context of demand forecasting. More recently, multicriterion clustering approaches for both settings have gained in popularity, as multiobjective evolutionary algorithms (MOEAs, [4]) have allowed for a more flexible and scalable implementation of the approach [10,9]. Here, we refrain from the use of a multiobjective evolutionary algorithm, as our aim is to provide a first proof-of-concept, but a more general implementation of our methodology will need to be based on these recent developments.

2.3   Forecasting of Analogous Time Series

The success of clustering methods is best assessed in the context of the overall success of a particular application [12], and this is the approach employed in our manuscript. Concretely, we consider the impact of a multicriterion clustering procedure within the context of an analytics pipeline developed to support the forecasting of analogous time series.

Analogous time series are time series that share similar levels in terms of their causal factors, giving rise to a similar behaviour of the time series over time [6,7]. Analogous time series are thought to exist in many settings, e.g. they have been studied in the context of income tax [6,7]. Traditional methods of forecasting of a single time series can be sensitive to outliers and are particularly inaccurate in the context of small-scale time series. Previous research [6,7] has demonstrated that, in such settings, the aggregation of information from groups of analogous time series can be valuable in improving the accuracy of forecasts. Essentially, by drawing information from several analogous time series, multiple data points are available per time period (one for each time series), and can yield valuable information on the reliability of observed demand fluctuations and trends. The success of the approach is dependent on the identification and grouping of analogous time series, which may be achieved using a number of approaches. The three approaches considered in the existing literature comprise clustering based on similarities in the time series points observed, clustering based on similarities in a set of underlying causal factors (e.g. demand drivers), and grouping of time series based on expert judgement.

3   Methods

In brief, we implement a clustering approach that segments a transportation market based on the time-dependent demand patterns of routes, as well as a second feature set representing potential demand drivers.

The demand patterns are described using time series, which are grouped using average link agglomerative hierarchical clustering. For the purpose of clustering, the Pearson correlation is employed to measure the distance $d_{TS}(i,j)$ between all pairs of time series i and j. Note, that more advanced choices of comparing time series would be possible here, such as the representation of each time series using a seasonal, auto-regressive integrated moving average (ARIMA, [2]) model, followed by distance calculations based on the linear predictive coding cepstrum [11].

Regarding the second feature set, we assume availability of a single demand driver only. This does not impact on the generality of our results, as our core analysis here concerns the availability of two different, incommensurable and noisy feature spaces. The distance $d_{CF}(i,j)$ between all pairs of time series i and j with respect to the second feature set is calculated using the Euclidean distance. Note that distance measures suitable to deal with categorical variables may be more appropriate in some practical settings.

To combine both information sources, we use a weighted sum between these two distance functions $d(i,j) = \omega \times d_{TS}(i,j) + (1 - \omega) \times d_{CF}(i,j)$, where the relative weight ω may vary from 0 to 1. This is a simplified analysis compared to the use of a full MOEA, as we may miss many of the Pareto optimal solutions (in particular those that are non-supported but also those that are missed due to the analysis of a limited number of weight settings only).

Hierarchical clustering returns a dendrogram that can be cut at different heights to give rise to different clustering solutions. In general, internal cluster validation techniques (such as the Silhouette Width [13]) may be used to identify the number of clusters to use in the following forecasting stage of the analysis. To eliminate this complicating factor from our analysis, we currently restrict our analysis to the segmentations obtained for the correct number of clusters only.

The segmentations are directly employed to inform demand forecasts for each individual route. To take advantage of the additional information provided by the route cluster associated with each route, a specialised forecasting methodology for analogous time series is employed. The approach uses cross-sectional pooling of time series data to inform forecasts for individual time series. Specifically, we use the Cross-Sectional Multi-State Kalman Filter (C_MSKF, [6]), which combines the Multi-State Kalman Filter (MSKF) with the Conditionally Independent Hierarchical Model (CIHM). In addition to the C_MSKF algorithm, we also applied forecasting to each time series in isolation, using the MSKF algorithm. This provides a baseline for our analysis, as we expect that an improved grouping should give rise to an increased performance gap between C_MSKF's and MSKF's forecasting performance.

In our final evaluation, we vary the number of data points used during training (time series length), and we forecast results for the next three points (years) in each series. To assess the accuracy of forecasts, we use the Mean Average Percentage Error (MAPE). To provide insight regarding the impact of these different aspects, we break up our results by segment, time series length and prediction horizon. Finally, in assessing the overall quality of the techniques, we consider the percentage of times that C_MSKF outperforms MSKF, and statistical significance of our findings is established using the Wilcoxon matched-pairs signed-rank tests with a significance threshold of $\alpha = 0.05$.

3.1   Generation of Synthetic Data Sets

To allow us to understand the impact of different factors in detail, our current analysis is focused on synthetic data where we have perfect control regarding the number of clusters, the noise levels, as well as full access to the ground truth.

To create a simple (but noisy) test case which meets the assumptions for analogous time series, a linear, logarithmic and quadratic function were used to simulate three groups of time series as a function of time t. Specifically, the following three models were used:

1. $f_1(t) = 0.8614t + 3.6348$
2. $f_2(t) = 4.3429ln(t) + 2$
3. $f_3(t) = -0.0883t^2 + 2.0087t + 0.9577$

Each of these models was used to generate a cluster of "analogous" time series where each group contains 10 separate time series, simulated for a series of 30 time steps. To create suitable diversity amongst the time series from the same model, random noise was added to each time point, with the level of noise set

to increase over time. Noise levels were different for the three models, with the
following setups used (listed by model):

1. $f'_1(t) = f_1(t) + N(0, (t/10)^2) + U(0,1)$
2. $f'_2(t) = f_2(t) + N(0, (t/13)^2) + U(0,1)$
3. $f'_3(t) = f_3(t) + N(0, (t/7)^2) + U(0,1)$

Here, $N(\mu, \sigma)$ describes a random variate drawn from a Normal Distribution
with mean μ and variance σ^2. Similarly, $U(a,b)$ describes a random variate
drawn from a Uniform Distribution with lower bound a and upper bound b. The
resulting set of 30 time series is displayed in Figure 1.

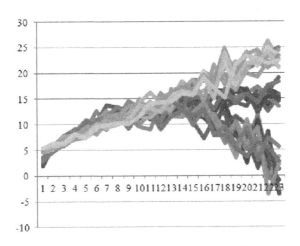

Fig. 1. Time series generated from the three mathematical models

We used a correspondingly simple setup to model the availability of known (but
noisy) causal factors. In this case, we knew the ground truth of the generating
model for all time series, so our causal factors could be directly linked to this by
introducing a different level of each factor, for each cluster. To add complexity,
random noise is used to perturb the appropriate levels for each individual time series. As discussed above, in this first analysis, we limited ourselves to the use of a
single causal factor. Its attribute value was drawn from the Normal Distribution
$N(m, 0.25)$, where $m \in [1,3]$ is an integer referring to the identifier of the mathematical model used for the generation of this particular time series (see above).

4   Results

Our aim here is to explore the potential benefits of a clustering approach that
is capable of considering several sources of incommensurable information and
explore different weight settings in the analysis. To achieve this aim, we present
results regarding the relative performance of C_MSFK and MSFK in three settings. The first two of these settings correspond to the ways in which clustering

has typically been conducted in the context of the forecasting of analogous time series [6], i.e. considering the points for the time series, or the causal factors only during cluster analysis. In these situations, we expect to obtain segments that are homogeneous with respect to the data considered, but not necessarily with respect to the second information source. The third setting corresponds to the use of a weighted sum to incorporate the information provided by both information sources. The time series identified under this approach are expected to be homogeneous with respect to both demand patterns and causal factors. It is expected that this clustering strategy may produce improved segments when both of the information sources are noisy and the noise of those information sources can be considered to be independent.

Despite the noisy nature of both information sources, the single-objective partitions capture valuable information about the structure of the data set, as evident from C_MSKF's superior performance (see Figure 2 and 3). As expected, the performance gap between C_MSFK and MSKF decreases with an increasing length of the time series, as the additional information gathered from analogous time series becomes less valuable in such settings. This is in agreement with the results previously presented in the literature [6].

From the single-objective results, it is also clear that the causal factor presents the noisier information source in this case and produces a poorer clustering overall. This can be seen from the smaller performance difference between C_MSKF and MSKF, and the fact that, for a prediction horizon of three years, C_MSKF is outperformed by MSKF for two out of three clusters. Note that the graphs for MSKF and C_MSKF are not directly comparable across figures, as MAPE results are presented by cluster and are therefore dependent on the particular cluster assignments employed.

Table 1 summarises the results obtained from exploring the simultaneous use of both information sources using a weighted-sum approach, for a selection of different weights. Shown is the percentage of times that C_{MSKF} improves performance relative to MSKF. It is evident that the weight has an impact on clustering quality. Good results are obtained here for a range of weight settings, but for more complex data we expect that it may become more difficult to set the weight without prior knowledge of the reliability of the two information sources. On the data considered here, the best results are obtained if a weight $\omega = 0.8$ or $\omega = 0.9$ is assigned to $d_{TS}(i, j,)$ (both weights give rise to the same clustering[1]). This is consistent with the findings of the single-objective analysis, which indicated that the time series data is more reliable than the information drawn from the casual factor. Full results for these particular weight settings are reported in Figure 4. They show that, compared to the single-objective settings, the performance gap between C_MSKF and MSKF increases significantly and C_MSFK maintains a significant performance advantage for all three years of predictions.

---

[1] For weight settings between 0.5 and 0.7, different clustering solutions are obtained and the performance difference becomes less pronounced.

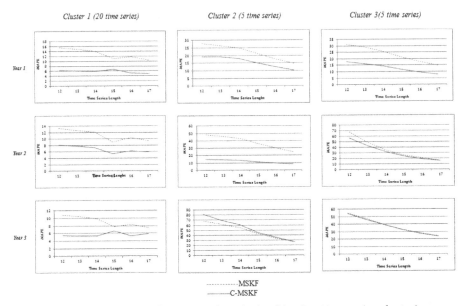

Fig. 2. Forecasting performance after single-objective time series clustering

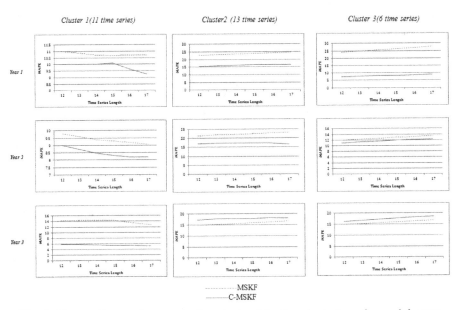

Fig. 3. Forecasting performance after single-objective clustering of causal factors

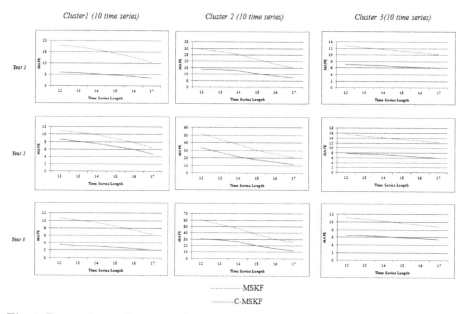

Fig. 4. Forecasting performance after multicriterion clustering with $\omega = 0.8$ or $\omega = 0.9$ (results are identical for both weights)

Table 1. Forecasting performance after multicriterion clustering with different weights ω used in the distance calculation $d(i, j) = \omega d_{TS}(i, j) + (1 - \omega)d_{CF}(i, j)$. Shown is the percentage of time that the forecasting performance (MAPE) of C_MSFK is bettern than that of MSFK.

Weight ω	Improvement of C_MSKF over MSKF (percentage of times)
1	77.8
0.9	100
0.8	100
0.7	100
0.6	100
0.5	100
0.4	88.9
0.0	66.7

5 Conclusion

In the context of demand forecasting, multicriterion clustering approaches present the opportunity to obtain segments that share similarities in terms of the demand patterns observed, but are also interpretable with respect to the type and strength of demand drivers between routes. Using experiments on synthetic data, we illustrate that our clustering strategy leads to actionable segments that are of immediate relevance to demand forecasting. Forecasting accuracy is increased through the agglomeration of available information across routes of the same segment, thus reducing the sensitivity to outliers and overcoming problems related to small-sample sizes. We present additional experiments to analyse how the length of the time series, and the weighting of descriptor variables impact on forecasting performance. The paper has presented a proof of concept using standard clustering methodology. Future work will focus on the integration of a fully flexible, multi-objective evolutionary clustering approach that will eliminate the need for a prior identification of suitable weight settings and allow for the identification of non-supported solutions.

References

1. Arup and Oxera, How has the market for rail passenger demand been segmented (2010) (accessed November 25, 2014)
2. Box, G.E.P., Jenkins, G.M.: Time Series Analysis: Forecasting and Control, Time Series and Digital Processing (1976)
3. Brusco, M.J., Cradit, J.D., Stahl, S.: A simulated annealing heuristic for a bi-criterion partitioning problem in market segmentation. Journal of Marketing Research 39(1), 99–109 (2002)
4. Deb, K.: Multi-objective optimization using evolutionary algorithms, vol. 16. John Wiley & Sons (2001)
5. Delattre, M., Hansen, P.: Bicriterion cluster analysis. IEEE Transactions on Pattern Analysis and Machine Intelligence 4, 277–291 (1980)
6. Duncan, G., Gorr, W., Szczypula, J.: Bayesian forecasting for seemingly unrelated time series: Application to local government revenue forecasting. Management Science 39(3), 275–293 (1993)
7. Lee, W.Y., Goodwin, P., Fildes, R., Nikolopoulos, K., Lawrence, M.: Providing support for the use of analogies in demand forecasting tasks. International Journal of Forecasting 23(3), 377–390 (2007)
8. Ferligoj, A., Batagelj, V.: Direct multicriteria clustering algorithms. Journal of Classification 9(1), 43–61 (1992)
9. Handl, J., Knowles, J.: On semi-supervised clustering via multiobjective optimization. In: Proceedings of the 8th Annual Conference on Genetic and Evolutionary Computation, pp. 1465–1472 (2006)
10. Handl, J., Knowles, J.: An evolutionary approach to multiobjective clustering. IEEE Transactions on Evolutionary Computation 11(1), 56–76 (2007)
11. Kalpakis, K., Gada, D., Puttagunta, V.: Distance measures for effective clustering of ARIMA time-series. In: Proceedings IEEE International Conference on Data-Mining, pp. 273–280 (2001)
12. Luxburg, U., Williamson, R., Guyon, I.: Clustering: Science or Art? In: MLR: Workshop and Conference Proceedings, vol. 27, pp. 65–79 (2012)
13. Rousseeuw, P.J.: Silhouettes: a graphical aid to the interpretation and validation of cluster analysis. Journal of Computational and Applied Mathematics 20, 53–65 (1987)

A Self-Organising Multi-Manifold Learning Algorithm

Hujun Yin$^{(\boxtimes)}$ and Shireen Mohd Zaki

School of Electrical and Electronic Engineering
University of Manchester, Manchester, UK
`h.yin@manchester.ac.uk`

Abstract. This paper presents a novel self-organising multi-manifold learning algorithm to extract multiple nonlinear manifolds from data. Extracting these sub-manifolds or manifold structure in the data can facilitate the analysis of large volume of data and discover their underlying patterns and generative causes. Many real data sets exhibit multiple sub-manifold structures due to multiple variations as well as multiple modalities. The proposed learning scheme can learn to establish the intrinsic manifold structure of the data. It can be used in either unsupervised or semi-supervised learning environment where ample unlabelled data can be effectively utilized. Experimental results on both synthetic and real-world data sets demonstrate its effectiveness, efficiency and promising potentials in many big data applications.

1 Introduction

Dimensionality reduction and manifold learning have been an active topic in machine learning and data analytics, as they can uncover the underlying subspace and hence aid functionality modeling[1][2][3]. They play an increasingly important role in the big data era where both complexity and volume of the data are enormous and/or increase rapidly. Extracting single manifold is becoming increasingly cumbersome if not impossible. It is also unrealistic to assume that a practical big data set sits in a single manifold.

Various dimensionality reduction and manifold methods have been proposed. Recent reviews can be found in [4][5]. Broadly speaking, there are four categories of methods [5]: multidimensional scaling (MDS) based, eigen-decomposition based, principal curve/surface based, and self-organising map (SOM) based. Typical exam-ple of the MDS based methods include the Sammon mapping, adaptive MDS (e.g. Isomap [6]), and curvilinear component analysis (CCA) [7]. Eigen-decomposition based has a large family of methods such as kernel PCA [8], LLE [9], Laplacian eigenmap [10] and spectral clustering [11]. SOM and its variant ViSOM [12][13] have been shown to be non-metric and metric scaling, respectively. They are also formally linked to the principal curves and surfaces [14].

There have been abundant of activities to extend single manifold methods to multi-manifolds. The potential value of multi-manifold technique has been

© Springer International Publishing Switzerland 2015
J.M. Ferrández Vicente et al. (Eds.): IWINAC 2015, Part II, LNCS 9108, pp. 389–398, 2015.
DOI: 10.1007/978-3-319-18833-1_41

demonstrated in many methods that are based on graph embedding framework and clustering analysis such as semi-supervised multi-manifold learning [15], K-manifold [16], multi-manifold clustering (mumCluster) [17], spectral multi-manifold clustering (SMMC) [18], graph-based k-means (GKM) [19] and multi-manifold discriminant analysis (MMDA) [20]. Furthermore, single manifold techniques also have been adapted into multi-manifold setting to address the complexity of mixture of manifolds such as multi-manifold ISOMAP (M-ISOMAP) [21] and nonnegative matrix factorization clustering (NMF-MM) [22]. These techniques have demonstrated their capabilities in handling multiple manifolds, which can be a combination of overlapping and widely separated manifolds, along with different intrinsic structures and dimensionalities.

The paper is divided into five sections. Section 2 briefly describes the related work. Section 3 presents the details of the proposed novel self-organising multi-manifold method. The experiments on both synthetic and real data sets, results and analysis are given in Section 4, followed by conclusions in Section 5.

2   Related Work

Manifold learning has attracted wide attention in many fields in pursuit to achieve faithful representation of high dimensional data structure in low dimensional. Most practical data sets such as face images consist of multiple manifolds to represent the variations of illumination, pose or face expression as well as subjects. A manifold of images of a person that represent poses could be on or near to a manifold, which may or may not intersect with the manifold of a different person. Therefore a useful multi manifold learning technique should be able to find these underlying structures and meaningfully present the variations in low dimensional spaces.

K-manifold clustering [16] aims to solve head-pose estimation problem by identi-fying and organising data points that lies on intersected manifolds. K-manifold starts with computing similarities between data points by estimating geodesic distance be-tween them such that,

$$s(x_i, x_j) = \sum_p -||x_p^i - x_p^j||^2 \tag{1}$$

where P is the number of poses for each subject. Then the affinity n-simplex is de-fined to seek the low dimensional embedding of intra-class compactness and inter-class separability in the pose subspace. The intra-class compactness embedding should minimise,

$$\sum_p \sum_{i,j} ||y_p^i - y_p^j||^2 \tag{2}$$

where y_p^i and y_p^j are the embedding of the head image x_p^i and x_p^j with the pose angle p. While inter-class separability should maximise,

$$\sum_s \sum_{i,j} ||y_i^s - y_j^s||^2 T_{ij} \tag{3}$$

where T_{ij} is the penalty for poses i and j of subject s if they are close together. Follow-ing the principle of K-means, K-manifold finds a new manifold embedding based on manifold embedding to manifold embedding distance metric. The data points are then classified based on the new manifold embedding. K-manifold performance was reported to exceed other manifold methods on a pose estimation experiment where manifolds were expected to be intersected, but failed to handle separated manifolds.

Multi-manifold clustering algorithm (mumCluster) [17] tackles single and intersecting manifolds separately and removes the requirement for a user to define number of clusters and intrinsic dimensionality of a data space. This method aims to define the above parameters automatically and divide given data samples into their individual manifold(s). However, mumCluster could give a misleading estimation of a manifold intrinsic structure especially when dealing with a noisy real-world data [18]. Therefore, spectral multi-manifold clustering (SMMC) algorithm was proposed to improve the robustness in handling data with outliers. This method takes a local approach to represent a global manifold, where it derived from a series of local linear manifolds. Each local manifold geometric structure is approximated using local tan-gent space at each data point. However, as classical local tangent space is based on Euclidean distance, two points that are close together but from different manifolds could lead to similar covariance matrices. Therefore, probabilistic principal component analyser (PPCA) is used to determine the local tangent space of each point using principal subspace of the local analyser. Then pairwise affinity between data points are calculated from the local tangent space such that

$$p_{ij} = p(\Theta_i, \Theta_j) = \left(\prod_{i=1}^{d} \cos(\theta_l) \right) \tag{4}$$

where Θ_i and Θ_j denotes the tangent space at x_i and $x_j (i, j = 1, , N)$ and θ denotes a series of principal angles between two tangent spaces Θ_i and Θ_j. An affinity matrix is then computed as

$$w_{ij} = p_{ij} q_{ij} = \begin{cases} \prod_{l=1}^{d} \cos(\theta_l), & \text{if } x_i \in Knn(x_j) \text{ or } x_j \in Knn(x_i). \\ 0, & \text{otherwise} \end{cases} \tag{5}$$

where Knn denotes K nearest neighbours of x, and W is projected onto a k-dimensional embedded space with a generalised eigenvector problem. The cluster labels for each manifold are then inferred using K-means algorithm. SMMC regularly achieved almost 100% accuracy on various toy datasets and had greater performance on object recognition experiment using COIL-20 database than K-manifold and mumCluster algorithms.

Graph-based K-means (GKM)[19] extends the classic K-means clustering to multi-manifold setting. Similar to the classical algorithm, GKM is implemented in two steps, adapting centroids and adapting cluster labels of data points. Taking a global approach in representing manifolds, GKM applies graph distance method instead of the usual Euclidean distance, where the graph is constructed using Dijkstras algo-rithm. Therefore GKM determine the centroids by optimising the following problem,

$$c_k = \arg \min_{x_j, j \in C_k} \frac{1}{2} \sum_{i \in C_k} d_G(x_i, x_j), k = 1, ..., K \tag{6}$$

where $d_G(x_i, x_j)$ denotes the graph distance given two vertices x_i and x_j. Meanwhile, tired random walk model is implemented in GKM to measure similarities on manifolds and determining the cluster membership of each point. Accumulated transition probability acts as a measurement of all possible paths on the graph between x_i and x_j where it measures an ability of a tired walker to walk continuously through the edges before it use up all its strength. GKM demonstrated a more meaningful placement of the centroids for each manifold compared to the classical K-means and this improves clustering representation of multi-manifold datasets.

3   Self-Organising Multi-Manifold Learning Algorithm (SOMM)

3.1   General Framework

The general framework for the proposed self-organising multi-manifold (SOMM) method is depicted in Fig. 1. The data is assumed to have multiple modalities, i.e. multiple variations, a typical case in practical large data sets. A number of SOM are used as discrete manifolds, each to represent a modality or typical variation. Such representations effectively confine data and their prototypes in Riemannian geometry. Information based learning algorithm is then derived to learn these sub-manifolds discretely with the SOMs. The learnt maps are the approximate to these sub-manifolds, up to a resolution of the map. Further validations are applied to ensure that the maps fit the data appropriately and optimally.

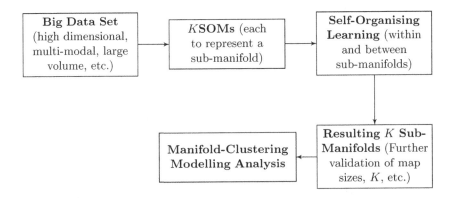

Fig. 1. General framework of the SOMM method

3.2 The SOMM Algorithm

The proposed SOMM algorithm uses K SOMs coherently to extract multi-manifold structure of the data. The details of the procedure are as follows.

1. Initialise the weight vectors w_{ij} of each SOM randomly, or to data first few principal subspace, where i denotes i-th neuron and j denotes the j-th SOM. Assume that the size of each SOM is fixed to M and the number of SOMs, K, is given a priori. When K is not known, a validation process can be applied to determine the best value. Similar principle can be applied to choose the best SOM sizes and different sizes can be used for the SOMs.
2. Update the weight vector of winning neuron, v, of the winning SOM, k, and the weight vectors of its neighbours.

$$\Delta w_{ik}(t) = \alpha(t)\eta_k(v, i, t)\left[x(t) - \Delta w_{ik}(t)\right] \tag{7}$$

   where α is the learning rate and η is the neighbourhood function, which is typically a Gaussian function of indexes of updating and winning neurons, i and v. Its width controls the effective neighbourhood.
3. Update the weight vectors of neurons of the neighbouring SOMs, by a small factor, β, which decreases with time,

$$\Delta w_{ij}(t) = \alpha(t)\beta(t)\left[x(t) - \Delta w_{ij}(t)\right] \tag{8}$$

4. Check whether K SOMs converge. If not, go back to step (2). Otherwise, stop.

The structure of the SOMs can be either 2D or 1D depending on the data. A growing structure, e.g. growing SOM or neural gas, can be used. In additions, metric scaling ViSOM can also be used to make the manifolds more metric, i.e. quantifiable.

4 Experiments

4.1 Synthetic Datasets

A number of synthetic data sets have been used to test the validity of the proposed method and as well as to compare with some existing methods. The first dataset is two noisy half circles where 3000 data points are distributed along each of two semi-circles of radius, 0.6, and centered at (1, 1) and (1.4, 1.6), respectively. Gaussian noise of standard deviation of 0.1 is added to both semi-circles. The second and third dataset are three- and five-intersected lines, centered at

Table 1. Manifold representation accuracy on MNIST handwritten digits

Algorithm	SOMM	K-Means	SMMC	K-Manifold
Representation Accuracy (%)	**96**	89	76	49

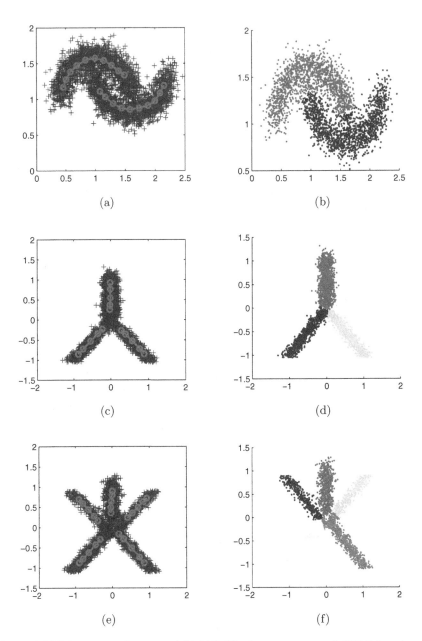

Fig. 2. SOMMs on toy datasets: (a)(c)(d) Final convergence of SOMM nodes and (b)(d)(e) Clustering results of the SOMMs (K=2, 3 and 5 respectively)

$(0, 0)$ with 0.1 standard deviation of Gaussian noise. Two 1D SOMs (K=2) with 10 nodes (M=10) each were trained on the two noisy half circles dataset. For the

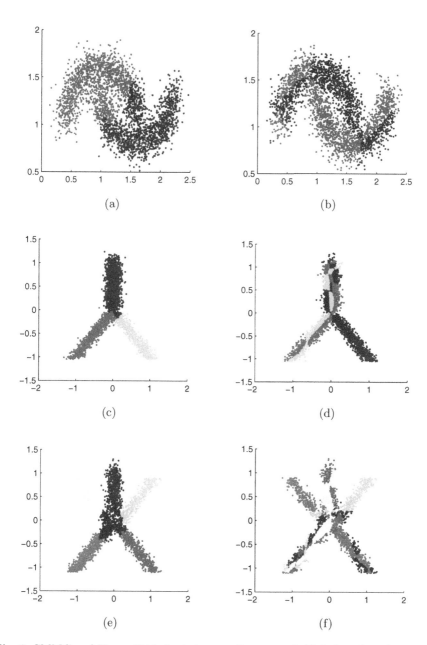

Fig. 3. SMMC and K-manifold clustering results on two half circles, three-intersected and five-intersected lines

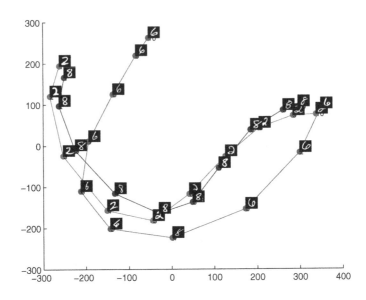

Fig. 4. SOMM on the MNIST handwritten digits database. MDS is further applied to resulting nodes of the SOMM for visualisation on 2D plot.

three- and five-intersected noisy lines datasets, three SOMs with 5 nodes each (K=3, M=5) and five SOMs with 5 nodes each (K=5, M=5) were trained on the datasets, respectively. As a comparison, all three datasets were also used by the SMMC [18] and K-manifold [16] algorithms. The results are shown in Fig. 2 and Fig. 3.

4.2   Real World Data

A real-world dataset was also used to demonstrate the validity and usefulness of the SOMM algorithm. The handwritten digits from MNIST [23] were used for this purpose. Three digits (2, 6 and 8) were used in the experiment, each digit having between 5917 and 5958 samples. SOMM with K=3 and M=10 were used for extracting the underlying sub-manifolds.

4.3   Results and Discussions

Fig. 2 demonstrates SOMM results on the synthetic datasets. Fig. 2 (a), (c) and (e) show final convergence of the SOMM on manifolds of the datasets, and Fig. 2 (b), (d) and (f) demonstrate clustering result of the data. Results of K-manifold and SMMC algorithms are shown in Fig. 3 as a comparison. The parameters for each method were tweaked to find the best result, however in most cases both K-manifold and SMMC failed to produce comparable clustering results to SOMM.

Fig. 4 shows a 2-dimensional representation of SOMM (with $K=3$, $M=10$) that were trained on three types of handwritten digits from the MNIST database. To produce a meaningful low-dimensional visualisation of SOMM result, metric multi-dimensional scaling (MDS) method was implemented to reduce the dimensionality of the nodes, and the nearest sample were mapped on each SOM node. From the figure, SOMM demonstrates that it was able to find the underlying sub-manifolds, where each SOM represents one kind of digit. At the same time, the nodes on each SOM capture the smooth changes that represent variations of the handwritten digits.

SOMM is also compared to the classic K-means, SMMC and K-manifold in terms of correct manifold representation on the three digits of the MNIST database. The results are shown in Table 1., indicating that the proposed SOMM significantly outperforms the other methods. In summary, SOMM does not only perform well on synthetic datasets, but also proved to be useful for real-world problems.

5   Conclusion

Many practical data sets exhibit multiple variations or lie in multiple sub-manifolds of the data space. Revealing these sub-manifolds is of great importance to data analytics. An adaptive self-organising multi-manifold method is proposed for extracting multiple non-linear sub-manifolds. The self-organising maps have been previously shown to be adaptive (single) manifold methods. This paper further extends them to coherent multi-manifold methods. The exact form of the submanifolds can vary in topology (1D, 2D or neural gas) and metric or non-metric learning (e.g. SOM or ViSOM) and can also be made dynamic or adaptive (e.g. growing structure). Experiments in synthetic data and real-world images show that the proposed method is an efficient and effective method for multi-manifolds.

Further work will involve dealing with sub-manifolds that are diverse, non uniform and un-balanced, as well as combining with classification for supervised or semi-supervising learning tasks. The advantages of using multi-manifolds for modeling variations in practical data such as pose and expression changes in face images will then be fully utilized.

References

1. Seung, H.S., Lee, D.D.: The manifold ways of perception. Science 290, 2268–2269 (2000)
2. Weinberger, K.Q., Saul, L.K.: Unsupervised learning of image manifolds by semidefinite programming. Int. J. Computer Vision 70(1), 77–90 (2006)
3. Huang, W., Yin, H.: On nonlinear dimensionality reduction for face recognition. Image and Vision Computing 30, 355–366 (2012)

4. Fodor, I.K.: A survey of dimension reduction techniques. Technical Report UCRL-ID-148494, Lawrence Livermore Nat Lab, Center for Applied Scientific Computing (2002)
5. Yin, H.: Advances in adaptive nonlinear manifolds and dimensionality reduction. Front. Electr. Electron. Eng. China 6(1), 72–85 (2011)
6. Tenenbaum, J.B., de Silva, V., Langford, J.C.: A global geometric framework for nonlinear dimensionality reduction. Science 290, 2319–2323 (2000)
7. Demartines, P., Herault, J.: Curvilinear component analysis: a self-organizing neural network for nonlinear mapping of data sets. IEEE Transactions on Neural Networks 8(1), 148–154 (1997)
8. Scholkopf, B., Smola, A., Muller, K.R.: Nonlinear component analysis as a kernel eigenvalue problem. Neural Computation 10(5), 1299–1319 (1998)
9. Roweis, S.T., Saul, L.K.: Nonlinear dimensionality reduction by locally linear embedding. Science 290, 2323–2326 (2000)
10. Belkin, M., Niyogi, P.: Laplacian eigenmaps for dimensionality reduction and data representation. Neural Computation 15(6), 1373–1396 (2003)
11. Weiss, Y.: Segmentation using eigenvectors: a unified view. In: Proceedings of IEEE International Conference on Computer Vision, pp. 975–982 (1999)
12. Yin, H.: ViSOM-A novel method for multivariate data projection and structure visualization. IEEE Transactions on Neural Networks 13(1), 237–243 (2002)
13. Yin, H.: Data visualization and manifold mapping using the ViSOM. Neural Networks 15(8-9), 1005–1016 (2002)
14. Yin, H.: On multidimensional scaling and the embedding of self-organizing maps. Neural Networks 21(2-3), 160–169 (2008)
15. Goldberg, A.B.: Multi-manifold semi-supervised learning. In: Proc. 12th International Conference on Artificial Intelligence and Statistics (AISTATS), pp. 169–176 (2009)
16. Liu, X., Lu, H., Li, W.: Multi-manifold modeling for head pose estimation. In: Proc. IEEE International Conference on Image Processing, pp. 3277–3280 (2010)
17. Wang, Y., Jiang, Y., Wu, Y., Zhou, Z.-H.: Multi-manifold clustering. In: Zhang, B.-T., Orgun, M.A. (eds.) PRICAI 2010. LNCS, vol. 6230, pp. 280–291. Springer, Heidelberg (2010)
18. Wang, Y., Jiang, Y., Wu, Y., Zhou, Z.-H.: Spectral clustering on multiple manifolds. IEEE Transactions on Neural Networks 22(7), 1149–1161 (2011)
19. Tu, E., Cao, L., Yang, J., Kasabov, N.: A novel graph-based k-means for nonlinear manifold clustering and representative selection. Neurocomputing 143, 1–14 (2014)
20. Yang, W., Sun, C., Zhang, L.: A multi-manifold discriminant analysis method for image feature extraction. Pattern Recognition 44(8), 1649–1657 (2011)
21. Fan, M., Qiao, H., Zhang, B., Zhang, X.: Isometric Multi-manifold Learning for Feature Extraction. In: Proc. IEEE 12th International Conference on Data Mining, pp. 241–250 (2012)
22. Shen, B., Si, L.: Nonnegative Matrix Factorization Clustering on Multiple Manifolds. In: Proc. 24th AAAI Conference on Artificial Intelligence, pp. 575–580 (2010)
23. LeCun, Y., Bottou, L., Bengio, Y., Haffner, P.: Gradient-based learning applied to document recognition. Proceedings of IEEE 86(11), 2278–2324 (1998)

Artificial Metaplasticity for Deep Learning: Application to WBCD Breast Cancer Database Classification

Juan Fombellida[1][✉], Santiago Torres-Alegre[1],
Juan Antonio Piñuela-Izquierdo[2], and Diego Andina[1]

[1] Group for Automation in Signals and Communications,
Technical University of Madrid, Madrid, 28040, Spain
jfv@alumnos.upm.es, d.andina@upm.es
[2] Universidad Europea de Madrid, Villaviciosa de Odón, Madrid, Spain

Abstract. Deep Learning is a new area of Machine Learning research that deals with learning different levels of representation and abstraction in order to move Machine Learning closer to Artificial Intelligence. Artificial Metaplasticity are Artificial Learning Algorithms based on modelling higher level properties of biological plasticity: the plasticity of plasticity itself, so called Biological Metaplasticity. Artificial Metaplasticity aims to obtain general improvements in Machine Learning based on the experts generally accepted hypothesis that the Metaplasticity of neurons in Biological Brains is of high relevance in Biological Learning. This paper presents and discuss the results of applying different Artificial Metaplasticity implementations in Multilayer Perceptrons at artificial neuron learning level. To illustrate their potential, a relevant application that is the objective of state-of-the-art research has been chosen: the diagnosis of breast cancer data from the Wisconsin Breast Cancer Database. It then concludes that Artificial Metaplasticity also may play a high relevant role in Deep Learning.

Keywords: Metaplasticity · Deep learning · Plasticity · MLP · MMLP · AMP · WBCD · Feature extraction · Machine learning · Artificial neural network

1 Introduction

In this research we progress on previous works [2], [3], [7], [8]. In the experiments that have been performed in the frame of this investigation, several neural networks belonging to the multiplayer perceptron type have been used to classify the patterns available in the Wisconsin Breast Cancer Database (WBCD)[10].

We will compare the results obtained for the classification of breast cancer data using several different implementations of the Artificial Metaplasticity Multilayer Perceptron (AMMLP) theory [7]. The experiments will use in different forms the AMMLP inherent distribution estimations of patterns in the learning sets.

© Springer International Publishing Switzerland 2015
J.M. Ferrández Vicente et al. (Eds.): IWINAC 2015, Part II, LNCS 9108, pp. 399–408, 2015.
DOI: 10.1007/978-3-319-18833-1_42

Inside each set of experiments several training methods for AMMLP have been used. The first step is to optimize the parameters used in the nominal Backpropagation (BPA) algorithm for well known Multilayer Perceptrons (MLPs) [4] to be sure that the results obtained with the modifications of the method are compared with the best performance possible using BPA. The following experiments use *a posteriori* probability estimation of the input distributions inside the MLP and AMMLP theory to implement the Metaplasticity Learning algorithm. Finally the main objective of the article is covered with the successful results presented, in comparison with state-of-the-art methods.

For assessing this algorithm's accuracy of classification, we used the most common performance measures: specificity, sensitivity and accuracy. The results obtained were validated using the 10-fold cross-validation method.

The remainder of this paper is organized as follows. Section 2 presents a detailed description of the database and the algorithms. In Section 3 the experimental results obtained are present. A brief discussion of these results is showed in Section 4. Finally section 5 summarizes the main conclusions.

2 Materials and Methods

2.1 WBCD Dataset

Breast cancer is a malignant tumor that develops from breast cells. Although research has identified some of the risk factors that increase a woman's chance of developing breast cancer, the inherent cause of most breast cancers remains unknown.

The correct pattern classification of breast cancer is an important worldwide medical problem. Cancer is one of the major causes of mortality around the world and research into cancer diagnosis and treatment has become an important issue for the scientific community. If the cancerous cells are detected before they spread to other organs, the survival rate is greater than 97%. For this reason, the use of classifier systems in medical diagnosis is increasing. Artificial intelligence classification techniques can enhance current research.

This study analyzed the Wisconsin Breast Cancer Database (WBCD). This data base has been used several times in the literature so is very useful in order to compare the results with the state of the art. This data base contains 699 patterns, each of this pattern is composed by 9 numerical attributes that corresponds to different physical characteristics that can be considered as markers of the possible presence of cancer in the sample.

2.2 Data Preparation

Numerically the attributes have been evaluated manually by an expert with values between 1 and 10, being value 1 the closest to an indicator of a benign nature of the sample and value 10 the closest to an indicator of a malicious nature of the sample. The database contains a field that indicates the final diagnosis of the nature of the sample. This value will be used as the ideal output of the networks during the supervised training.

In the original data base there are 16 samples whose attributes are not completely filled. In order to work with a homogeneous set of patterns with all the numerical attributes filled, incomplete elements have been eliminated from the experiment. Finally we will use 683 patterns that are divided in 444 benign samples (65%) and 239 malicious samples (35%).

To obtain results statistically independent of the distribution of the patterns a 10 fold cross validation evaluation method has been considered. Using this method the possible dependence of the results with the distribution of the samples in the training or performance evaluation sets is eliminated: all the samples are used to train the networks and all the samples are used to evaluate the performance of the results in different executions of the experiment for the same initial neural networks, mean values are calculated to establish the final performance results.

It it has empirically been proved that the classifiers based on neural networks produce better results if the training sets are equilibrated presenting the same number of patterns belonging to each one of the possible classes. In order to achieve this situation in the creation of the sets used to train and to evaluate the system some malicious patterns will be repeated instead of eliminating some benign patterns to get these equilibrated sets. It has been considered better to duplicate a small number of malicious elements as inputs for the networks instead of losing the potential information present in some of the benign elements.

For these experiment we have used ten data sets with the following distribution of patterns:

- G1: 90 total patterns: 45 benign and 45 malicious
- G2: 90 total patterns: 45 benign and 45 malicious
- G3: 90 total patterns: 45 benign and 45 malicious
- G4: 88 total patterns: 44 benign and 44 malicious
- G5: 88 total patterns: 44 benign and 44 malicious
- G6: 88 total patterns: 44 benign and 44 malicious
- G7: 88 total patterns: 44 benign and 44 malicious
- G8: 88 total patterns: 44 benign and 44 malicious
- G9: 88 total patterns: 44 benign and 44 malicious
- G10: 90 total patterns: 45 benign and 45 malicious

Using these elements as initial sets we will create 10 different final folders. In each one of the training sets that will be used as inputs to the networks for training the system and evaluating the evolution of the error will consists in 9 of the previous 10 groups. The final evaluation of the performance of the network will use the other element. The 10 folders will be created with the variation of the initial set that is used for evaluation and not for training.

The networks are trained from the same initial conditions presenting the data corresponding to each of the 10 folders. Finally the mean values of the results will be calculated to eliminate the possible statistical influence in the results due to the concrete fixed selection of some patterns to train the system and the fixed selection of other patterns to evaluate the results.

2.3 Artificial Metaplasticity Neural Network Model

ANNs, widely used in pattern classification within medical fields, are biologically inspired distributed parallel processing networks based on the neuron organization and decision-making process of the human brain [8]. In this paper we continue with our previous work [7] applying metaplasticity to the MLP for classifying breast cancer patterns.

The concept of biological metaplasticity was defined in 1996 by Abraham [1] and now is widely applied in the fields of biology, neuroscience, physiology, neurology and others [1], [9]. Recently, Ropero-Peláez [8], Andina [2] and Marcano-Cerdeño [5] have introduced and modeled the biological property metaplasticity in the field of ANNs, obtaining excellent results.

For these experiments Multiplayer Perceptron (MLP) neural networks have been used with a input composed by 9 attributes contained in each single pattern, a hidden layer composed by 8 neurons (previous experiments proved that 8 neurons is enough to get the flexibility needed), and an output layer with just 1 neuron (result of the classification).

The activation function used in all the neurons of the system is sigmoidal, input patterns set normalized and the initialization of the weights of the neurons is random but included in an interval $[-0.5, +0.5]$, parameter value σ in the sigmoidal activation function is constant and equal to 1. Doing this so the range of inputs to the activation function $\sigma \sum \omega_i x_i$ will be limited to the interval $[-4.5, +4.5]$. Then the initial part of the training is compliant with the premise of not saturating the output of the neurons.

To introduce AMP in an arbitrary MLP training [7], all that has to be done is to introduce a weighting function $\frac{1}{f_X^*}$ in the MLP learning equation that has the properties of a probability density function [3]. Then, it is up to the designer to find a function that improves MLP learning. Several have been already proposed [2], [3], [7], [8], and we introduce a new one in this paper.

$$w_{ij}^{(s)}(t+1) = w_{ij}^{(s)}(t) - \eta \frac{\partial E^*[W(t)]}{\partial w_{ij}^{(s)}} = w_{ij}^{(s)}(t) - \eta \frac{1}{f_X^*} \frac{\partial E[W(t)]}{\partial w_{ij}^{(s)}} \qquad (1)$$

where $s, j, i \in N$ are the MLP layer, node and input counters, respectively, for each $W(t)$ component, $w_{ij}^{(s)}(t) \in R$ and being $\eta \in R^+$ the learning rate.

The Backpropagation (BP) algorithm presents some limitations and problems during the MLP training [6]. The Artificial Metaplasticity on MLP algorithm (AMMLP) aims to improve BP algorithm by including a variable learning rate $\eta(x) = \frac{\eta}{f_X^*}$ in the training phase affecting the weights in each iteration step based on an estimation of the inherent distribution of training patterns.

Considering this basic methods some alternative weighting functions or alternative AMMLP implementations are considered and studied in this paper. For each modification the same initial networks have been used.

Artificial Metaplasticity by Gaussian Weighting Function or by Inputs *a Posteriori* Distributions. Two cases are considered:

– A given probability distribution known or assumed: One suboptimal solution [7] is:

$$f_X^*(x) = \frac{A}{\sqrt{(2\pi)^N} \cdot e^{B \sum_{i=1}^{N} x_i^2}} = \frac{1}{w_X^*(x)} \quad (2)$$

where $w_X^*(x)$ is defined as $1/f_X^*(x)$, N is the number of neurons in the MLP input, and parameters A and $B \in R^+$ are algorithm optimization values which depend on the specific application of the AMLP algorithm.

– Considering the estimation of *a posteriori* probability density function. In this case:

$$\widehat{y}_L \cong P(H_l/x) = f_X^*(x) \quad (3)$$

where $\widehat{y}_L$ is the output of the neuron that estimate the *a posteriori* probability. Equation 3 takes advantage of the inherent *a posteriori* probability estimation for each input class of MLP outputs. Note that if this is not the case, as it happens in first steps of BPA training algorithm, the training may not converge. In this first steps, the outputs of the MLP does not provide yet any valid estimation of the probabilities, but rather random values corresponding to initial guess of the MLP weights. In these first steps of training is better either to apply ordinary BPA training or to use another valid weighting function till BPA starts to minimize the error objective.

For two classes in the training set we only need one output (for N classes classification problems only N-1 output neurons are needed). If the desired output activation correspond to input vectors $x \in H_1$, then the AMP is implemented by:

$$f_X^*(x) = \widehat{y}_L \quad (4)$$

and for the complementary class patterns $x \in H_0$:

$$f_X^*(x) = (1 - \widehat{y}_L) \quad (5)$$

to swap the roles of "1" and "0" is a choice of the designer.

3 Results

3.1 Network Characteristics

In this section we present the results obtained in this research. All the models used in this study were trained and tested with the same data and validated using 10-fold cross-validation. The MLP and AMMLP proposed as classifiers implemented in MATLAB (software MATLAB version R2012a).

Structure of the network

- Number of inputs: equal to the number of attributes of the pattern (9).
- Number of hidden layers: 1
- Number of neurons included in the hidden layer: Based on previous experience 8 neurons are considered ideal for a tradeoff between the flexibility in the definition of the decision regions and the complexity of the system.
- Number of neurons in the output layer: 1 to classify in two classes.
- Activation function: Sigmoidal with output included in the interval $(0, 1)$.

Conditions considered to finalize the network training:

- Reach a defined number of inputs presented to the network to have enough iterations to reach a stable output without overspecializing the network.

3.2   Evaluation Method

In each one of the experiments 50 networks have been considered and trained. Using the 10 fold cross validation method the results are not dependant of the concrete patterns used for training and for performance evaluation. Using 50 different initial networks and calculating mean values we assure that the results are independent of the initial random value in the creation of the networks.

The following hypothesis are defined:

- $H(1/1)$: The pattern is malicious and has been classified as malicious.
- $H(1/0)$: The pattern is benign and has been classified as malicious.
- $H(0/1)$: The pattern is malicious and has been classified as benign.
- $H(0/0)$: The pattern is benign and has been classified as benign.

Using these hypothesis it is possible to build the confusion matrix:

Table 1. Confussion matrix model

True Positive $H(1/1)$	False Positive $H(1/0)$
False Negative $H(0/1)$	True Negative $H(0/0)$

To evaluate the performance of the classifiers two measures are used and defined as follows:

$$Sensitivity(SE) = \frac{TP}{TP + FN}(\%) \tag{6}$$

$$Accuracy(AC) = \frac{TP + TN}{TP + TN + FP + FN}(\%) \tag{7}$$

Where TP, TN, FP, and FN stand for true positive, true negative, false positive and false negative, respectively.

50 initial networks have been trained and evaluated with the 10 folder cross validation algorithm. From the results obtained for the same network with each one of the folders the mean confusion matrix is obtained for each network. Once these 50 mean values are calculated an additional calculation is made and the final mean value is obtained as the final result of the experiment.

The most important figure in these experiments in the sensitivity (considered as true positive percentage), these is due to the intrinsic nature of the experiment (it is much more important to detect all the malicious patterns than classifying as malicious a benign input).

Nominal Backpropagation Algorithm. Results for this case have been presented in [7] but it is now relevant to add the graphical evolution of learning, shown in Figure 1.

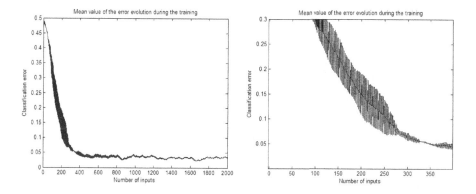

Fig. 1. Evolution of the evolution of the classification error (detail in right figure) $\eta = 1$ - Nominal Backpropagation - 2000 input patterns

Gaussian AMP Function. Results for this case have also been presented in [7] but, as in the case of nominal BPA, it is now relevant to add the graphical evolution of learning, shown in Figure 2

AMP Based on the Output of the Network. In order to check the theoretical approach of applying the *a posteriori* estimation of the probability distribution as AMP function it is necessary to reach a point when the network has started to learn. This experiment will be divided in two parts:

- The initial part of the training will use the backpropagation classic algorithm using learning rate $\eta = 25$ (empirically determined, as in usual BPA) until classification error reach 0.3.

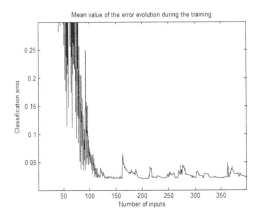

Fig. 2. Detail of the evolution of the classification error $A = 10$ $B = 0.45$ - Gaussian AMP - 2000 input patterns

- The second and principal part of the training will use equations 1, 3, 4 and 5 from error 0.3 to the finalization of the training.
- % Accuracy = 97.9009
- % Sensitivity = 98.8739

Confusion matrix for these parameters rate is shown in table 2.

Table 2. Confusion matrix $\eta = 25$ - Output as probability estimation - 2000 input patterns

98.87 %	3.07 %
1.13 %	96.93 %

The evolution of the classification error during the training phase is shown in Figure 3.

The ROC of this experiment can be found on Figure 4, the area under the curve associated is 0.995.

4   Discussion

- The best sensitivity result is obtained by the Gaussian AMP and the best accuracy result is obtained by AMP based on the output of the network.
- If we compare the results obtained by the nominal BPA, with both AMP training methods we can observe a considerable improvement in the quality of the performance (for both accuracy and sensitivity). We can also observe a considerable quicker learning for AMP, measured in iterations to achieve the final error.

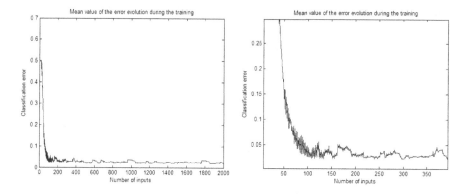

Fig. 3. Evolution of the classification error $\eta = 25$ (detail in right figure) - AMP based on the output of the network - 2000 input patterns

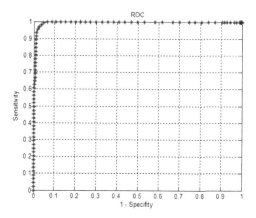

Fig. 4. ROC $\eta = 25$ - AMP based on the output of the network - 2000 input patterns

– Observing the evolution of the classification error, in the first two experiments (BPA and Gaussian AMP) there are a lot of peaks in the learning evolution. In the last AMP where the output of the network is used, the evolution of the error is more natural.

5   Conclusions

In this paper, several alternative Artificial Metaplasticity Learning implementations on MLPs show improved learning over the classical MLPs that corresponds to a uniform plasticity model. Not only in final performance, but also in the evolution of learning. Applied to a challenge and impact relevant classification problem, cancer detection, the learning has been more natural (regular), taken less training steps and provided better performance. Due to the general nature of the Metaplasticity concept, that is as wide as plasticity itself, this results are coherent with the hypothesis that modelling Metaplasticity in Machine Learning must be a relevant issue for Deep Learning Algorithms, as it is plasticity.

References

1. Abraham, W.C.: Activity-dependent regulation of synaptic plasticity (metaplasticity) in the hippocampus. In: The Hippocampus: Functions and Clinical Relevance, pp. 15–26. Elsevier, Amsterdam (1996)
2. Andina, D., Ropero-Pelaez, J.: On the biological plausibility of artificial metaplasticity learning algorithm. Neurocomputing (2012), http://dx.doi.org/10.1016/j.neucom.2012.09.028
3. Andina, D., Alvarez-Vellisco, A., Jevtic, A., Fombellida, J.: Artificial metaplasticity can improve artificial neural network learning. Intelligent Automation and Soft Computing; Special Issue in Signal Processing and Soft Computing 15(4), 681–694 (2009)
4. Andina, D., Pham, D.: Computational Intelligence for Engineering and Manufacturing. Springer, The Nederlands (2007)
5. Benchaib, Y., Marcano-Cedeño, A., Torres-Alegre, S., Andina, D.: Application of Artificial Metaplasticity Neural Networks to Cardiac Arrhythmias Classification. In: Ferrández Vicente, J.M., Álvarez Sánchez, J.R., de la Paz López, F., Toledo Moreo, F. J. (eds.) IWINAC 2013, Part I. LNCS, vol. 7930, pp. 181–190. Springer, Heidelberg (2013)
6. Leung, H., Haykin, S.: The complex backpropagation algorithm. IEEE Transactions on Signal Processing 39(9), 2101–2104 (1991)
7. Marcano-Cedeño, A., Quintanilla-Dominguez, J., Andina, D.: Breast cancer classification applying artificial metaplasticity algorithm. Neurocomputing 74(8), 1243–1250 (2011)
8. Ropero-Pelaez, J., Andina, D.: Do biological synapses perform probabilistic computations? Neurocomputing (2012), http://dx.doi.org/10.1016/j.neucom.2012.08.042
9. Kinto, E.A., Del Moral Hernandez, E., Marcano, A., Ropero Peláez, J.: A preliminary neural model for movement direction recognition based on biologically plausible plasticity rules. In: Mira, J., Álvarez, J.R. (eds.) IWINAC 2007. LNCS, vol. 4528, pp. 628–636. Springer, Heidelberg (2007)
10. http://archive.ics.uci.edu/ml/datasets.html

Comparing ELM Against MLP
for Electrical Power Prediction in Buildings

Gonzalo Vergara[1(✉)], Javier Cózar[2], Cristina Romero-González[2],
José A. Gámez[2], and Emilio Soria-Olivas[1]

[1] IDAL Research Group, E.T.S.E, University of Valencia, Valencia, Burjassot, Spain
gonzalo.vergara@uv.es
[2] SIMD Research Group, I^3A, University of Castilla-La Mancha, Albacete, Spain

Abstract. The study of energy efficiency in buildings is an active field
of research. Modelling and predicting energy related magnitudes leads to
analyse electric power consumption and can achieve economical benefits.
In this study, two machine learning techniques are applied to predict
active power in buildings. The real data acquired corresponds to time,
environmental and electrical data of 30 buildings belonging to the University of León (Spain). Firstly, we segmented buildings in terms of their
energy consumption using principal component analysis. Afterwards we
applied ELM and MLP methods to compare their performance. Models
were studied for different variable selections. Our analysis shows that the
MLP obtains the lowest error but also higher learning time than ELM.

Keywords: Extreme learning machines · Multilayer perceptron · Power
prediction · Buildings electrical power

1 Introduction

Nowadays, building consumption is estimated to be about 40% from total energy
consumed in developed countries [15]. The International Energy Agency (IEA)
claims that building sector is one of the most interesting sectors to invest in
energy efficiency. From the economical point of view, it is possible to reduce the
energy consumption in 1509 Mtoe [1] in year 2050. This fact also implies important
environmental benefits: if the energy demand is reduced from buildings, the
carbon dioxide (CO_2) emissions could be dramatically reduced too. It can be
estimated that 12.6 Gt of CO_2 could not be emitted to the atmosphere in 2050
[8].

In the literature, there is a great variety of data mining techniques that can be
used in prediction of energy demanding tasks, among them linear regression (LR)
is applied to predict monthly electrical consumption in large public buildings [10],
autoregressive integrated moving average (ARIMA) models let failure detection
in electronic equipment [2], and building occupancy is considered to improve
predictions [12]. Artificial neural networks (ANN) have been widely applied to

---

[1] 1 toe = 1 tonne oil equivalent = 11.63 MWh.

© Springer International Publishing Switzerland 2015
J.M. Ferrández Vicente et al. (Eds.): IWINAC 2015, Part II, LNCS 9108, pp. 409–418, 2015.
DOI: 10.1007/978-3-319-18833-1_43

the study of energy consumption in buildings: minimization of the energy to air condition an office-type facility [9], energy of office buildings with daylighting [19] and air condition heating dependence on electrical consumption [5] are some examples. Support vector regression (SVR) is utilized in [21] to predict electrical consumption in large buildings using simulated dataset.

In this study we used real data from 30 different buildings of the University of León (Spain) [1]. These buildings have different uses and can be classified into three types: academic (e.g. faculties), support (e.g. dormitory building, cafeteria, etc.) and research (e.g. technology center).

From previous research [17] we know that MLP yields the better results when using multivariate data, however they are quite time consuming. In this paper we study the performance of other neural net-based network, ELM and compare it with MLP. Our objective is to obtain models with the two techniques and different variable selections to predict power consumption for the next hour. Furthermore we compare different models showing their differences in terms of errors and computing times and finally we select the best one for each building.

The rest of the paper is organized as follows. In Section 2 the datasets are detailed. Afterwards, in Section 3, MLP and ELM are explained. The experiment setup and results are shown in Section 4. Finally, Section 5 concludes this paper.

2   Analysed Datasets

From the technical code of spanish buildings [11], the city of León (Spain) is placed in E1 climate zone at 914.7 m above sea level. This zone has severe climate conditions in winter as well as in summer, hence the electrical consumption due to Heating, Ventilating and Air Conditioning (HVAC) equipment is high (in 2010 the minimum temperature was $-10°C$ and the maximum $34°C$).

The available datasets contain information about environmental, electric and time information variables. The environmental variables were acquired from a single external meteorological station, meanwhile, the rest of variables were separately acquired from each building [1]. The variables involved in this study are: hour, day, month, year, ambient temperature ($°C$), relative humidity (%), solar radiation (W/m^2), working day, active power (kW) and standard deviation of hourly active power (kW).

2.1   Data Processing

The data have been registered each two minutes during 13 months (from first of March 2010 to thirty-first of March 2011), reaching to 285120 instances for each building. In order to obtain an hourly measure per variable, we have calculated the mean for each hour. However, instances having missing values are deleted and only hours containing 30 samples are considered.

To group buildings according to their power behaviour we carried out a principal component analysis (PCA) using a set of additional power related variables:

minimum, maximum, mean, range, standard deviation and coefficient of variation of active power for all buildings. Over the obtained PCA space we perform k-medoids clustering. We found that the best choice for the number of clusters, in terms of mean silhouette index, was three (Fig. 1). These three groups have 4, 10 and 16 buildings respectively. From a descriptive analysis, we found that the first group of buildings has high hourly power (100 kW) and low power variation (30%), the second group has medium power (30 kW) and high power variation (90%) and the third group has low power (10 kW) and high power variation (100%).

In order to test the two different techniques we have selected three buildings per group. These buildings are chosen in such a way that they are dispersed all over each cluster (Fig. 1). We can see how this selection scans the whole range of first and second principal components which explain the 95.73% of the variation of power related variables. The nine selected buildings with their codes are:

Group 1: Veterinary (11), Biology (13) and Agriculture (24).
Group 2: Technology (5), Philosophy (6) and Sports (17).
Group 3: Cafeteria II (22), Molecular (23) and Dormitory (25).

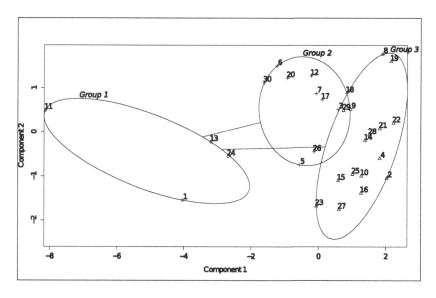

Fig. 1. Grouping of the 30 buildings in terms of first and second principal components

Finally, hour variable is transformed to improve the performance of the applied data mining techniques, avoiding the discontinuity between 23 and 0 h. In order to do that, it is common to unfold the hour in two new variables [4]:

$$h_x = \sin\left(\frac{2\pi h}{24}\right) \quad \text{and} \quad h_y = \cos\left(\frac{2\pi h}{24}\right)$$

where h is the hour. In addition, datasets have been randomly arranged to avoid causal dependency between consecutive instances. Inputs for all variables were standardised.

3 Utilized Techniques

In certain cases, linear techniques, as AR or LR, could be satisfactory [14]. These methods are easy to program and their computational cost is low. When linear techniques are not enough, it is recommended to use non-linear techniques as MLP [3,6]. MLP has proven robustness and good performance in a broad variety of applications [13,16] but its main inconvenient is long learning time. On the other hand ELM has proven good performance in artificial regression problems as well as in some real-world regression problems [7].

Below, the aforementioned techniques are briefly described.

3.1 Multilayer Perceptron

The neural network architecture used in this study is the multilayer perceptron. This model consists in a layered arrangement of individual computation units known as artificial neurons [6]. The neurons of a given layer feed with their outputs the neurons of the next layer. Each input $\mathbf{x}_i$ to a neuron j is multiplied by an adaptive coefficient $\mathbf{w}_{ij}$, called synaptic weight, which represents the strength of the connectivity between neurons. Then, a non-linear function (named activation function) is applied; usually the function is a sigmoid-shaped (sigmoid or hyperbolic tangent function) ϕ. Therefore, the output of the j-th neuron, o_j, is given by:

$$o_j = \phi\left(\sum_{i=0}^{n} \mathbf{w}_{ij} \cdot \mathbf{x}_i\right)$$

where n is the number of variables of the dataset. Neurons from a specific network are grouped together in layers that form a fully connected network. The training process consists of adjusting the coefficients (synaptic weights) of the MLP. We have utilised the *Levenberg-Marquardt* learning algorithm which is a combination of steepest descendent and *Gauss-Newton* algorithms. This algorithm produces a stable and fast convergence [20].

3.2 Extreme Learning Machines

The algorithm known as extreme learning machines is due to Huang [7] and it is based on a one single layer MLP with H hidden neurons with input weights randomly initialised. The MLP can be considered as a linear system and the output weights can be obtained analytically with the pseudoinverse matrix of H hidden outputs neurons for a training set. So, given a set of N input vectors, there exist the parameters β_i, $\mathbf{w}_i$ and b_i with:

$$\sum_{i=1}^{H} \beta_i \phi(\mathbf{w}_i \mathbf{x}_j + b_i) = \mathbf{t}_j,$$

with $j = 1, \ldots, N$ and $\mathbf{t}_i$ is the i-th target vector. This equation can be expressed in matrix form as:

$$\mathbf{H}\beta = \mathbf{T},$$

where $\mathbf{H}$ is the $N \times H$ matrix of outputs hidden layer, β is a $H \times n$ matrix of outputs weights and $\mathbf{T}$ is a $N \times n$ matrix of N targets. The MLP training consists on a minimum squares problem with optimal weights $\hat{\beta} = \mathbf{H}^\dagger \mathbf{T}$, where the dagger indicates pseudoinverse of *Moore-Penrose* [7].

4   Experimental Evaluation

In this section, we carry out a study divided in two steps. First of all, we analyse how useful are the variables of the datasets for each method, that is ELM and MLP (we need to point out that input and output variables are the same for all the problems). The idea of this first step is to check if some variables can induce noise to the models, or if they simply do not provide useful information to the learning methods (but increase the computational time to run the algorithms). After that, we compare ELM versus MLP, after applying its corresponding variable selections.

In order to evaluate each configuration a 10 fold cross-validation has been carried out for the nine buildings selected (three buildings per group). The same

Table 1. Train and test MAE of ELM models for different variable selection. First row indicates train set and second row test set.

Group	Building	P	P1	P2	E	EP	EP1	EP2
1	Veterinary	8.825	7.195	6.507	4.650	4.431	5.129	5.129
		8.977	7.638	7.155	6.552	**5.890**	6.515	6.515
	Biology	6.305	5.077	4.586	3.534	3.455	3.783	3.579
		6.518	5.268	4.997	4.783	**4.329**	4.715	4.767
	Agriculture	5.914	5.200	4.828	3.860	4.284	4.412	4.393
		5.954	5.367	5.147	5.401	**4.792**	5.083	5.066
2	Technology	3.557	3.364	2.938	2.020	2.336	2.651	2.722
		3.609	3.567	3.252	3.298	**2.963**	3.149	3.192
	Philosophy	4.865	3.875	3.205	2.233	2.056	2.341	2.345
		5.076	4.084	3.671	3.354	**3.080**	3.319	3.331
	Sports	3.620	3.440	3.429	2.201	2.351	2.696	2.660
		3.646	3.609	3.587	3.350	**2.961**	3.228	3.226
3	Cafeteria II	0.160	0.313	0.317	0.288	0.251	0.253	0.256
		0.169	0.329	0.326	0.304	0.273	0.288	0.289
	Molecular	2.037	1.799	1.669	1.501	1.538	1.576	1.536
		2.042	1.887	1.802	1.791	**1.682**	1.780	1.783
	Dormitory	2.466	2.030	1.907	1.586	1.684	1.732	1.700
		2.489	2.121	2.044	2.049	**1.941**	1.988	1.994

partitions have been used to test all the evaluated configurations (variables and algorithms).

The error measure utilised is the *Mean Absolute Error* (MAE) commonly used in electric consumption data (usually having outlayers), and more convenient than others measures as the Root Mean Squared Error (RMSE) [18]:

$$\text{MAE} = \frac{\sum_{i=1}^{N} |y_i - \widehat{y}_i|}{N}$$

where y_i are real data, $\widehat{y}_i$ are predictions and N is the number of instances. Models are constructed for each fold and mean MAE for each test set is calculated and utilized for the study.

The following parameters has been used for the evaluated techniques: ELM models have been trained using from 100 to 2000 neurons in steps of 100 and a fine tuning in steps of 10 neurons have also been done. MLP is a one hidden layer perceptron with a sigmoid activation function in the hidden layer and a linear activation function in the output layer, the number of neurons goes from 2 to 20 neurons in steps of 2 neurons. The configuration with lower mean MAE is chosen to select the models for each variable selection. The measurements of computational times correspond to an Intel®Core™i7 3770 processor at 3.4 GHz running OS X.

Table 2. Train and test MAE of one hidden layer MLP models for different variable selection. First row indicates train set and second row test set.

Group	Building	P	P1	P2	E	EP	EP1	EP2
1	Veterinary	8.840	5.752	5.413	8.413	3.810	3.792	3.772
		8.828	5.832	5.551	8.701	3.988	**3.960**	3.970
	Biology	6.326	4.359	4.039	4.952	2.892	2.812	2.824
		6.364	4.394	4.139	5.137	2.979	**2.941**	2.955
	Agriculture	5.903	5.043	4.843	6.707	3.930	3.840	3.768
		5.902	5.073	4.904	7.021	3.979	3.920	**3.897**
2	Technology	3.546	2.655	2.402	4.360	1.583	1.597	1.584
		3.533	2.690	2.474	4.467	**1.645**	1.663	1.672
	Philosophy	4.894	2.696	2.555	4.370	1.888	1.857	1.864
		4.920	2.725	2.640	4.550	**1.960**	1.961	1.978
	Sports	3.592	2.905	2.878	3.764	2.036	2.042	2.004
		3.590	2.922	2.898	3.898	**2.099**	2.114	2.115
3	Cafeteria II	0.157	0.134	0.134	0.415	0.146	0.131	0.132
		0.155	**0.137**	0.141	0.437	0.161	0.148	0.150
	Molecular	2.027	1.792	1.749	1.899	1.341	1.336	1.317
		2.025	1.798	1.767	1.953	1.370	1.359	**1.357**
	Dormitory	2.469	2.141	2.021	2.072	1.511	1.505	1.523
		2.472	2.153	2.059	2.136	**1.566**	1.567	1.581

4.1   Variable Selection Method

As it was pointed out previously, all the datasets involve the same variables. However, not all these variables are equally useful for the learning methods. In fact, it is possible that some variables lead to the problem of overfitting, or induce noise to the models. Moreover, even if useless variables do not hurt the performance, the computational time required to build these models are increased indeed. To test if it is worthy to deal with the whole set of input variables we have classified them into two groups: environmental variables (h_x and h_y, day, working day, month, ambient temperature, relative humidity, solar radiation) and power variables (a pair of power and standard deviation for a specific hour $\{p_t, sd_t\}$, one $\{p_{t-1}, sd_{t-1}\}$, and two hours before $\{p_{t-2}, sd_{t-2}\}$).

First of all, we want to test if environmental variables can be used without the power variables and vice versa. With this approach we can see if power consumption is mostly dependant on the environmental circumstances, or on the contrary this kind of information is irrelevant. In this case, the variable selection consists on the eight environmental variables (E), and the three combination of power variables: p_t and sd_t (P); p_t, sd_t, p_{t-1} and sd_{t-1} (P1); p_t, sd_t, p_{t-1}, sd_{t-1}, p_{t-2} and sd_{t-2} (P2). Finally, to test the behaviour of both environmental and power variables together, the combination of the three last cases with E has been tested: E ∪ P (EP), E ∪ P1 (EP1) and E ∪ P2 (EP2).

Training and test error for each dataset are shown in Table 1 for ELM models and in Table 2 for MLP models. In the case of ELM, the best variable selection is clearly EP reaching the best results in 8 out of 9 datasets. The Cafeteria II is an

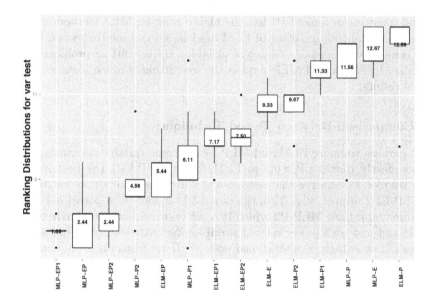

Fig. 2. Ranking distribution for ELM and MLP fot different variables selections

Table 3. Computing times with standard deviation and number of neurons of best models for different buildings. MAE corresponds to test set.

Group	Building	Method	MAE	Neurons	Time (s)
	Veterinary	ELM-EP	5.890	1590	43.484 ± 2.435
		MLP-EP1	3.960	16	164.974 ± 12.770
1	Biology	ELM-EP	4.329	1390	28.310 ± 0.754
		MLP-EP1	2.941	20	181.567 ± 14.123
	Agriculture	ELM-EP	4.792	760	6.909 ± 0.399
		MLP-EP2	3.897	10	154.569 ± 31.801
	Technology	ELM-EP	2.963	1120	16.700 ± 0.348
		MLP-EP	1.645	14	179.203 ± 4.174
2	Philosophy	ELM-EP	3.080	2000	81.570 ± 0.763
		MLP-EP	1.960	18	201.908 ± 0.715
	Sports	ELM-EP	2.961	1270	23.697 ± 0.509
		MLP-EP	2.099	14	201.700 ± 0.365
	Cafeteria II	ELM-P	0.169	40	0.202 ± 0.020
		MLP-P1	0.137	8	200.703 ± 0.197
3	Molecular	ELM-EP	1.682	580	3.290 ± 0.205
		MLP-EP2	1.357	12	209.827 ± 23.396
	Dormitory	ELM-EP	1.941	870	9.170 ± 0.263
		MLP-EP	1.566	20	229.051 ± 33.953

outlayer and its best model just uses the variables power and deviation at current hour (P variable selection). In the case of MLP, EP gives the best results in four buildings and EP1 or EP2 are the best selections for two buildings. Nevertheless, if we calculate mean ranks EP1 has the higher rank for MLP methodology.

Related to the training error of ELM models, we can see that when we take into account environmental and power variables, the overfitting problem is more noticeable. In the case of MLP models the overfitting is much lower compared to ELM results.

4.2 Comparison Between Tested Techniques

In this part we compare ELM and MLP for different variable selections. Figure 2 shows clearly that MLP wins to ELM, being MLP-EP1 the best selection. In the boxplot we observe that second and third best ranks goes for MLP-EP and MLP-EP2 respectively. Also, the best ELM model (fifth place) is ELM-EP with lower rank than MLP-P2 which is a selection without any environmental variable and just with power related variables. Nevertheless, MLP has a known drawback that is high computational cost [7]. If we focus on mean computing time necessary to train models, in Table 3 we can see that for the nine buildings (excepting the outlayer Cafeteria II) training times for ELM models goes from 3.290 to 43.484 s but MLP models needs from 164.974 to 229.051 s. The number of neurons for ELM models is high going from 580 to 2000 neurons meanwhile MLP models just need 10 to 20 neurons to reach minimum error.

5   Conclusions

In this paper we use data mining techniques to build models to make one hour predictions for the power consumption in different buildings that belongs to the University of León. Each model is built through a learning method, which uses data about a certain building as input. This data contains information about several variables related to environment and electrical power (temperature, humidity, actual power consumption, etc.).

We have evaluated the behaviour of ELM and MLP independently for this problem and both methodologies were chosen by their similar structure based in one hidden layer of neurons. First of all, we have analysed how useful are input variables for each learning method, in order to discard some of them which could induce the models to the problem of overfitting or just increase the computational time without improving the prediction performance. Here we show that the best models includes power related variables as well as environmental ones. Afterwards, we have compared ELM versus MLP for different variable selections. Results show that MLP with one hidden layer overcomes clearly ELM. However we can observe a clear difference, as on the contrary to MLP, ELM does not take advantage of using power consumption in previous hours, it just use the data from the current hour to predict the next one.

Nevertheless, we show that in our problem ELM is clearly faster than MLP with one hidden layer of neurons. Thus, we think ELM could be an advisable method when more data (variables and/or instances) is available, as it scales better. Furthermore, the use of MLP in a like-incremental approach is prohibitive, so if we want to take into account the last actual data to retrain our model, ELM could be used but not MLP.

Acknowledgements. This work has been partially funded by FEDER funds and JCCM through the project PEII-2014-049-P. Javier Cózar and Cristina Romero-González are also funded by the MECD grants FPU12 / 05102 and FPU12 / 04387 respectively. Authors want to thank Manuel Domínguez of the SUPRESS research group of the University of León for his collaboration.

References

1. Alonso, S.: Supervisión de la energía eléctrica en edificios públicos de uso docente basada en técnicas de minería de datos visual. Ph.D. thesis, Departamento de Ingeniería Eléctrica, Electrónica, de Computadores y Sistemas. Universidad de Oviedo (2012)
2. Bian, X., Xu, Q., Li, B., Xu, L.: Equipment fault forecasting based on a two-level hierarchical model. In: 2007 IEEE International Conference on Automation and Logistics, pp. 2095–2099 (2007)
3. Bishop, C.: Pattern Recognition and Machine Learning (Information Science and Statistics). Springer-Verlag New York, Inc. (2006)
4. Carpinteiro, O., Alves da Silva, A., Feichas, C.: A hierarchical neural model in short-term load forecasting. In: IJCNN (6), pp. 241–248 (2000), http://doi.ieeecomputersociety.org/10.1109/IJCNN.2000.859403

5. Ekici, B., Aksoy, U.: Prediction of building energy consumption by using artificial neural networks. Advances in Engineering Software 40(5), 356–362 (2009)
6. Haykin, S.: Neural Networks and Learning Machines, 3rd edn. Prentice-Hall (2009)
7. Huang, G., Zhu, Q., Siew, C.: Extreme learning machine: Theory and applications. Neurocomputing (70), 489–501 (2006)
8. I.E.A. International Energy Agency: Energy Performance Certification of Buildings (2013)
9. Kusiak, A., Li, M., Tang, F.: Modeling and optimization of HVAC energy consumption. Applied Energy 87(10), 3092–3102 (2010)
10. Ma, Y., Yu, J., Yang, C., Wang, L.: Study on power energy consumption model for large-scale public building. In: Proceedings of the 2nd International Workshop on Intelligent Systems and Applications, pp. 1–4 (2010)
11. Ministerio de Fomento, Gobierno de España: Código Técnico de la Edificación (2010), http://www.codigotecnico.org
12. Newsham, G., Birt, B.: Building-level occupancy data to improve arima-based electricity use forecasts. In: Proceedings of the 2nd ACM Workshop on Embedded Sensing Systems for Energy-efficiency in Building, BuildSys 2010, pp. 13–18. ACM (2010)
13. Paliwal, M., Kumar, U.: Neural networks and statistical techniques: A review of applications. Expert Systems with Applications 36(1), 2–17 (2009)
14. Soliman, S., Al-Kandari, A.: Electric load modeling for long-term forecasting. In: Electrical Load Forecasting, pp. 353–406. Butterworth-Heinemann, Boston (2010)
15. U.S. Department of Energy: Buildings Energy Data Book (2010), http://buildingsdatabook.eren.doe.gov/DataBooks.aspx
16. Vellido, A., Lisboa, P., Vaughan, J.: Neural networks in business: a survey of applications (1992–1998). Expert Systems with Applications 17(1), 51–70 (1999)
17. Vergara, G., Carrasco, J., Martínez-Gómez, J., Domínguez, M., Gámez, J., Soria-Olivas, E.: Machine learning models to forecast daily power consumption profiles in buildings. Journal of Electrical Power and Energy Systems (2014) (submitted)
18. Willmott, C., Matsuura, K.: Advantages of the mean absolute error (mae) over the root mean square error (rmse) in assessing average model performance. Climate Research 30(1), 79 (2005)
19. Wong, S., Wan, K., Lam, T.: Artificial neural networks for energy analysis of office buildings with daylighting. Applied Energy 87(2), 551–557 (2010)
20. Yu, H., Wilamowski, B.: The Industrial Electronics Handbook, vol. 5. CRC (2011)
21. Zhao, H., Magoulès, F.: Parallel support vector machines applied to the prediction of multiple buildings energy consumption. Journal of Algorithms & Computational Technology 4(2), 231–249 (2010)

Fish Monitoring and Sizing Using Computer Vision

Alvaro Rodriguez[1(✉)], Angel J. Rico-Diaz[1,2],
Juan R. Rabuñal[1,2], Jeronimo Puertas[3], and Luis Pena[3]

[1] Department of Information and Communications Technologies,
[2] Centre of Technological Innovation in Construction and Civil Engineering
(CITEEC),
[3] Department of Hydraulic Engineering (ETSECCP),
University of A Coruña, Campus Elviña s/n 15071, A Coruña, Spain
{arodriguezta,angel.rico,juanra,jpuertas,lpena}@udc.es

Abstract. This paper proposes an image processing algorithm, based in a non invasive 3D optical stereo system and the use of computer vision techniques, to study fish in fish tanks or pools.

The proposed technique will allow to study biological variables of different fish species in underwater environments.

This knowledge, may be used to replace traditional techniques such as direct observation, which are impractical or affect the fish behavior, in task such as aquarium and fish farm management or fishway evaluation.

The accuracy and performance of the proposed technique has been tested, conducting different assays with living fishes, where promising results were obtained.

Keywords: Segmentation · Computer vision

1 Introduction

Inspection of the appearance or behavior of fishes can provide with a variety of information regarding the fish species, including their health and development, and its relations with the ecosystem. In particular it can be used to provide early indications of health problems, estimating grown rates and predicting the optimal development stage to an eventual commercial exploitation.

Non invasive fish inspection is an important question in fields related with fish studies; such as marine biology, oceanography and others. Where it is common to maintain specimens of different fish species in closed and controlled ecosystems, which need to be managed.

Additionally, fish size inspection is a critical question in fish farming applications consisting in raising fish in tanks or enclosures, usually to be used in the food industry. In this applications, the optimum fish size and the parameters concerning fish grown have to be studied and monitored [1].

In these scenarios, fish are usually contained in more or less artificial structures, and the usual process to obtain the required information about fish is direct observation.

© Springer International Publishing Switzerland 2015
J.M. Ferrández Vicente et al. (Eds.): IWINAC 2015, Part II, LNCS 9108, pp. 419–428, 2015.
DOI: 10.1007/978-3-319-18833-1_44

In this context, Computer Vision is a rapid, economic, consistent, objective, and non-destructive inspection technique which may constitute an more efficient alternative [2]. It has been already widely used in the agricultural and food industry, and its potential in this field have analysed in works such as [3,4].

Computer Vision techniques have also been applied with success in underwater scenarios and in research related with fish. An early examples of these applications are the use of acoustic transmitters and a video camera for observing the behavior of various species [5]. More recently, different computer vision techniques have been used to study fish swimming trajectories [7], classify fish species [8,9], fish sizing [10,11] or monitoring fish behavior [12].

The imaging techniques to detect and study fish in images are usually based on color segmentation such as in [13] where fluorescent marks are used for the identification of fishes in a tank or in [14] where color properties and background subtraction are used to recognize live fish in clear water.

Other used techniques are stereo vision [10], background models [15], shape priors [16], local thresholding [17], moving average algorithms [18] or pattern classifiers applied to the changes measured in the background [19].

The works previously described, use color features, background or fish models, and stereo vision systems in different applications. However, most of the published techniques rely on manual marking features on the images [10,11] or on fish marking [20], monitoring the fish passage through special devices [7] or in the location of special fish features dependent of the particular selected fish species [21] or are used in out of water operations [22].

The proposed technique, uses a stereo camera system and uses different computer vision techniques to estimate without human supervision, distribution of fish sizes in a small water environment (Fig. 1).

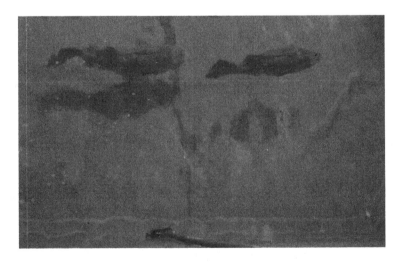

Fig. 1. Experimental conditions during one of the assays conducted in the Finisterrae Aquarium of A Coruña

2   Fish Detection

The first stage in the proposed algorithm is the detection of fish. This step will be based on the use of stereopsis, which allows the system to obtain measurements in the real world of objects located at different distances from the cameras.

Principles of stereopsis can be explained considering a point P in the real 3D space, being projected simultaneously in two image points p and p' through two camera projection centers C and C'. The points P, p, p', C and C' lie in a plane called epipolar plane. The epipolar plane, intersects each image in a line. This lines correspond to the projection of the rays through p and P, and p' and P. They are called the epipolar lines. This projection is described by epipolar geometry [23] (Fig. 2).

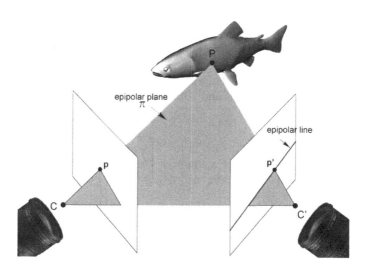

Fig. 2. Epipolar Geometry

In practise this process is done in two steps. First, the estimation of the intrinsic parameters of each camera is performed. The intrinsec parameters, determine the process how a point in the real space is projected on the image plane in an single optical system. This projection process can be precisely described using *Pin-hole* mathematical model [24,25].

$$\begin{bmatrix} p_x \\ p_y \\ 1 \end{bmatrix} = \underbrace{\begin{bmatrix} f_x & 0 & c_x \\ 0 & f_y & c_y \\ 0 & 0 & 0 \end{bmatrix}}_{M} * \begin{bmatrix} P_x/P_z \\ P_y/P_z \\ 1 \end{bmatrix} \tag{1}$$

Where p is a point in the image with coordinates p_x, p_y, f represents the focal length of the camera (which can be view as the distance from the lens to the camera sensor) and c determines the image coordinates where a point is projected through the center of the lens C. Therefore, the matrix M models the projection from a point in the image, through a camera system to the real 3D space, obtaining a projection line pP, as shown in (Fig. 2) with plane coordinates $(P_x/P_z, P_y/P_z)$.

Additionally, the pin-hole model describes the projection of a point in a perfect optical system. In practice, the real position of the point differs from the ideal one because imperfections in camera lenses introduce some distortions in the image. These distortions follow simple mathematical formulas (2) depending on a reduced number of parameters D [26].

$$
\begin{aligned}
dr_x &= xk_1r^2 + xk_2r^4 \\
dr_y &= yk_1r^2 + yk_2r^4 \\
dt_x &= k_3(r^2 + 2x^2) + 2k_4xy \\
dt_y &= sk_3xy + k_4(r^2 + 2y^2) \\
D &= \begin{bmatrix} k_1 & k_2 & k_3 & k_4 \end{bmatrix}
\end{aligned}
\tag{2}
$$

Where a distortion for a point in the image, in cooditates (x,y) is modelled, and being $r = \sqrt{x^2 + y^2}$

The final step to estimate the epipolar geometry of the camera system is the estimation of the the translation and rotation of the second camera relative to the first one.

$$
\begin{bmatrix} C'_x \\ C'_y \\ C'_z \end{bmatrix} = \begin{bmatrix} r_{11} & r_{12} & r_{13} \\ r_{21} & r_{22} & r_{23} \\ r_{31} & r_{32} & r_{33} \end{bmatrix}_R * \begin{bmatrix} C_x \\ C_y \\ C_z \end{bmatrix} + \begin{bmatrix} t_1 \\ t_2 \\ t_3 \end{bmatrix}_T
\tag{3}
$$

Taking this into account, the complete camera model depends on a limited number of parameters which can be solved using a set of known correspondences in both cameras. This is achieved using a calibration pattern of known geometry and easily detectable feature points, and then minimizing the total re-projection error for all the points in the available views from both cameras using the Levenberg-Marquardt optimization algorithm.

Once the camera model has been obtained, to estimate a 3D map from an observed scene, it is necessary to perform a matching process, with the aim of solving the correspondence of the images obtained in the same moment by the two cameras. The result of this process will be the estimation of a disparity map, referred to the difference in image location the pixels seen by the left and right cameras.

Disparity is inversely related with the distance from the cameras. Furthermore, using the previously estimated camera model, it can be used to obtain the 3D position of the objects observed in the scene.

To estimate the disparity map, a parallelized implementation of the Block-Matching algoritmh, avaliable in OpenCV [27] has been used. This technique analyses the statistical similarity of pixel values in each region (block) of the image. The purpose is to solve the correspondence problem for each block, finding in the next image the region representing the most likely displacement.

Numerically, the point *(i', j')* corresponds to *(i,j)* after applying the displacement $d(i,j)=(dx,dy)$, which may be described as described in (1).

$$(i', j') = (i + x, j + y) \tag{4}$$

This approach is based on the assumption expressed in (2).

$$I(i, j) + N = I'(i + x, j + y) \tag{5}$$

N being a noise factor following a specific statistical distribution, I the image obtained from camera C and I' the one obtained from camera C'.

In this point, a segmentation based on disparity thresholding of the scene, will provide us with the objects located in the field of view of the cameras and located at a distance of interest.

3   Estimating Fish Size

In the previous stage, objects in the image have been detected in the water by filtering the distance map obtained with the stereo image processing.

However, the real time obtained disparity map may not be accurate enough to estimate the fish size properly and may not possess enough information to distinguish fish from other objects. The proposed technique uses a segmentation technique to detect fish in the region of the RGB space corresponding to the fish location in the disparity map, and combining both analysis, fish are then properly detected and sized.

The fish segmentation technique consists in a background subtraction process using a dynamic background modelling technique. According to this, every pixel of the scene must be matched to the background or foreground category. To this end a widely used model based in estimate the RGB color space probability distribution for every pixel in the image has been chosen. The segmentation algorithm works using Bayesian probability to calculate the likelihood of a pixel $x_{ij}(t)$, at time t in coordinates *(i,j)*, being classified as background *(B)* or foreground *(F)*. This is expressed as follows:

$$p(B|x) = \frac{p(x|B)p(B)}{p(x|B)p(B) + p(x|F)p(F)} \tag{6}$$

$$p(B|x) \sim x_{ij}(t)^T H_{ij}(t) \tag{7}$$

Where $H_{ij}(t)$ is an histogram estimation in time t for *(i,j)*. This estimation is based in a set of observations $\chi_{ij} = \{x_{ij}(t - T), , x_{ij}(t)\}$ where T is a time period used to adapt to changes. According to this, new samples are added to the set and old ones are discarded, while the set size does not exceed a certain value. Therefore, the estimated probability distribution will be adapted to changes in the scene. This is performed by using weighting values depending a on decaying factor that is used to limit the influence of the old data.

Background subtraction through the estimation of pixel distributions have been studied in several works. Most of these works are based in the use of parametric models to estimate the probability distribution of pixel values [28,29]. These techniques are based commonly on the use of single Gaussian distribution models or mixture of Gaussian models, according to the complexity of the scene [30,31].

In this work, a nonparametric algorithm proposed in [32] has been used. Therefore, the distribution itself is estimated and updated, assuming a priori probabilities $p(B)=p(F)=0.5$. Thus, a pixel is classified as a background if $p(x|B) >0.8$. The weights of the model are updated according to the following equation:

$$H_{ij}(t+1) = (1 - \alpha)H_{ij}(t) + \alpha x_{ij}(t) \tag{8}$$

Where $x_{ij}(t)$ is the current sample and α is a constant value, equivalent to $1/T$, as expressed above. In this work, α was set by default to 0.025. Given that, the new weighting value in the histogram for a sample y observed k frames in the past and which had a weight w, will be $w(1 - \alpha)^k$.

At this point, a thresholding will be performed on the result, obtaining a binary representation of the candidate foreground objects. Finally a representation and interpretation phase is conducted, using the edges of the detected objects, whose aim is determining if they can be considered as fish.

In this stage objects that do not achieve certain restrictions are eliminated, these restrictions are defined as valid ranges on the object size, its area, and its length-height ratio, defined empirically according to the fish species.

Finally, if the segmented object is accepted as a fish the camera calibration and the disparity map are used to measure the fish size.

4   Results

To measure the performance of the proposed system. a set of experiments were conducted with living fishes in different fish tanks.

To this end, five experiments were conducted in a fish tank located in the Centre of Technological Innovation in Construction and Civil Engineering (CITEEC) of A Coruña, using specimens of European perch (*Perca fluviatilis*) and brown trout (*Salmo trutta*) Fig. 3; and two experiments were conducted in a pool located in the Finisterrae Aquarium of A Coruña, using specimens of Atlantic wreckfish (*Polyprion americanus*) Fig. 4.

Near 2000 images were used in the experiments, and the obtained results were achieved comparing the measurements obtained by the system with a manual labeling of the different fish specimens conducted by experts.

Obtained results are summarized in Table 1.

False positive, were automatically marked as anomalous detections which estimated length L is outside the limits $[\mu - 2\sigma, \mu + 2\sigma]$. Being μ, σ the average and standard deviation respectively of the estimated sizes in the assay.

Fig. 3. Example of obtained results. (a) Image of a detected wreckfish recorded in Centre of Technological Innovation in Construction and Civil Engineering (CITEEC) of A Coruña. (b) Segmented and processed Image.

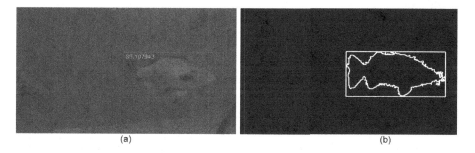

Fig. 4. Example of obtained results. (a) Image of a detected European perch recorded in the Finisterrae Aquarium of A Coruña. (b) Segmented and processed Image.

Table 1. The average results obtained with the different techniques

Results	European perch	Brown trount	Atlantic wreckfish
Avg. Measured Size	8.8	6.5	92.6
Std.Dev. Measured Size	0.8	0.8	8.4
Avg. Absolute Error (cm)	0.6	0.9	6.5
Std.Dev. Absolute Error (cm)	0.6	0.7	5.2
Avg. Relative Error	0.07	0.12	0.07
Std.Dev. Relative Error	0.09	0.09	0.06
True Positives	620	600	182
Detected False Positives	44	31	47
Precision	0.9	1.0	0.8
False Positive Ratio	0.1	0.0	0.2

Analyzing the results it may be observed that, in general, a 10% of error in estimated fish size was achieved. Obtaining a low standard deviation of error and a precision rate near the 90%.

The absolute error in measurements of European perch and brown trout is less than 1cm, which constitute a very promising result. However a higher absolute error is observed when measuring Atlantic wreckfish, this is motivated to the big size of the used specimens (some bigger than 1m), which caused some detections when a part of the fish was outside the visual frame. Although this situation, will be managed in future versions of this work, obtained results are still accurate enough to study this kind of fishes.

Summarizing, it can be stated that the proposed technique can obtain a reliable measurement of fish size regarding the variability in luminosity changes, environmental conditions and fish species.

A higher absolute error is observed when measuring Atlantic wreckfish than with other species, this is motivated

In general, it can be seen that a very low false positive rate is achieved, so obtained results are very reliable, and detected fishes represent real fishes with a high probability.

Finally, the obtained error in the measurements is very low, achieving also a low standard deviation of the error. These results indicate that the proposed technique may used in real conditions.

5   Conclusions

In this work, a non invasive solution to estimate fish size in underwater environments is proposed, aimed to the optimization of aquarium and fish farm management.

The proposed technique combines stereo imaging with computer vision techniques to detect and to measure fish in images, without need of fish marking techniques or direct fish observation.

The accuracy and performance of the proposed technique has been tested, conducting different assays with living fishes.

The results obtained with this system have been very promising, as they allowed us to obtain the fish size in the image with a low error rate.

In future stages of this work, the segmentation and detection algorithms will be optimized, to improve the reliability of the system and a tracking algorithm will be defined to manage measurements according to each specimen.

Additionally, further experiments will be conducted and the functionality of the proposed technique will be extended to estimate the distribution of fish weights and sizes.

Acknowledgments. This work was supported by FEDER funds and Spanish Ministry of Economy and Competitiveness (Ministerio de Economa y Competitividad) (Ref. CGL2012-34688). The authors would also like to thank the Spanish Ministry of Education (FPI grant Ref. BES-2013-063444). The authors also wish to thank the managers and personnel of the Finisterrae Aquarium of A Coruña for their support, technical assistance and for allowing the unrestricted use of the Finisterrae facilities.

References

1. Leon-Santana, M., Hernandez, J.M.: Optimum management and environmental protection in the aquaculture industry. Ecological Economics 64, 849–857 (2008)
2. Cappo, M., Harvey, E., Malcolm, H., Speare, P.: Potential of video techniques to monitor diversity, abundance and size of fish in studies of marine protected areas. In: Aquatic Protected Areas-What Works Best and How do We Know, pp. 455–464 (2003)
3. Brosnan, T., Sun, D.-W.: Inspection and grading of agricultural and food products by computer vision systems a review. Computers and Electronics in Agriculture 36, 193–213 (2002)
4. Costa, C., Antonucci, F., Pallottino, F., Aguzzi, J., Sun, D.-W., Menesatti, P.: Shape Analysis of Agricultural Products: A Review of Recent Research Advances and Potential Application to Computer Vision. Food and Bioprocess Technology 4, 673–692 (2011)
5. Armstrong, J.D., Bagley, P.M., Priede, I.G.: Photographic and acoustic tracking observations of the behavior of the grenadier Coryphaenoides (Nematonorus) armatus, the eel Synaphobranchus bathybius, and other abyssal demersal fish in the North Atlantic Ocean. Marine Biology 112, 1432–1793 (1992)
6. Steig, T.W., Iverson, T.K.: Acoustic monitoring of salmonid density, target strength, and trajectories at two dams on the Columbia River, using a split-beam scaning system. Fisheries Research 35, 43–53 (1998)
7. Rodriguez, A., Bermudez, M., Rabuñal, J., Puertas, J.: Fish tracking in vertical slot fishways using computer vision techniques. Journal of Hydroinformatics (2014)
8. Zion, B., Shklyar, A., Karplus, I.: Sorting fish by computer vision. Computers and Electronics in Agriculture 23, 175–187 (1999)
9. Zion, B., Shklyar, A., Karplus, I.: In-vivo fish sorting by computer vision. Aquacultural Engineering 22, 165–179 (2000)
10. Petrell, R.J., Shi, X., Ward, R.K., Naiberg, A., Savage, C.R.: Determining fish size and swimming speed in cages and tanks using simple video techniques. Aquacultural Engineering 16, 63–84 (1997)
11. Israeli, D., Kimmel, E.: Monitoring the behavior of hypoxia-stressed Carassius auratus using computer vision. Aquacultural Engineering 15, 423–440 (1996)
12. Ruff, B.P., Marchant, J.A., Frost, A.R.: Fish sizing and monitoring using a stereo image analysis system applied to fish farming. Aquacultural Engineering 14, 155–173 (1995)
13. Duarte, S., Reig, L., Oca, J., Flos, R.: Computerized imaging techniques for fish tracking in behavioral studies. European Aquaculture Society (2004)
14. Chambah, M., Semani, D., Renouf, A., Courtellemont, P., Rizzi, A.: Underwater color constancy enhancement of automatic live fish recognition. In: IS&T Electronic Imaging (SPIE) (2004)
15. Morais, E.F., Campos, M.F.M., Padua, F.L.C., Carceroni, R.L.: Particle filter-based predictive tracking for robust fish count. In: Brazilian Symposium on Computer Graphics and Image Processing (SIBGRAPI) (2005)
16. Clausen, S., Greiner, K., Andersen, O., Lie, K.-A., Schulerud, H., Kavli, T.: Automatic segmentation of overlapping fish using shape priors. In: Ersbøll, B.K., Pedersen, K.S. (eds.) SCIA 2007. LNCS, vol. 4522, pp. 11–20. Springer, Heidelberg (2007)

17. Chuang, M.-C., Hwang, J.-N., Williams, K., Towler, R.: Automatic fish segmentation via double local thresholding for trawl-based underwater camera systems. In: IEEE International Conference on Image Processing (ICIP) (2011)
18. Spampinato, C., Chen-Burger, Y.-H., Nadarajan, G., Fisher, R.: Detecting, Tracking and Counting Fish in Low Quality Unconstrained Underwater Videos. In: Int. Conf. on Computer Vision Theory and Applications (VISAPP) (2008)
19. Lines, J.A., Tillett, R.D., Ross, L.G., Chan, D., Hockaday, S., McFarlane, N.J.B.: An automatic image-based system for estimating the mass of free-swimming fish. Computers and Electronics in Agriculture 31, 151–168 (2001)
20. Frenkel, V., Kindschi, G., Zohar, Y.: Noninvasive, mass marking of fish by immersion in calcein: evaluation of fish size and ultrasound exposure on mark endurance. Aquaculture 214, 169–183 (2002)
21. Martinez-de Dios, J., Serna, C., Ollero, A.: Computer vision and robotics techniques in fish farms. Robotica 21, 233–243 (2003)
22. White, D.J., Svellingen, C., Strachan, N.J.C.: Automated measurement of species and length of fish by computer vision. Fisheries Research 80, 203–210 (2006)
23. Hartley, R.I., Zisserman, A.: Multiple View Geometry. Cambridge University Press (2004)
24. Abad, F.H., Abad, V.H., Andreu, J.F., Vives, M.O.: Application of Projective Geometry to Synthetic Cameras. In: XIV International Conference of Graphic Engineering (2002)
25. Martin, N., Perez, B.A., Aguilera, D.G., Lahoz, J.G.: Applied Analysis of Camera Calibration Methods for Photometric Uses. In: VII National Conference of Topography and Cartography (2004)
26. Zhang, Z.: Flexible Camera Calibration By Viewing a Plane From Unknown Orientations. In: International Conference on Computer Vision (ICCV) (1999)
27. OPENCV: Open Source Computer Vision, http://opencv.org (Visited: February 2015)
28. Coifman, B., Beymer, D., McLauchlan, P., Malik, J.: A real-time computer vision system for vehicle tracking and traffic surveillance. Transportation Research Part C: Emerging Technologies 6, 271–288 (1998)
29. Horprasert, T., Harwood, D., Davis, L.S.: A statistical approach for real-time robust background subtraction and shadow detection. In: IEEE ICCV, pp. 1–19 (1999)
30. KaewTraKulPong, P., Bowden, R.: An improved adaptive background mixture model for real-time tracking with shadow detection. In: Video-Based Surveillance Systems, pp. 135–144. Springer (2002)
31. Zivkovic, Z.: Improved adaptive Gaussian mixture model for background subtraction. In: International Conference on Patern Recognition (ICPR 2004), pp. 28–31 (2004)
32. Godbehere, A.B., Matsukawa, A., Goldberg, K.: Visual tracking of human visitors under variable-lighting conditions for a responsive audio art installation. In: American Control Conference (ACC), pp. 4305–4312 (2012)

Computer-Aided Development of Thermo-Mechanical Laser Surface Treatments for the Fatigue Life Extension of Bio-Mechanical Components

José-Luis Ocaña, Carlos Correa, Ángel García-Beltrán[✉],
Juan-Antonio Porro, Leonardo Ruiz-de-Lara, and Marcos Díaz

Centro Láser U.P.M., Edificio Tecnológico La Arboleda, Campus Sur U.P.M.,
Carretera de Valencia, km. 7,300, 28031 Madrid, Spain
jlocana@etsii.upm.es
http://www.upmlaser.upm.es

Abstract. Bio-mechanical components (i.e. spinal, knee and hip prostheses) are key elements definitely improving the quality of life of human beings. These components development has been traditionally subject to mechanical and functional designs based primarily on intuitive medical approaches, not always optimized from an engineering point of view, what in turn has been responsible for undesirable cases of mechanical failure implying the need for additional surgical interventions and its associate life risk for aged patients. Laser Shock Processing (LSP) uses the high peak power of short pulse lasers to generate an intense shock wave into the material finally leading to the generation of a compressive residual stresses field definitely protecting the component against crack initiation and propagation, thus improving its mechanical response and in-service fatigue life. Developments in the field of the predictive assessment of LSP are presented along with practical examples of the design-motivated improvements in prostheses achievable by LSP.

Keywords: Computer-aided development · Laser shock processing · Bio-mechanical components

1 Introduction

Bio-mechanical components (i.e. human replacement prostheses) are key elements definitely improving the quality of life of human beings. Major advancements in the development of spinal, knee and hip implants have, among others, allowed the extension and improvement of life conditions of an increasingly huge number of aged persons, providing them with life expectations remarkably higher than ever before.

However, the development of these bio-mechanical components has been traditionally subject to mechanical and functional designs based primarily on intuitive medical approaches, not always optimized from a mechanical engineering point

© Springer International Publishing Switzerland 2015
J.M. Ferrández Vicente et al. (Eds.): IWINAC 2015, Part II, LNCS 9108, pp. 429–438, 2015.
DOI: 10.1007/978-3-319-18833-1_45

of view, what in turn has been responsible for a certainly undesirable number of cases of prosthesis mechanical failure implying the need for additional surgical interventions and its associate life risk for aged patients.

Consistently with the development of different types of prostheses in different materials and with continuously improved mechanical designs, several types of surface treatments for the improvement of their final surface condition and mechanical fatigue life have been developed of are under development.

One of the most recent and promising techniques for the application of such kind of thermo-mechanical surface treatments is Laser Shock Processing (LSP), a technique using the high peak power of short pulse lasers to generate an intense shock wave into the treated material finally leading to the generation of a compressive residual stresses field definitely protecting the component against crack initiation and propagation, thus improving its mechanical response and its in-service fatigue life.

Laser Shock Processing has revealed itself in the last years as a powerful technique able to substitute with advantage classical surface conditioning techniques used for analogous purposes but implying important unpredictability and environmental drawbacks, as i.e. shot peening [1, 2, 3, 4].

A conceptually relevant feature of Laser Shock Processing is its clear ability for a close control of the effects induced in the treated materials as a consequence of the capability of robotized systems to guide the incident laser beam within accuracies in the range of tenths of a millimeter and of the prospects for the numerical predictability of the laser effect by means of experimentally consistent models.

The development of these experimentally supported prediction models is nowadays considered as a key assignment for the development and applicability of the technique to real components and is, in turn, the key for the truly adjusted design of the prosthesis elements with maximum warranty of quality and durability. The development is, in summary, an excellent example on how computer-aided design is providing enhanced solutions to real life problems on the track to improved human quality life.

In the present paper, a summary of the developments achieved by the authors in the field of the predictive assessment of Laser Shock Processing is presented along with some practical examples of the design-motivated improvements in replacement prostheses achievable by application of the technique.

2   Model Description

Laser Shock Processing (LSP) is based on the application of a high intensity pulsed Laser beam (I >1 GW/cm^2; $\tau < 50$ ns) on a metallic target forcing a sudden vaporization of its surface into a high temperature and density plasma that immediately develops inducing a shock wave propagating into the material.

During a first step (during which the laser beam is active on the piece), the laser energy is deposited at the interface between the target and a transparent confining material (normally water). This transparent layer increases (around ten times) the pressure in the generated plasma even with a very small thickness [5]. This pressure induces two shock waves propagating in opposite directions (on in the target and another in the confining material). When the laser is switched off, the plasma continues to maintain a pressure, which decreases during its expansion as a consequence of the increase of the plasma volume.

The description of the relevant laser absorption phenomena becomes hardly complicated because of the non-linear effects appearing along the interaction process and which significantly alter the shocking dynamics.

The most popular methods reported in the literature for the analysis of the LSP phenomenology [6, 7, 8] try to induce the intensity and temporal profile of the shock wave launched into the treated solid material by means of the analysis of the impulse conservation between the external interface of such material and the frontier of the confining material without any reference to the detailed physics of the plasma formation process taking place in the outermost layers of the solid target: this plasma is assumed to be built up to certain degree as a consequence of the initial laser energy deposition, but no analysis is provided about its real dynamics.

From the point of view of the acceleration of the confining medium as a consequence of the excess pressure due to the expansion of the generated plasma, the simplified models proposed in the cited references can provide adequate results that can even be experimentally contrasted. However, from the point of view of the actual compression dynamics of the solid target (the main objective of the study and for which the contrast to experimental results is far more complicated), such simplified models are presumably not able to provide a correct estimate of the pressure/shock waves effectively launched, at least in the initial moments of the laser interaction, in which the complex physics related to plasma ionization dynamics can substantially modify the target state in view of the subsequent process development.

The modeling of LSP process requires a three level description with adequate interconnection of the data obtained in each phase in a self-consistent way.

The referred three-level description includes:

1. Analysis of the plasma dynamics, including breakdown phenomenology in dielectric media
2. Simulation of the hydrodynamic phenomenology arising from plasma expansion between the confinement layer and the base material
3. Analysis of the propagation and induction of permanent material changes by shock wave evolution

The authors have developed a simulation model (SHOCKLAS) that deals with the main aspects of LSP modeling in a coupled way [9],[5],[10]. A scheme with the different codes and their relations with input and output variables is shown in Fig. 1 (adapted from [11]).

2.1 Description of Laser Plasma Interaction

For the description at this level, the maximum detail is provided by the HELIOS code [12]. HELIOS is a 1-D radiation-hydrodynamics code that is used to simulate the dynamic evolution of laser created plasmas.

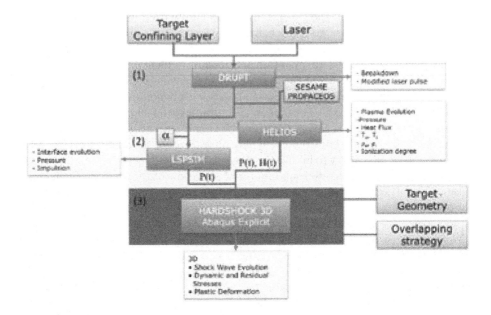

Fig. 1. Calculational scheme of the developed system SHOCKLAS

With the aid of HELIOS, The plasma is described by its thermodynamic state and by the velocity of flow. The state variables are related by the material equation of state (E.O.S.). The Navier-Stokes equations of fluid mechanics yield the needed five relations expressing the basic conservation laws of physics applied to the moving fluid. The code uses a two (differentiated for ions and electrons due to the weak energy coupling between the two populations) temperature-fluid scheme. Electrons and ions are assumed to flow as one fluid what implies no charge separation.

Material equation of state (EOS) properties is based on either SESAME tables [13] or PROPACEOS tables [12]. Opacities are based on tabulated multi-group PROPACEOS data. Radiation emission and absorption terms are coupled to the electron temperature equation.

Laser energy deposition is computed using an inverse Bremsstrahlung model, with the natural physical restriction that no energy in the beam passes beyond the critical surface.

2.2   Target Thermo-mechanical Behavior Description

The study LSP processes without coating needs a coupled treatment of thermal and mechanical transient processes. The target material subject to LSP is heated due to direct laser interaction heating (input from HELIOS simulations). To study these 3D problems, the authors have developed a model based in the FEM commercial code ABAQUS [14]. It solves the shock propagation problem into the solid material, with specific consideration of the material response to alterations induced by the thermal and mechanical interaction (i.e. effects as elastic-plastic behavior). The resulting temperature rise is computed and produces a local thermal expansion of the target material whose subsequent thermal strains have to be consistently calculated.

From the point of view of time differencing, the usual strategy of explicit differencing for the initial fast shock propagation phase followed by standard implicit differencing for the analysis of the final residual stresses equilibrium is not used. Instead, only explicit differencing has been used with long time evolutions in order to reach thermal and stress equilibration.

From the geometrical point of view, two kinds of configurations have been considered: an axis-symmetric one for studies of the effect of one single pulse or several pulses in the same point and a full 3D configuration for those geometric overlapping of pulses in which a full 3D process dependence has to be considered.

The FEM element used for the mechanical simulation is a 4-node brick reduced integration with hourglass control bilinear, namely CAX4RT, and an 8-node brick reduced integration with hourglass control trilinear, namely C3D8RT, for fully 3D simulations.

In LSP processes the material is stressed and deformed in a dynamic way, with strain rates exceeding 104s-1. The material behavior of Al2024 and Ti6Al4V is modeled using Johnson-Cook theory and parameters [15, 16] and the simulation of practical scale components is accomplished using explicit time differencing and the eigenstrain reconstruction method [17, 18, 19]

3   Results: Study of Compressive Residual Stresses Induced by LSP in Hip Replacement Parts

Titanium alloys such as Ti6Al4V are well known for their superior mechanical properties as well as for their good biocompatibility, making them desirable as surgical implant materials [20]. Total hip replacement composed of Ti6Al4V is successfully applied to patients afflicted by hip diseases [21, 22, 23, 24].

Basically, the purpose of an artificial hip joint is to restore as much as possible the kinematics of the locomotor system and allows it to withstand physiological loads while minimizing wear and friction and avoiding harmful reactions in the body [25].

The interest of the present analysis is particularly focused in the prevention of the stem fatigue failure, related to the last described failure mechanism. Despite the fact that the fatigue failures of hip prostheses have been significantly reduced in the past two decades [26, 27], every new implant design aspires to be more resistant against fatigue failure [28].

Accordingly, traditional shot peening (SP) has been proposed to induce compressive residual stresses in order to improve fatigue life in hip replacements [29]. However, recent investigations (carried out with Ti6Al4V turbofan blades) have shown that LSP provides deeper compressive residual stresses and a greater fatigue life enhancement than SP [30, 31].

Consequently, in this paper we propose to treat Ti6Al4V hip replacements by LSP. A typical Charnley stem-head shape has been considered (Fig. 2 in [32]). Stem shapes have significant influence on the performance of prostheses. Stem shapes with smooth surfaces like the Charnley stem shape generally reduce stress concentrations and lead to longer fatigue life of the prosthesis.

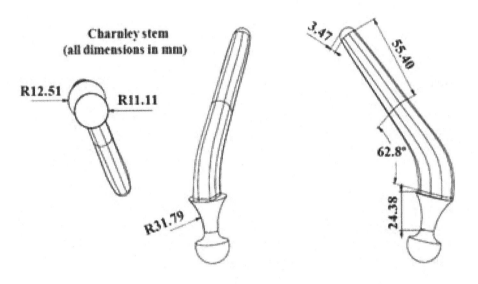

Fig. 2. Typical geometry of a Charnley hip replacement prosthesis (adapted from [32])

The numerical simulation of the LSP treatment of one of these typical prosthesis under the typical parameters of laser pulse diameter $\emptyset = 1{,}5$ mm and overlapping pitch 0,33 mm (equivalent to a pulse overlapping density of 918 pulses/cm^2) has been performed over the areas shown in Fig. 3.

This figure displays in colour scale the residual minimum principal stress in the hip replacement from in different views showing the superficial homogeneity of the induced stresses due to high overlapping density (pulses/cm^2). Superficial compressive residual stress values are greater than 600 MPa in the whole treated patch. There is no observable influence of curvature in residual stresses field. The reason is definitely related to the big difference between the applied spot diameter and the radius of curvature.

In Fig. 4, two different cross-sections are represented in order to show the internal distribution of the residual stresses field. The obtained FEM data predict that the compressed depth is 1.64 mm. This value is consistent with the outcome experimentally obtained by Shepard et al. for Ti6Al4V blades (1.27-1.90 mm).

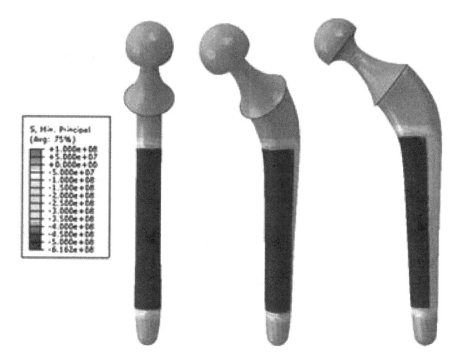

Fig. 3. Treatment geometry and colour scale presentation of the minimum principal superficial residual stresses induced in the considered hip replacement by LSP

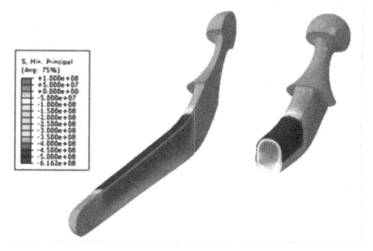

Fig. 4. Sample result showing the internal residual stresses fields induced in a hip replacement prosthesis by LSP

4   Discussion and Outlook

The numerical model developed by the authors for the prediction of residual stress fields induced in real components by LSP has been successfully applied to the simulation of residual stresses induced in real scale components subject to this fatigue life improvement technique.

One of the most significant applications of the technique, however not yet sufficiently developed, is the treatment of hip replacement prostheses, which can benefit from an relevant enhancement in their fatigue life expectations, thus improving the life quality and expectation of the surgically implanted patients.

The development of the reported experimentally supported prediction models is to be considered as a relevant improvement for the truly adjusted design of the prosthesis elements with maximum warranty of quality and durability as well as an excellent example on how computer-aided design is providing enhanced solutions to real life problems on the track to improved human quality life.

Acknowledgements. Work partly supported by Spanish MINECO (Project MAT2012-37782).

References

1. Sano, Y., Mukai, N., Okazaki, K., Obata, M.: Residual stress improvement in metal surface by underwater laser irradiation. Nuclear Instruments and Methods in Physics Research B 121, 432–436 (1997)
2. Rubio-González, C., Ocaña, J.L., Gómez-Rosas, G., Molpeceres, C., Paredes, M., Banderas, A., Porro, J.A., Morales, M.: Effect of laser shock processing on fatigue crack growth and fracture toughness of 6061-T6 aluminum alloy. Mat. Sci. Eng. A 386, 291–295 (2004)
3. Ocaña, J.L., Morales, M., Molpeceres, C., Porro, J.A.: Laser Shock Processing as a Method for Surface Properties Modification of Metallic Materials. In: Schulze, V., Niku-Lari, A. (eds.) Shot Peening and other Mechanical Surface Treatments, pp. 466–471. I.I.T.T, Paris (2005)
4. Sánchez-Santana, U., Rubio-González, C., Ocaña, J.L., Gómez-Rosas, G., Molpeceres, C., Porro, J.A., Morales, M.: Wear and friction of 6061-T6 aluminum alloy treated by laser shock processing. Wear 260, 847–854 (2006)
5. Morales, M., Porro, J.A., Blasco, M., Molpeceres, C., Ocaña, J.L.: Numerical simulation of plasma dynamics in laser shock processing experiments. Appl. Surf. Sci. 255, 5181–5185 (2009)
6. Griffin, R.D., Justus, B.L., Campillo, A.J., Goldberg, L.S.: Interferometric studies of the pressure of a confined laser-heated plasma. Journal of Applied Physics 59, 1968–1971 (1986)
7. Fabbro, R., Fournier, J., Ballard, P., Devaux, D., Virmont, J.: Physical study of laser produced plasma in confined geometry. Journal of Applied Physics 68, 775–784 (1990)
8. Ocaña, J.L., Morales, M., Molpeceres, C., Torres, J.: Numerical simulation of surface deformation and residual stresses fields in laser shock processing experiments. Appl. Surf. Sci. 238, 242–248 (2004)

9. Morales, M., Ocaña, J.L., Molpeceres, C., Porro, J.A., García-Beltrán, A.: Model based optimization criteria for the generation of deep compressive residual stress fields in high elastic limit metallic alloys by ns-laser shock processing. Surface & Coatings Technology 202, 2257–2262 (2008)

10. Morales, M., Porro, J.A., Molpeceres, C., Holgado, M., Ocaña, J.L.: Analysis of plasma thermal surface effects on the residual stress field induced by LSP in Al2024-T351. Journal of Optoelectronics and Advanced Materials 12, 718–722 (2010)

11. Morales, M., Correa, C., Porro, J.A., Molpeceres, C., Ocaña, J.L.: Thermomechanical modelling of stress fields in metallic targets subject to laser shock processing. International Journal of Structural Integrity 2, 51–61 (2011)

12. MacFarlane, J.J., Golovkin, I.E., Woodruff, P.R.: HELIOS-CR A 1-D radiation-magnetohydrodynamics code with inline atomic kinetics modeling. Journal of Quantitative Spectroscopy & Radiative Transfer 99, 381–397 (2006)

13. Lyon, S.P., Johnson, J.D.: SESAME: Los Alamos National Laboratory Equation of State Database. Technical report LA-UR-92-3407, Los Alamos National Laboratory, Los Alamos (1992)

14. ABAQUS, Inc.: ABAQUS Users Manual, Pawtucket (2009)

15. Johnson, G.R., Cook, W.H.: A constitutive model and data for metals subjected to large strains, high strain rates and high temperatures. In: Proceedings of the 7th International Symposium on Ballistics, pp. 541–547 (1983)

16. Kay, G.: Failure Modeling of Titanium 6Al-4V and Aluminum 2024-T3 with the Johnson-Cook material model. Technical Report U.S. Department of Transportation DOT-FAA-AR-97-88 (2003)

17. Korsunsky, A.M.: Residual elastic strain due to laser shock peening: Modelling by eigenstrain distribution. The Journal of Strain Analysis for Engineering Design 41, 195–204 (2006)

18. Jun, T., Korsunsky, A.M.: Evaluation of residual stresses and strains using the Eigenstrain Reconstruction Method. International Journal of Solids and Structures 47, 1678–1686 (2010)

19. Achintha, M., Nowell, D.: Eigenstrain modelling of residual stresses generated by laser shock peening. Journal of Materials Processing Technology 211, 1091–1101 (2011)

20. Bordji, K., Jouzeau, J.Y., Mainard, D., Payan, E., Netter, P., Rie, K.T., Stucky, T., Hage-Ali, M.: Cytocompatibility of Ti-6Al-4V and Ti-5Al-2.5Fe alloys according to three surface treatments, using human fibroblasts and osteoblasts. Biomaterials 17, 929–940 (1996)

21. Savilahti, S., Myllyneva, I., Pajamaki, K.J.J., Lindholm, T.S.: Survival of Lubinus straight (IP) and curved (SP) total hip prostheses in 543 patients after 413 years. Arch. Orthop. Trauma Surg. 116, 1013 (1997)

22. Marston, R.A., Cobb, A.G., Bentley, G.: Stanmore compared with Charnley total hip replacement: a prospective study of 413 arthroplasties. J. Bone Joint Surg. 78-B, 178–184 (1996)

23. Neumann, L., Freund, K.G., Sorenson, K.H.: Long-term results of Charnley total hip replacement. Review of 92 patients at 15 to 20 years. J. Bone Joint Surg. 76-B, 245–251 (1994)

24. Delaunay, C.P., Kapandji, A.I.: Primary total hip arthroplasty with the Karl Zweymuller first-generation cementless prosthesis: a 5 to 9 year retrospective study. J. Arthroplasty 11, 643–652 (1996)

25. Campioni, I., Notarangelo, G., Andreaus, U., Ventura, A., Giacomozzi, C.: Hip prosthesis computational modeling: FEM simulations integrated with fatigue mechanical tests, Biomechanics imaging and computational modeling in biomechanics. Lecture Notes in Computational Vision and Biomechanics 4, 81–109 (2013)
26. Sotereanos, N.G., Engh, C.A., Glassman, A.H., Macalino, G.E.: Cementless femoral components should be made from cobalt chrome. Clin. Orthop. 313, 146–153 (1995)
27. Wroblewski, B.M., Sidney, P.D.: Charnley low friction arthroplasty of the hip Long term result. Clin. Orthop. 292, 191–201 (1993)
28. Zafer Senalp, A., Kayabasi, O., Kurtaran, H.: Static, dynamic and fatigue behavior of newly designed stem shapes for hip prosthesis using finite element analysis. Materials & Design 28, 1577–1583 (2007)
29. Champaigne, J.: Shot peening of orthopaedic implants for tissue adhesion. US Patent 7,131,303 (2006)
30. Shepard, M.J.: Laser shock processing induced residual compression: impact on predicted crack growth threshold performance. Journal of Materials Engineering and Performance 14, 495–502 (2005)
31. Hammersley, G., Hackel, L.A., Harris, F.: Surface prestressing to improve fatigue strength of components by laser shock peening. Optics and Lasers in Engineering 34, 327–337 (2000)
32. Charnley, J.: Femoral prosthesis, US Patent 4,021,865 (1977)

From Smart Grids to Business Intelligence, a Challenge for Bioinspired Systems

Irene Martín-Rubio[1,2(✉)], Antonio E. Florence-Sandoval[2],
Juan Jiménez-Trillo[1], and Diego Andina[1]

[1] Group for Automation in Signals and Communications,
Technical University of Madrid, 28040 Madrid, Spain
[2] ETSIDI, Technical University of Madrid - UPM, Madrid, Spain
{irene.mrubio,d.andina}@upm.es

Abstract. Interconnected networks for delivering electricity are in the need of powerful Information Technology Systems to successfully process information about the behaviours of suppliers and consumers. They are becoming Smart Grids, increasingly complex infrastructures that require the automated intelligent management of multi-tier services and utility's business, improving the efficiency, reliability, economics, and sustainability of the production and distribution from suppliers to consumers. This paper makes a review of the State-of-the-art of this technological challenge, where Big Data from Smart Grids empowers Business Intelligence. Bioinspired computing that models adaptive, reactive, and distributed intelligent processing is candidate to play an important role in tackling this complex problems.

Keywords: Smart grid · Business intelligence · Big data · Bioinspired systems · Multi utilities

1 Introduction

A new generation of fully interactive Information and Communication Technologies (ICT) infrastructure is being developed to support the optimal exploitation of energy sector. Big Data (BD) from smart meters empower a Business Intelligence (BI) from advanced metering infrastructure which has the potential to react almost in real time as well as production than in consumption behaviours. Utilities must be able to plan for and manage an exponential increase in network connections. The growth of the Smart Grid (SG) will create opportunities up and down the industry value chain, as well as in related industries. This growth occur across the value chain, from customer-side applications to grid-wide automation upgrades and will vary regionally around the world. Opening up a closed infrastructure as that of energy networks will reshape the energy business sector [4,22,8] . Utilities are deploying Advanced Metering Infrastructure (AMI) and intelligent grid device across their Transmission and Distribution (T&D) service areas. These increasingly complex infrastructures require management of multi-tier services and new technologies that span Information Tecnology (IT)

© Springer International Publishing Switzerland 2015
J.M. Ferrández Vicente et al. (Eds.): IWINAC 2015, Part II, LNCS 9108, pp. 439–450, 2015.
DOI: 10.1007/978-3-319-18833-1_46

Systems and Applications and the utility's business. Due to its challenging complex and distributed nature, Bionspired systems as those of Swarms or massive parallel processing bioispired systems have the opportunity of playing a relevant role. This paper presents the state-of-the art of the problem to serve as a reference to specialized researchers.

2   State-of-the-Art

In the last decade energy utility sector has undergone major changes in terms of liberalization, increased competition, efforts in improving energy efficiency, and in new technological solutions such as smart meter and grid operations. There are new information technology solutions that will not only introduce new technical and organizational concepts, but have a very strong potential to radically change *modus operandi* of utility companies [18,4]. Utility companies are redefining their supply chains in terms of multi-commodities, delivering several products or services to the customer across multiple jurisdictions and as result of dynamic pricing. Utility companies are switching from energy procurement, distribution and sales into IT-driven efficient management of complex information [18,8].

2.1   Smart Grids

Smart Grids are interconnected networks for delivering electricity that use Information and Communications Technologies (ICTs) to process information about the behaviours of suppliers and consumers, improving the efficiency, reliability, economics, and sustainability of the production and distribution The smart grid embodies the intersection of power engineering and information technology. Building a culture that encourages rich dialog and compelling solutions across these competencies is difficult but required in order to success [23]. Few companies can or should develop the full suite of capabilities in-house. The ability to identify, convene and manage partnerships, alliances and more informal consortia is growing in importance. Three main business segments for smart grid technologies: customer applications, Advanced Metering Infrastructure (AMI)/smart meters, and grid applications [7,5,8].

Deployment of smart grid technologies is accelerating, particularly in the United States [7,8]. The technological revision and management methods of the electrical infrastructure will require redesign of the roles of current electrical companies and also entry new players into the power generation industry [31].

The size of the smart gird opportunity for solutions providers is still unclear. Across business segments, growth and value will be determined by the technology, the level of competition, and emerging regulation and policy. Growth will likely be slower where existing energy infrastructure is highly developed, such as in Europe [7]. Growth on a percentage basis in China and other areas where the transmission and distribution (T&D) infrastructure is still developing, will be faster as smart grid technologies will be used be leapfrog traditional infrastructure investments. For López et al. [22], smart grid concept represents one of the most challenging engineering projects, as long as it means a revolution at every domain of such a complex and critical system as the electrical grid.

2.2 Multiutilities

New technologies and concepts are emerging as we move towards a more dynamic, service-based, market-driven infrastructure, where energy efficiency and savings can be better addressed though interactive distribution networks [20]. Highly distributed business processes will need to be established to accommodate these technology and market evolutions. The traditional static customer process will increasingly be superseded by a very dynamic, decentralized and market-oriented process where a growing number of providers and consumers interact. The utility must first understand the end-to-end network across the value chain -from generation to the substation to the home. A true understanding of the connections, relationships, and dependencies between the smart meters, backhaul network, AMI network, meter data management system, business applications, and the back-office IT system is vital to keep the utiliy's business highly available and running at peak performance [33].

These applications require a range of components across the value chain - transmission and distribution infrastructure, a communication network, and a computing platform [8]-. Implementing these elements, will also require utilities to change their business processes, as the added level of technology will require new management and oversight procedures [17]. In order to capture the smart grid business opportunity, smart grid players must build a deep understanding of where the value is in the evolving smart grid, and they must also develop a compelling business model to pursue this value. Utilities are demanding more clearly articulated value propositions, and their direct linking to underlying value drivers (e.g. reduced operating expenses, increased grid efficiency, improved capital productivity). A compelling business model is the one that aligns the utility goals by creating rate-based opportunities and reducing operating expenses, while also reducing risk [33]. Utilities now face the challenge of packaging solutions, which will require a more consultative approach to sales and marketing and coordination across traditionally siloed business units within a utility (e.g. distribution operations and IT). From the utility perspective, getting secure, future-proofed products has become a top priority [5]. Concerns about risk emanate from a number of areas, including uncertainty around standards and interoperability as well as the capability of utilities to deploy the new technologies successfully. Another problem is the widespread use of renewable energy resources in Distributed Generation. It will have a significant impact in the energy market.

2.3 Interoperability of Multi Utility

Interoperability is a critical parameter in Smart Grids [22,5]. Interoperability can result in real cost savings for the utility and the vendors, who no longer have to write custom integrations for every new technology that's brought in [11]. Interoperability is a challenge for many industries, not just the utility sector.

The medical and financial industries, for example, are in the midst of tackling the challenges of interoperability, as are their peers in the utility sector. The increase connectivity and bandwidth in the utility industry, just like other industries, are driving the need for greater interoperability. Interoperability requires market momentum resulting in investments in the kind of cross-functional bodies that provide for verification, testing, modelling and design.

López et al. [22] consider interoperability as the communication infrastructure that allow equipment from different manufacturers to interact seamlessly through standardization; their research focussed on M2M (Machine-to-Machine) communications without human intervention because smart meters are not just sensors any longer, but they become part of the core of the power distribution network itself.

3D interoperability concept adds application dimension to the more common 2D models that foucus on devices and communications with one set of head end and software. In a 2D interoperability model -the historical approach that utilities have followed- there has been a strong focus on devices and communications [11].

3D interoperability models aspire to a full three dimensions of freedom, meaning that technology can move along any axis without disrupting existing technology along the other axis. So a technology change to the billing department would not impact technology supporting customer service. This approach promises lower system maintenance costs since it eliminates the rippling upgrade costs commonly associated with past 2D implementations. When considering interoperability, standards and alliances are crucial as well as design flexibility. Design is crucial, especially a collaborative, flexible design when considering all stakeholders' and preferences. Utilities adopt a standard network design, and a standard integration or interoperation patterns and approaches, including standard for data sharing, data governance and information sharing. Utilities have endorsed a standard infrastructure model, so that not every application runs on its own set of infrastructure.

3D interoperability can simplify information sharing. Interoperability helps to drive down costs by providing a more simple, reusable information sharing platform. It gives utilities the ability to provide enhanced security at the right points, and enables a more reliable and monitoring and management solution when it comes to the use of data and interactions. One big goal achieved is to minimize user intervention and minimize time of implementation by making things simpler. With the right interoperability architecture, business can bring together data and information from application systems across and enterprise, irrespective of vendors, to serve analytic and business needs. So its about big data: taking that data, analysing and doing even more with it, such as improving performance. Many utilities see a dramatic uptick in customer service [11], which is the top priority. Streamlining and automating processes relieves some of the stress and some of the training that is required for staff.

2.4   Big Data and Information Systems

Information Systems can be classified according to their function, but must work in a coordinated and integrated manner:

- AGC: Automatic Generation Control
- AMS: Asset Management System
- WMS: Work Management System
- OMS: Outage Management System
- GIS: Geographic Information System
- AGC: Automatic Generation Control
- DSM: Demand Side Management
- DMS: Distribution Management System
- CIS: Customer Information System
- NMS: Network Management System
- Billing, Planning and Advanced Simulation.

Today, the ubiquity and proliferation of data and analytics is profoundly altering the business landscape. A number of articles have noted the demand for data scientists is racing ahead of supply [13,14,12]. The power of analytics is racing while costs are falling. Data visualization, wireless communications and cloud infrastructure are extending the power and reach of information, creating a new foundation for competitiveness. With abundant data from multiple touch points, and new analytics tools companies are getting better and better in creating transparency, enabling experimentation to discover actionable insights, exposing variability and improving performance, segmenting populations to customize actions, and innovating/customizing products and services. Companies are learning to test and experiment with big data [14]. As companies collect more data from operations, some are looking to create new information-based products that add new revenue streams. In summary, big data and analytics is about effecting transformational evidence based decision making and process, organizations need to change from a lean continuous improvement mindset to an experimentation driven learning organization. Despite the widespread recognition of big data's potential, organizational and technological complexities as well as the desire for perfection often slow progress.

2.5   Business Intelligence (BI)

To meet these challenge, utilities need a way to collect, correlate, and analyze events from multiple sources, leveraging existing tools while gaining end-to-end management of the entire network. Business are increasingly monitoring and tracking data about what it takes to keep themselves running. Business Intelligence (BI) is the practice of interpreting and visualizing data to make useful business-oriented decisions. BI tasks occur in offices, universities, and data centers and provide data-oriented lifeblood to research and business organizations worldwide. BI systems must often appeal to broad audiences, from knowledge workers to Chief Executive Officers (CEOs) to stockholders [15]. They must allow for rapid analysis for decision making, developing insights, and communication those insights' results . BI tasks fall into three categories: - Exploring and analyzing data - Monitoring ongoing dataflows through dashboards, and - Communicating insights to others, both in and outside a company. A BI system is made of seven layers: IT and related infrastructure, data acquisition, data integration, data storage, data organizing, data analytics and data presentation [16]. Much BI work happens around data that's stored specifically for BI analysis, often called a data warehouse. Data in those warehouses might be stored in a precomputed structure called an OLAP (Online Analytical Processing) cube. OLAP cubes make it easy to look at various data aggregates, filtering on a number of selected fields. Fisher et al. [15] look to learn about the processes that practitioners currently follow in this area and how new BI techniques and capabilities will help users understand and act on widely disparate types of data. BI is an area of inquiry and explore beyond the current standard practices. In the Information System (IS) and BI (Business Intelligence) literature Business Intelligence Systems (BIS) are well recognized as contributing to decision-making, especially when firms operate in highly competitive environments [29,30]. BIS are most commonly identified as technological solutions holding quality information in well-designed data stores, connected with business friendly tools that provide users -incumbents of executives, managers, business analytsts and other roles within a firm utilizing BIS -enabled information for analytical decision making- with timely access to as well as effective analysis and insightful presentation of the information generated by enterprise- wide applications, enabling them to make the right decisions or take the right actions.

2.6   (R)evolution in the Utility Supply Chain Structure

The rapid evolution of new IT (smart grids), combined with AMI promise to revolutionize the way utility companies deliver their products to the customers. The utility supply chain structure is being reorganized [18] in all segments, as schematically shown in Figure 1.

Development of the smart grid will create opportunities for traditional infrastructure distributors while opening the market to new players. Traditional vendors will benefit from large-scale renewal of utility assets as customer and grid applications are deployed and will be able to differentiate their product lines through increased functionality and integration with other smart technologies. New players - IT hardware providers, software firms, networking and telecommunications companies, semiconductor manufacturers, and systems integrators- will benefit from major technology investments. Capturing these opportunities, however, will not be easy.

Competition will be stiff, and solutions providers will need to develop a market-entry approach that reflects the long sales cycles, technology obsolescence concerns and regulatory constraints that characterize utility procurement processes [26].

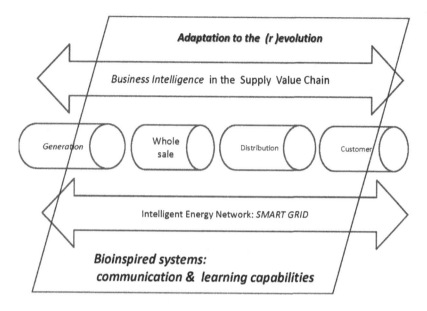

Fig. 1. From Smart Grids to BI: A challenge for Bioinspired Systems

Technology alone will not make energy supply more intelligent. Crucially important is the intelligent of the service company that runs the business in the grid [4]. With Smart Grids there is not only a revolution in the management of electricity networks but also in the business arena.

2.7   Key Performance Indicators

A smart grid seeks to improve the current power delivery system, increasing the interaction between different stakeholders and providing an easy connection to their elements. However, the main characteristics of a smart grid is its multi-disciplinary character [35,3,28]. It combines different systems that require and advanced communications systems that provide interoperability amongst them [2,19]. As evidence of the Smart Grid progress, several demonstrations projects or initiatives have been developed in the last years by various agencies, in the USA and Europe.

- In USA, RDSI (Renewable and Distributed Systems Integration) projects involve different Smart Grid technologies, devices and systems, seeking to reduce peak demand.
- EPRI (Electric Power Research Institute) focused on the integration of different technologies to form a VPP (Virtual Power Plant), employing integrated

control of Distributed Generation (DG), storage, and demand response technology. Another EPRI initiative is the IntelliGrid program which is focused on the development of new ICT (Information and Communication Technologies) for Smart Grids.

- Other North American initiatives are also noted because of their intense activities, such as the Edison Electric Institute, the GridWise Alliance, and the Galvin Electricity Initiative.
- On the other hand, in Europe, it is normal to find a lot of projects mainly oriented towards environmental protection. These projects are typically funded by the FP7 (7th Framework Programme) and their initiatives pursue the goals known as 20-20-20 targets. Thus, these initiatives also have continuity within the next European Framework Programme for Research and Innovation (Horizon 2020).

The presence of Smart Grid systems is increasing. Nevertheless, their complexity and multidisciplinary character make their evaluation difficult. Due to this, it is essential to apply a metric to help this process. There exist some metrics in the literature. Personal et al. [28] suggest a new hierarchical metric with a novel approach based on BI to evaluate a Smart Grid project. This hierarchy seeks a dual purpose. It serves not merely to seek to facilitate the project evaluation, this hierarchy also allows the system to adapt their information for different audiences (with different technical knowledge). Personal et al. adopt a PlanPotT (a powerful planning support tool), which can estimate network behaviour under different virtual scenarios. [28] implemented their model on a real Smart Grid project, called "Smartcity Malaga". The importance of selecting a proper metric, which accommodates the different initiatives of the evaluated project must be underscored. These tools, translate the company visions and targets into indicators; thus these elements reflect the process or assessment of each project initiative.

Finally, it is also important to highlight several projects in Brazil, Oman and Southeast Asia, where initiatives such as NEDO (New Energy and Industrial Technology Development Organization), Alliance (in Japan) and SGCC (Stage Grid Corporation of Chicha) are very dynamic in the last few years. As previously mentioned, a Smart Grid project involves a lot of technologies, each being a complex research in and of itself.

However, the progress so far has been conducted individually, without analysing the interaction amongst them. Due to this, some authors conclude that the next step in Smart Grid networks is to develop projects which integrates and analyse the systems as a whole [21]. However, this poses several problems, because of its complex and multidisciplinary nature, it is very difficult to become to classical engineering solutions that assess the overall success. Here is where Bio-inspired artificial intelligent systems that implement adaptive, reactive, complex and distributed processing may lay the technological solution. Some Artificial Intelligence solutions based on Artificial Neural Networks [1], [10] or Swarm Intelligence [6] are already contributing to the analysis of "big data", routing methods and improving supply chain execution and effective supplier relationship management (SRM).

3   Discussion

The existence of smart meters that can be also accessed in a seamless and uniform way via standardized methods is a must for the service oriented infrastructure. Assuming that smart meters will be accessed via web-services, has far reaching implications, since now business processes can actively integrate them in their execution. Instead of providing only their data (one-way communication), which limited their usage, they can be active and host business intelligence (bidirectional communication) which does not have to rely only on the back-end systems. Since the meter do have computing capabilities, are able to process locally their data and take local decisions, this data does not need to be sent to the backend systems. The business process could be even more distributed since the meter may trigger an external Internet service which will do advance the business process itself. There are great advantages i.e more lightweight business process which can outsource or parallelize specific execution steps, it has been reduced communication overhead since the data do not have to be transferred to the backend but stay at their original source, and more sophisticated business process data are highly distributed.

Business Intelligence has to be able to automatically calculate the impact of infrastructure and application issues on business services and is the only solution that can model business processes and tie them to underlying infrastructure. With precise impact analysis, utilities can take swift to protect the services and process most critical to the business. Business Analytics must calculate the impact of network, system, and application problems by totalling the values of all affected customer, service, and infrastructure components. Administrators supported by Automated Decision Systems will be able to assign the "weights" according to business importance and can also customize the impact analysis to leverage external data, such as importing penalty rates from service contracts to automatically calculate business value. BI supports mapping of business processes. It can represent an extended set of business entities -customers, lines of business, business units, and more. BI can calculate how problems in one business process impact a related process. Different IT Performance Reporter graphically represents performance metrics across the environment in one consolidated view, enabling better business decisions.

Production generation and wholesale will be changed due to innovative solar and wind generators as well as technological innovations of automotive industry. Downstream supply chain structure will be restructured mainly due to technological innovations represented by smart grids and AMI. Smart grids will form a distribution backbone that will enable integration of traditional and innovative upstream and downstream segments of future utility supply chain. AMI, on the other hand, will closely link consumers to utility supply chain since it will enable consumer to closely monitor energy consumption in real time through electronic meter solutions. Apart from consumer involvement, AMI will affect downstream utility supply chain in delivering new services (telecom services, financial services, security services.

These services will evolve due to AMI penetration to each consumer and apart from metering function, on the basis of highly developed distribution network, consequently providing playground for added value services (telecom services, financial services, security services). Technological (r)evolution is therefore offering numerous business opportunities to utility organizations. In order to gain business value, utility organizations have to meet several challenges.

In recent years, Business Intelligence is being implemented in several organizations [34,9] to operate as efficiently as possible [27], but some projects fail [32] due to cost overruns, technical obstacles and net-generation information challenges stemming from pervasive computing [32] and furthermore, motivational factors from managers acceptances and use [9]. At the technical side, it is needed a rationalization framework to facilitate the management of business intelligence systems geared towards a more efficient and effective use of explicit knowledge. At the motivational side, it is needed to clarify benefit factors (organizational rewards, reputation, and reciprocity) to exchange reports with other industry managers [9].

In the last few decades, computer scientist have invented some new computing paradigms and algorithms such as evolutionary computing and swarm intelligence, but they are far from capturing the nuances of evolutionary theory and behavioural biology. Transformative bio-inspirations for building computing and communication models will play an increasingly important role in the efforts to innovatively solve some real-world complex and difficult problems, which can be hard to tackle with existing computing algorithms [24],[25]. BI and Smart Grids can benefit from the adaptation possibilities that brings Bioinspired Systems.

4   Conclusions

Smart grids are empowering Business Intelligence based on the successful application of IT to a growingly complex Big Data in Multi Utilities systems environment. The challenge is to find engineering solutions that assess the overall success integrating and analysing the systems as a wholein. Such solutions must have the potential to perform complex, evolving and multidisciplinary tasks. Those are the characteristics of Bioinspired Systems, a discipline that starts to develop solutions in a field where the demand of solutions is racing ahead the supply.

References

1. Abbott, D.: Applied Predictive Analytics: Principles and Techniques for the Professional Data Analyst. John Wiley & Sons (2014)
2. Ancillotti, E., Bruno, R., Conti, M.: The role of communication systems in smart grids: Architectures, technical solutions and research challenges. Computer Communications 36(17-18), 1665–1697 (2013)

3. Ardito, L., Procaccianti, G., Menga, G., Morisio, M.: Smart grid technologies in europe: An overview. IEEE Power and Energy Magazine 10(4), 22–34 (2014)
4. Arends, M., Hendriks, P.H.: Smart grids, smart network companies. Utilities Policy 28, 1–11 (2014)
5. Bhatt, J., Shah, V., Jani, O.: An instrumentation engineer's review on smart grid: Critical applications and parameters. Renewable and Sustainable Energy Reviews 40, 1217–1239 (2014)
6. Bonabeau, E., Meyer, C.: Swarm intelligence: A whole new way to think about business. Harvard Business Review 79(5), 106–115 (2001)
7. Booth, A., Demirdoven, N., Tai, H.: The smart grid opportunity for solutions providers. McKinsey on Smart Grid (2010)
8. Cardenas, J.A., Gemoets, L., Rosas, J.H.A., Sarfi, R.: A literature survey on smart grid distribution: an analytical approach. Journal of Cleaner Production 65, 202–216 (2014)
9. Chang, Y.-W., Hsu, P.-Y., Wu, Z.-Y.: Exploring managers' intention to use business intelligence: the role of motivations. Behaviour & Information Technology, no. ahead-of-print, 1–13 (2014)
10. Choy, K.L., Lee, W., Lo, V.: Design of an intelligent supplier relationship management system: a hybrid case based neural network approach. Expert Systems with Applications 24(2), 225–237 (2003)
11. Connections, E.V.: Getting in synch: Here comes 3d: The next generation of interoperability. Elster Vital Connections Special Report (2014)
12. Davenport, T.H., Patil, D.: Data scientist, the sexiest job of the 21st century. Harvard Business Review 90, 70–76 (2012)
13. Dhar, V.: Data science and prediction. Communications of the ACM 56(12), 64–73 (2013)
14. Dhar, V., Jarke, M., Laartz, J.: Big data. Communications of the ACM 56(12), 64–73 (2013)
15. Fisher, D., Drucker, S., Czerwinski, M.: Business intelligence analytics. IEEE Computer Graphics and Applications 34(5), 22–24 (2014)
16. Glaser, J., Stone, J.: Effective use of business intelligence. Healthc. Financ. Manage. 62(2), 68–72 (2008)
17. Groznik, A.: Potentials and challenges in multi utility management (2010)
18. Groznik, A.: Towards multi utility management in europe 2(8), 195 (2014)
19. Gungor, V., Bin, L., Hancke, G.: Opportunities and challenges of wireless sensor networks in smart grid. IEEE Transactions on Industrial Electronics 57(10), 3557–3564 (2010)
20. Karnouskos, S., Terzidis, O., Karnouskos, P.: An advanced metering infrastructure for future energy networks. In: New Technologies, Mobility and Security, pp. 597–606. Springer (2007)
21. Kezunovic, M., Vittal, V., Meliopoulos, S., Mount, T.: The big picture: Smart research for large-scale integrated smart grid solutions. IEEE Power and Energy Magazine 10(4), 22–34 (2014)
22. López, G., Moreno, J., Amarís, H., Salazar, F.: Paving the road toward smart grids through large-scale advanced metering infrastructures. In: Electric Power Systems Research, vol. 120, pp. 194–205 (2015); smart Grids: World's Actual Implementations
23. Luthra, S., Kumar, S., Kharb, R., Ansari, M.F., Shimmi, S.: Adoption of smart grid technologies: An analysis of interactions among barriers. Renewable and Sustainable Energy Reviews 33, 554–565 (2014)

24. Ma, Z.S.: Towards computational models of animal cognition, an introduction for computer scientists. Cognitive Systems Research 33, 42–69 (2015)
25. Ma, Z.S.: Towards computational models of animal communications, an introduction for computer scientists. Cognitive Systems Research 33, 70–99 (2015)
26. Martín-Rubio, I., Florence-Sandoval, A., González-Sánchez, E.: Agency and learning relationships against energy-efficiency barriers. In: The Handbook of Environmental Chemistry, pp. 1–34. Springer, Heidelberg (2014), http://dx.doi.org/10.1007/698_2014_305
27. Olexová, C.: Business intelligence adoption: a case study in the retail chain. World Scientific and Engineering Academy and Soceity Transactions on Business and Economics 11, 95–106 (2014)
28. Personal, E., Guerrero, J.I., Garcia, A., Pea, M., Leon, C.: Key performance indicators: A useful tool to assess smart grid goals. Energy 76, 976–988 (2014)
29. Popovič, A., Hackney, R., Coelho, P.S., Jaklič, J.: Towards business intelligence systems success: Effects of maturity and culture on analytical decision making. Decision Support Systems 54(1), 729–739 (2012)
30. Popovič, A., Hackney, R., Coelho, P.S., Jaklič, J.: How information-sharing values influence the use of information systems: An investigation in the business intelligence systems context. The Journal of Strategic Information Systems 23(4), 270–283 (2014)
31. Salem, A.-A.: Electricity agents in smart grid markets. Computers in Industry 64(3), 235–241 (2013)
32. Sapp, C.E., Mazzuchi, T., Sarkani, S.: Rationalising business intelligence systems and explicit knowledge objects: Improving evidence-based management in government programs. Journal of Information & Knowledge Management 13(02) (2014)
33. Starace, F.: The utility industry in 2020. In: Handbook Utility Management, pp. 147–167. Springer (2009)
34. Tarallo, M.: Analytics for everyone. Security Management, November 2014 Print Issue. ASIS Intl. (2014)
35. Wissner, M.: The smart grid a saucerful of secrets? Applied Energy 88(7), 2509–2518 (2011)

Neural Recognition of Real and Computer-Designed Architectural Images

M.D. Grima Murcia[1]([✉]), Maria J. Ortíz[2], M.A. López-Gordo[3,4],
J.M. Ferrández-Vicente[5], and Eduardo Fernández[1]

[1] Institute of Bioengineering, University Miguel Hernández and
CIBER BBN Avenida de la Universidad, 03202, Elche, Spain
maria.grima@alu.umh.es, e.fernandez@umh.es
[2] Department of Communication and Social Psycology,
University of Alicante, Alicante, Spain
[3] Nicolo Association, Churriana de la Vega, Granada, Spain
[4] Department of Signal Theory, Communications and Networking,
University of Granada, 18071, Granada, Spain
[5] Department of Electronics and Computer Technologyt,
University of Cartagena, Cartagena, Spain

Abstract. Neuro-architecture seeks to define and understand the relationship between our psychological state and the artificial structures in which we spend most of our time and incorporate that insight into the design. However, little is known about the subjective judgment of real architectural models and about the cognitive processes in aesthetic appreciation applied to architecture. In the present study, we used real and computer-designed architectural images of bedrooms, in order to compare both types of images. Participants were asked to judge the arousal and valence of their own emotional experience after viewing each image. Furthermore, we used EEG recording to gain a better understanding of the regions of the brain involved in the processing of these images. Our results show that there are significant differences in the early stages of processing of both types of images and emphasize the importance of generating familiar and recognizable images for the acceptance by people.

Keywords: Neuro-architecture · EEG · Aesthetic appreciation

1 Introduction

Since 1962 architects have been using 3D modeling instead of hand crafted designs and drawings [1]. This technology is very powerful and today it is possible to create precise 3D models for different purposes [2]. This method is normally employed to create 3D interior and 3D exterior models. Furthermore, different business firms are taking interest in creation of 3D product models, which demonstrate the significance of 3D architectural modeling.

By applying architectural 3D models it is possible to envision the final results on a computer screen, which in general generates positive effects on the potential cutomers [3]. Thus, we can visualize the entire building, or a given room, using 3D

© Springer International Publishing Switzerland 2015
J.M. Ferrández Vicente et al. (Eds.): IWINAC 2015, Part II, LNCS 9108, pp. 451–458, 2015.
DOI: 10.1007/978-3-319-18833-1_47

rendering before the building is complete. This saves time and helps to modify details that require to be changed before the actual construction, which is extremely cost effective and saves ample time. However, little is known about the subjective judgment of real photographs and 3D architectural models and about the cognitive processes in aesthetic appreciation applied to architecture [4].

Neuroscientists and neuropsychologists have recently approached the traditionally philosophical field of aesthetics aiming to characterize the neural and evolutionary foundations of our species' capacity to appreciate beauty and art. This approach, known as neuroaesthetics, has begun to provide some insights into the neurobiological bases of aesthetic appreciation [5][6], and could also be useful for advising on architecture.

In the present study we used real and computer-designed architectural images of bedrooms, in order to compare both types of visual stimuli [7][8]. Participants were asked to judge the arousal and valence of their own emotional experience after viewing each image and electroencephalography recordings (EEG) were used to compare the descriptive and judgment processes [9]. Our goal was to investigate the processing characteristics and temporal time courses associated to each type of images [10]. Our preliminary results show that there are significant differences in the early stages of processing of both types of images and provide some insights regarding the emotional response to a given image and how aesthetic judgments are made.

2   Methods

Participants

Twenty two participants (mean age: 24.7; range: 19.7–33; eleven men, eleven women) participated in the study. All participants had no personal history of neurological or psychiatric illness, drug or alcohol abuse, or current medication, and they had normal or corrected to normal vision. All were right handed with a laterality quotient of at least + 0.4 (mean 0.8, SD: 0.2) on the Edinburgh Inventory [11]. All subjects were informed about the aim and design of the study and gave their written consent for participation.

Stimuli

We used twelve different images. Seven images were real photographs and five were computer-designed images performed with the help of a 3D modeling software (SketchUp). The photographs were real rooms, with different decoration and arrangements of furniture, windows etc. The computer-designed rooms were similar and involved several room types with different arrangements of the door, windows and bed. All images were presented in color, equated for luminance and contrast using a commercial stimulus presentation and experimental design system (STIM2, Compumedics, Charlotte, NC, USA).

REAL IMAGE

COMPUTER-GENERATED IMAGES

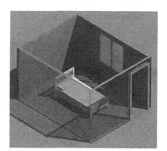

Figure 1.A: Stimulus type representative

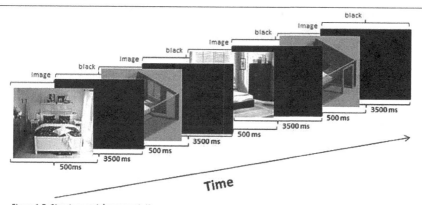

Figure 1.B: Structure serial representative

Fig. 1. Representative stimulus examples

Main Experiment

Figure 1 summarizes the experimental design. Each image was presented for 500ms and followed by a black screen for 3500ms. The images appeared in randomized order and only once. The participants were told that they would be asked to judge the arousal and valence of their own emotional experience after viewing each image (during the black period). Scales ranged from 1 (very unpleasant) to 9 (very pleasant).

EEG Recordings

We instructed subjects to remain as immobile as possible, avoiding blinking during image exposure and trying to keep the gaze toward the monitor center. EEG data was continuously recorded at a sampling rate of 1000 Hz from 64 locations (FP1, FPZ, FP2, AF3, GND, AF4, F7, F5, F3, F1, FZ, F2, F4, F6, F8, FT7, FC5, FC3, FC1, FCZ, FC2, FC4, FC6, FT8, T7, C5, C3, C1, CZ, C2,

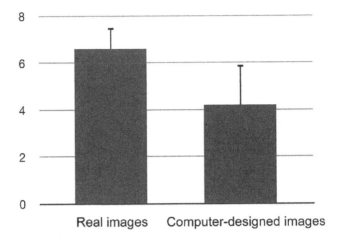

Fig. 2. Subjective scores of real photographs and computer−designed images (mean and standard deviation)

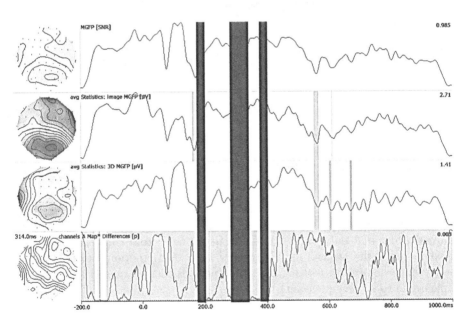

Fig. 3. Time points of significant differences in EEG activity for the 2 contrasts (photographs and computer designed images) as indicated by the T ANOVA analysis, depicting 1 minus p-value across time. Significant p values are plotted (α <0.01). The three vertical rectangles contain significant differences.

REAL IMAGES

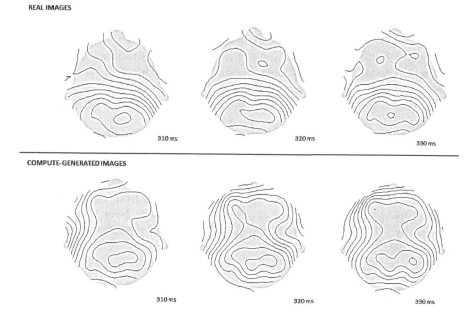

COMPUTE-GENERATED IMAGES

Fig. 4. Isopotential values of the EEG activity at different time windows

C4, C6, T8, REF, TP7, CP5, CP3, CP1, CPZ, CP2, CP4, CP6, TP8, P7, P5, P3, P1, PZ, P2, P4, P6, P8, PO7, PO5, PO3, POZ, PO4, PO6, PO8, CB1, O1, OZ, O2, CB2) using the international 10/20 system [12]. EEG was recorded via cap-mounted Ag-AgCl electrodes. A 64-channel NeuroScan SynAmps EEG amplifier (Compumedics, Charlotte, NC, USA) was used for EEG signal amplification and digitizing. The electrodes were filtered using a 45 Hz low-pass filter and a high-pass filter of 0.5 Hz. The impedance of recording electrodes was monitored for each subject prior to data collection and the thresholds were kept below 25 KΩ. All recordings were performed in a dimly lit and silent room.

Data Analysis

Experimental stimuli constisted of real photographs and 3D computer-designed images. To study the event-related EEG dynamics, we extracted data from an epoch 100 ms before stimulus onset to 1000 ms after stimulus. Signal processing was performed using the Curry 7 software (Compumedics, Charlotte, NC, USA). Data were re-referenced to a Common Average Reference (CAR), because the subsequently applied statistical and source analysis methods required CAR data. Furthermore, we used the frontal electrodes (FP1, FPZ, FP2, AF3 and AF4) to eliminate flicker effects. We chose these electrodes because they were located in regions that can be easily affected by possible involuntary movements. The pre and post times (-200ms, +500ms) defined the interval from the point at which the artifact was detected. The detected artifacts were corrected following standard procedures using Principal Component Analysis (PCA) [13]. Furthermore, to constrain our analysis, we used the examination of topographic changes in EEG activity [14][15]. This approach considers whole-scalp EEG activity elicited by a stimulus as a finite set of alternating spatially stable activation patterns, which reflect a succession of information processing stages. Therefore, the evolution of whole-scalp activity can be assessed over time in order to see how it differs between experimental conditions that impose different information processing demands.

Boundary Element Method (BEM) was used in the head reconstruction [16]. Topographical differences were tested through a non-parametric randomization test known as TANOVA (Topographic ANOVA), that allows to quantify differences in global dissimilarity of EEG activity between two conditions. This allows us to assess if the topographies are significantly different from each other on a timepoint-by-timepoint basis. For this paper, significance level is $\alpha=0.01$. As suggested by [17], the corresponding required number of repetitions was chosen to be p>1000.

3   Results

Subjective Scores

Images were scored by the participants during the experiment. The average score for the group of real photographs was 6.6 (SD: 0.9). By contrast, the average score for the group of images designed by computer was significantly lower: 4.2 (SD: 1.7); p<0.01 (Figure 2).

EEG

We found significant differences in stimulus-elicited activity between both types of images, photographs and computer designed images (α <0.01). These differences started approximately at 189 ms after picture onset. The time intervals with the larger differences were: 189-203 ms, 296-344 ms and 424-474 ms (Figure 3).

Figure 4 shows the isopotential areas during these periods, what provide information about the spatial configuration of the activated brain areas when looking real photographs or computer-designed images.

4   Conclusions

In this study we found significant differences in subjective perception of real photographs and computer-designed images. The subjective scoring revealed that the participants preferred real images (photographs) rather than computer-designed images. Moreover, our preliminary results show that there are significant differences in the brain processing of both types of images, specially in the early stages of processing, what could help to better understand how aesthetic judgments are made.

These differences in cognitive processing could be justified by effects related with the familiarity of the images. Thus, a number of psychological studies have shown that people usually prefer familiar stimuli, an effect currently explained under the umbrella of the processing fluency theory. In this context, Reber [18] suggested that objects vary with regards to the fluency with which they are processed. Given that fluent processing is experienced as hedonically pleasurable, and that aesthetic experiences are strongly influenced by affective states, it follows that positive aesthetic experiences arise especially from confident processing, such as that afforded by prototypical exemplars of a category [19]. In this framework the photographs are more familiar to most of the people and therefore are perceived as being more pleasurable than the 3D images created by a computer.

One would assume that if the images were more realistic 3D models, not only representing walls, bed, window, door, etc; the differences would be lower. In future work we will conduct more realistic computer models; we will study if there are still differences in the processing or if instead, the brain is not able to distinguish among very realistic computer generated images and real photographs.

Acknowledgement. This work has been supported in part by grant MAT2012-39290-C02-01, by the Bidons Egara Research Chair of the University Miguel Hernández and by a research grant of the Spanish Blind Organization (ONCE).

References

1. Szalapaj, P.: CAD principles for architectural design: analytical approaches to computational representation of architectural form. Architectural Press
2. Bijl, A.: Computer aided architectural design. In: Parslow, R.D., Green, R.E. (eds.) Advanced Computer Graphics, pp. 433–448. Springer, US
3. Lang, P.J., Bradley, M.M., Cuthbert, B.N.: Emotion, attention, and the startle reflex 97(3), 377–395
4. Vartanian, O., Navarrete, G., Chatterjee, A., Fich, L.B., Leder, H., Modrono, C., Nadal, M., Rostrup, N., Skov, M.: Impact of contour on aesthetic judgments and approach-avoidance decisions in architecture 110, 10446–10453, http://www.pnas.org/cgi/doi/10.1073/pnas.1301227110, doi:10.1073/pnas.1301227110
5. Watson, C.E., Chatterjee, A.: The functional neuroanatomy of actions 76(16), 1428–1434, http://www.neurology.org/cgi/doi/10.1212/WNL.0b013e3182166e2c, doi:10.1212/WNL.0b013e3182166e2c
6. Jacobsen, T.: Beauty and the brain: culture, history and individual differences in aesthetic appreciation 216(2), 184–191, http://doi.wiley.com/10.1111/j.1469-7580.2009.01164.x, doi:10.1111/j.1469-7580.2009.01164.x
7. Vartanian, O., Goel, V.: Neuroanatomical correlates of aesthetic preference for paintings 15(5), 893–897
8. Kedia, G., Mussweiler, T., Mullins, P., Linden, D.E.J.: The neural correlates of beauty comparison 9(5), 681–688, doi:10.1093/scan/nst026

9. Jacobsen, T., Höfel, L.: Descriptive and evaluative judgment processes: Behavioral and electrophysiological indices of processing symmetry and aesthetics 3(4), 289–299, doi:10.3758/CABN.3.4.289
10. Schupp, H.T., Cuthbert, B.N., Bradley, M.M., Cacioppo, J.T., Ito, T., Lang, P.J.: Affective picture processing: The late positive potential is modulated by motivational relevance 37(2), 257–261, http://doi.wiley.com/10.1111/1469-8986.3720257, doi:10.1111/1469-8986.3720257
11. Oldfield, R.: The assessment and analysis of handedness: The edinburgh inventory 9(1), 97–113, http://linkinghub.elsevier.com/retrieve/pii/0028393271900674, doi:10.1016/0028-3932(71)90067-4
12. Klem, G.H., Lders, H.O., Jasper, H.H., Elger, C.: The ten-twenty electrode system of the international federation. The International Federation of Clinical Neurophysiology 52, 3–6
13. Meghdadi, A.H., Fazel-Rezai, R., Aghakhani, Y.: Detecting determinism in EEG signals using principal component analysis and surrogate data testing, IEEE, 6209–6212, http://ieeexplore.ieee.org/lpdocs/epic03/wrapper.htm?arnumber=4463227, doi:10.1109/IEMBS.2006.260679
14. Murray, M.M., Brunet, D., Michel, C.M.: Topographic ERP analyses: A step-by-step tutorial review 20(4), 249–264, http://link.springer.com/10.1007/s10548-008-0054-5, doi:10.1007/s10548-008-0054-5
15. Martinovic, J., Jones, A., Christiansen, P., Rose, A.K., Hogarth, L., Field, M.: Electrophysiological responses to alcohol cues are not associated with pavlovian-to-instrumental transfer in social drinkers 9(4), e94605, http://dx.plos.org/10.1371/journal.pone.0094605, doi:10.1371/journal.pone.0094605
16. Vatta, F., Meneghini, F., Esposito, F., Mininel, S., Di Salle, F.: Realistic and spherical head modeling for EEG forward problem solution: A comparative cortex-based analysis 2010, 1–11, http://www.hindawi.com/journals/cin/2010/972060/, doi:10.1155/2010/972060
17. Rosenblad, A.: B. f. j. manly: Randomization, bootstrap and monte carlo methods in biology, 3rd edn., vol. 24(2), pp. 371–372. Chapman & hall/CRC, Boca raton, pp., $79.95 (2007), http://link.springer.com/10.1007/s00180-009-0150-3, doi:10.1007/s00180-009-0150-3 ISBN: 1-58488-541-6
18. Reber, R., Schwarz, N., Winkielman, P.: Processing fluency and aesthetic pleasure: Is beauty in the perceiver's processing experience? 8(4), 364–382, http://psr.sagepub.com/cgi/doi/10.1207/s15327957pspr0804_3, doi:10.1207/s15327957pspr0804_3
19. Winkielman, P., Halberstadt, J., Fazendeiro, T., Catty, S.: Prototypes are attractive because they are easy on the mind. 17(9), 799–806, http://pss.sagepub.com/lookup/doi/10.1111/j.1467-9280.2006.01785.x, doi:10.1111/j.1467-9280.2006.01785.x

Author Index

·

Printed in the United States
By Bookmasters